The Biophysical Chemistry of

Nucleic Acids

The Biophysical Chemistry of

Nucleic Acids

Thomas E. Creighton

HP

Helvetian Press

Published by Helvetian Press 2011

www.HelvetianPress.com

HelvetianPress@gmail.com

ISBN 978-0-9564781-8-4

CONTENTS

~ PREFACE ~

The field of molecular biology continues to be the most exciting and dynamic area of science and is predicted to dominate the 21st century. Only by investigating biological phenomena at the molecular level is it possible to understand them in detail. Such understanding is vital for advances in medicine, and the pharmaceutical industry that produces new drugs and cures is greatly dependent upon molecular biology. But molecular biology also contributes to the understanding of what human beings are and how they fit into this universe.

This volume builds on its companion volume, *The Physical and Chemical Basis of Molecular Biology* (Helvetian Press, 2010), and will be most intelligible and useful if the reader is aware of the information in that volume. That volume is also available in five sub-volumes: *Fundamentals*; *Visualizing Biological Structures and Molecules*; *Spectroscopy in Molecular Biology*; *Hydrodynamic Properties of Biological Macromolecules*; and *Interactions between Biological Molecules* (all Helvetian Press, 2011). They are available in printed form from www.Lulu.com.

Proteins and nucleic acids are the primary subjects of molecular biology. They carry, transmit, and express the genetic information that defines each living organism. It is vital to understand how these molecules function. The present volume is dedicated primarily to nucleic acids, whereas its companion volume, *The Biophysical Chemistry of Proteins* (Helvetian Press, 2011), focuses on proteins. Both are available in a single volume, *The Biophysical Chemistry of Nucleic Acids & Proteins* (Helvetian Press, 2010).

The first chapter is a general introduction to the covalent structures and conformations of macromolecules. The remaining chapters deal with the nucleic acids. The structural and chemical properties of DNA are the basis of its central role in storing and transmitting the genetic information (Chapter 2). DNA molecules tend to be immensely long, equivalent to a rope that is many kilometers long, which gives them special topological properties that must be accommodated (Chapter 3). The structure of RNA differs from DNA only very slightly, but this gives it remarkably different properties and functions (Chapter 4). The abilities of individual strands of DNA and RNA to base-pair with other strands with complementary nucleotide sequences are central to many techniques of molecular biology and increasingly to molecular medicine (Chapter 5). The ability to manipulate nucleic acids is central to molecular biology and described in Chapter 6.

Central to the functions of proteins and nucleic acids is their interactions with other molecules, and some of the physiologically most important interactions are those between proteins and nucleic acids (Chapter 7). The most impressive and important property of proteins is their ability of catalyze the rates of chemical reactions by many orders of magnitude, and usually incredibly specifically; nucleic acids also share in this ability (Chapter 8).

The references listed were chosen to be those that would best provide the interested reader with entry to the literature. They should not be assumed to be those most important for the subject. The latest references on any topic can be found by searching www.ncbi.nlm.nih.gov.

No one person can be expert in all the areas of molecular biology, so I have made ample use of the work of many others more expert than me, but too numerous to specify. Very special thanks are due to Eric Martz of the University of Massachusetts for making available the program Firstglance in Jmol (http://firstglance.jmol.org). It is incredibly useful for examining nucleic acid and protein structures and, at least as important, is very easy to use.

Of course, shortcomings and errors in this volume are totally my responsibility, for which I apologize in advance. Criticisms and suggestions would be welcome and can be sent to me at HelvetianPress@gmail.com.

Thomas E. Creighton

~ CHAPTER 1 ~

CONFIGURATIONS AND CONFORMATIONS

Molecules are generated by the formation of covalent bonds between pairs of atoms, in which the two atoms share electrons. A covalent bond forms when atoms such as carbon individually do not have enough electrons for a complete octet: if two atoms can complete their octets by sharing electrons, they can do so by forming a covalent bond. Covalent bonds can be explained only by quantum mechanics, but here it is necessary simply to recognize that covalent bonds are generally not broken in isolation under most conditions experienced in molecular biology. When a covalent bond is broken, as in a chemical or enzymatic reaction, it is generally exchanged with another covalent bond to a different atom. Consequently, covalent bonds define the structures and properties of small molecules, and those of large molecules are determined by the covalent structures of the smaller substituents from which they are made.

The most important molecules in biology are proteins and the nucleic acids deoxyribonucleic acid (DNA) and ribonucleic acid (RNA); all are macromolecules characterized by their very large sizes and high molecular weights. These giant molecules can contain many thousands, millions, even billions, of atoms. Fortunately, these macromolecules are **polymers**, produced by linking together in a linear fashion only a few relatively simple monomers of known structure: four nucleotides in the case of DNA and RNA, and 20 amino acids in the case of proteins. In each of these cases, each residue, *i*, of the chain consists of two parts: group X comprises the backbone and is the constant, repeating part of the polymer, while the side-chains (A, B, C, …) attached to the backbone are variable:

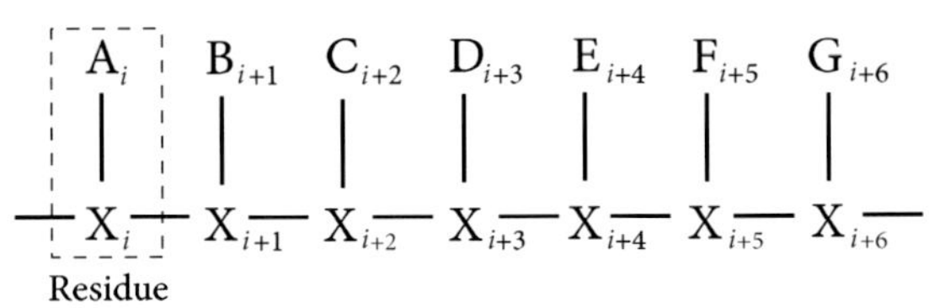

(1.1)

The side-chains connected to the backbone are all the same in **homopolymers**, as in carbohydrates or polymers made chemically, but they are variable in **copolymers**; the natural proteins and nucleic acids are extreme examples with several different types. Normally the individual residues are indexed from 1 to n, starting from one end of the polymer chain and finishing at the other. The bonds between the residues are numbered similarly, with bond i joining residues i and $i + 1$; there are then $n - 1$

bonds linking the *n* residues. Normally the backbone primarily has a structural role, while the side-chains contain the functional groups. In spite of the enormous sizes of proteins and nucleic acids, it is possible to determine their detailed covalent structures because, knowing the detailed structures of all the possible monomers, it is necessary only to determine their linear sequence in the polymer.

The detailed structures of the monomers are extremely important, because they determine the global properties of the macromolecule. They occur many times in the polymer, and their structures are multiplied many times over. Many of the monomers occur in only one of several possible isomers; for example natural proteins are composed solely of L-amino acids and nucleic acids of D-ribose or D-deoxyribose. While these details of the structure might seem very minor and mundane, they have extremely important consequences for the three-dimensional (3-D) structures of biopolymers and their functions. These consequences even extend to the macroscopic level; for example, the left/right asymmetry of all but the simplest microorganisms is believed to result from asymmetry at the atomic level of certain molecules.

Biopolymers. A. G. Walton & J. Blackwell (1973) Academic Press, NY.

An Introduction to Macromolecules. L. Mandelkern (1983) Springer-Verlag, NY.

Introduction to Macromolecular Science. P. Munk (1989) Wiley-Interscience, NY.

Advanced Organic Chemistry, 2nd edn. J. March (2000) Wiley-Interscience, NY.

Virtual exploration of the chemical universe up to 11 atoms of C, N, O, F: assembly of 26.4 million structures (110.9 million stereoisomers) and analysis for new ring systems, stereochemistry, physicochemical properties, compound classes, and drug discovery. T. Fink & J. L. Reymond (2007) *J. Chem. Inf. Model.* **47**, 342–353.

1.1. STEREOCHEMISTRY

Isomers are two molecules that share the same elemental formula but have different structures. A simple example is ethanol and dimethyl ether:

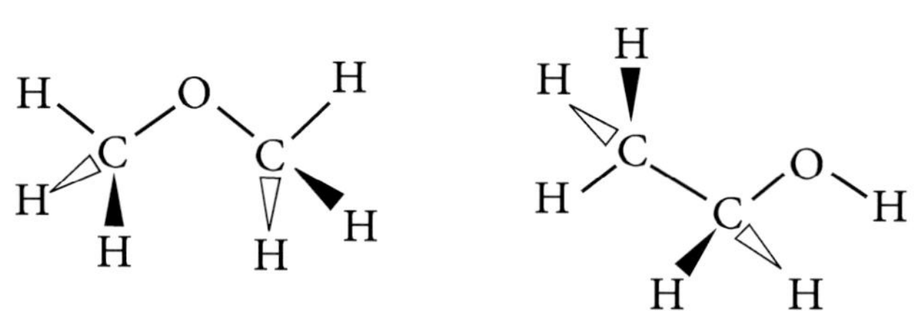

Dimethyl ether Ethanol (1.2)

They have the same elemental formula C_2H_6O but different covalent structures. Their 3-D structures are indicated here by solid tapered bonds that project above the plane of the paper, while open tapered bonds project below the plane. **Structural isomers like these cannot be interconverted without breaking chemical bonds.** They can differ dramatically in their 3-D structures and in their chemical, physical and biological properties. The potential for different isomers increases dramatically with the size of the molecule and is especially acute with macromolecules.

Isomers can differ in various ways. Ethanol and dimethyl ether (Equation 1.2) are **structural isomers** because the number and type of bonds linking the atoms are different. **Geometric isomers** differ in their geometrical arrangement of bonds:

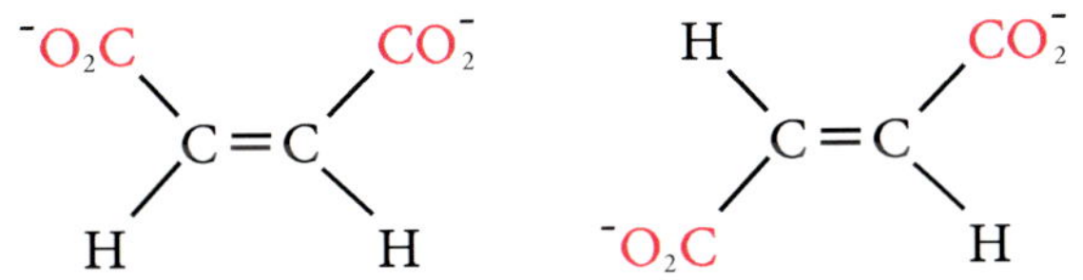

Maleate Fumarate (1.3)

Fumarate and maleate are geometric isomers because they differ only in the rotation about the double bond. Double and triple bonds are not readily rotated, so such geometric isomers are not readily interconverted. The double bond makes these molecules planar, with all the C and H atoms in the plane of the paper. The two carboxyl groups are highlighted to emphasize that they are on the same side in maleate but on opposite sides in fumarate; they can be said to be *cis* and *trans*, respectively (Section 1.1.A.5).

Two molecules are **stereoisomers** if they differ only in the spatial orientation of those atoms that cannot be rapidly interconverted by rotation about single bonds. Stereoisomers contain the same number and type of bonds and have the same chemical name, except for a prefix (e.g. D or L) that is sometimes used to discriminate between them. Stereoisomers are divided into **enantiomers**, molecules with nonsuperimposable mirror images (Section 1.1.A.1), and **diastereomers**, which comprise all other types of stereoisomers (Section 1.1.A.3). **Tautomers** are a specialized class of isomers that are distinguished by their ability to equilibrate rapidly (Section 1.1.C).

Stereoisomers, diastereomers and enantiomers are specialized types of isomers, in which the differences in the molecules are due solely to the **configuration**, which identifies the spatial arrangement of atoms within the structure of a molecule. Configurations are interconverted only by altering the chemical bonds between atoms, and they should be distinguished from **conformations** (Section 1.2), which maintain all covalent bonds and differ only in rotations about single bonds. The two terms configuration and conformation should not be confused.

Complete relative stereochemistry of multiple stereocenters using only residual dipolar couplings. J. Yan *et al.* (2004) *J. Am. Chem. Soc.* **126**, 5008–5017.

Isolation of isomers based on hydrogen/deuterium exchange in the gas phase. U. Mazurek *et al.* (2004) *Eur. J. Mass Spectrom.* **10**, 755–758.

The evolution of stereochemistry. H. D. Arndt (2006) *Angew. Chem. Int. Ed. Engl.* **45**, 4542–4543.

Mechanistic inferences from stereochemistry. I. A. Rose (2006) *J. Biol. Chem.* **281**, 6117–6119.

1.1.A. Chirality

A structure is **chiral** if it cannot be superimposed on its mirror image; examples are our left and right hands. The nonsuperimposable mirror-image isomers are **enantiomers**. The most common structural feature of chiral molecules is the presence of a tetrahedral atom, such as C, with four different substituents, which is by definition a **chiral center**. There are two different ways to arrange four different substituents around a tetrahedral atom, and they are mirror images. Chiral centers include the hydroxymethylene (–HCOH–) carbons of carbohydrates (Equation 1.6) and the C^{α} atoms of the α-amino acids:

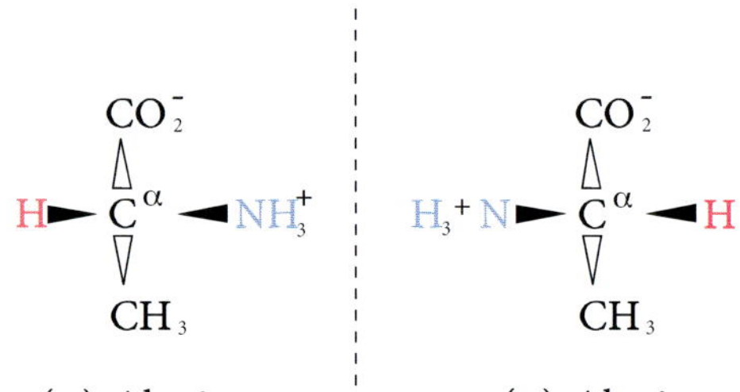

D-(-)-Alanine L-(+)-Alanine (1.4)

Mirror plane

The exceptional amino acid is glycine, which has two H atoms bonded to the C^{α} atom. The L and D designate the absolute configuration of the chiral center, while the + and – indicate in which direction an aqueous solution of the enantiomer will rotate the plane of polarized light. Enantiomers can be distinguished because they interact differently with polarized light; this is the basis of optical rotation and circular dichroism.

A tetrahedral chiral center is not required for chirality, neither does the presence of a chiral center require that the molecule be chiral. A molecule with an internal mirror plane may have chiral centers but not be chiral; these are **meso** compounds. Because of their internal symmetry, these molecules can be superimposed on their mirror image. For example, the meso form of tartaric acid, HO_2C–CHOH–HOCH–CO_2H, has two chiral centers, at each of the two middle C atoms, but there is a mirror plane between them:

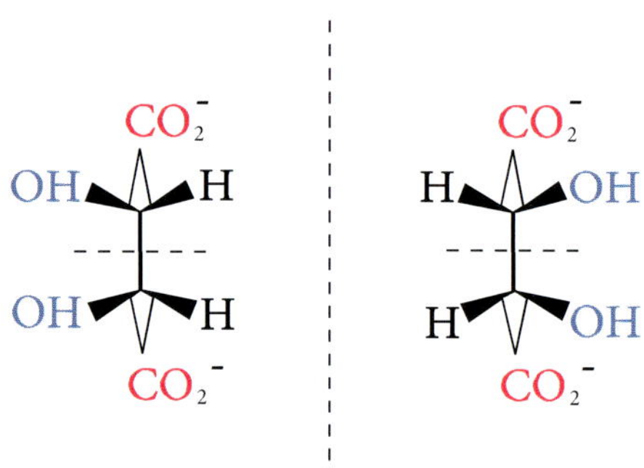

(1.5)

Mirror planes

Consequently, these mirror images can be superimposed by a rotation of 180°, unlike the two isomers of alanine (Equation 1.4).

Supramolecular chirality of self-assembled systems in solution. M. A. Mateos-Timoneda *et al.* (2004) *Chem. Soc. Rev.* **33**, 363–372.

Stereolabile chiral compounds: analysis by dynamic chromatography and stopped-flow methods. C. Wolf (2005) *Chem. Soc. Rev.* **34**, 595–608.

Nonlinear optical spectroscopy of chiral molecules. P. Fischer & F. Hache (2005) *Chirality* **17**, 421–437.

A novel spectroscopic probe for molecular chirality. N. Ji & Y. R. Shen (2006) *Chirality* **18**, 146–158.

Absolute configuration of chirally deuterated neopentane. J. Haesler *et al.* (2007) *Nature* **446**, 526–529.

1. Enantiomers

Enantiomers are recognized most readily by determining whether each chiral center could have an opposite configuration (Equation 1.4). Enantiomers interact identically with achiral compounds, but they can interact differently with other chiral objects. The physical property that differentiates enantiomers is the direction in which they rotate plane-polarized light, as in optical rotatory dispersion and circular dichroism. Thus, two enantiomers are differentiated as either dextrorotatory (+) or levorotatory (–), depending on whether the rotation of the polarized light is clockwise or counterclockwise, respectively (Equation 1.4). Solutions containing an excess of either of the enantiomers rotate polarized light and are said to be **optically active**.

Fischer introduced a general procedure that designated enantiomers as either D- or L- based on whether the nonhydrogen substituent was on the right or left when the molecule was drawn as a **Fischer projection**:

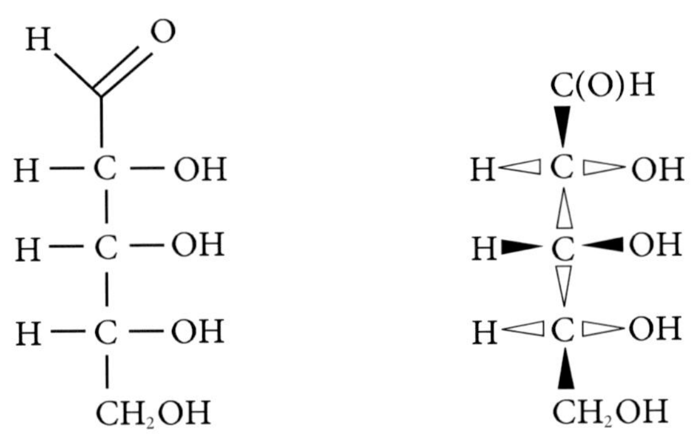

Fischer projection   Stereochemical drawing   (1.6)

By convention the Fischer projection has the vertical bonds directed away from the viewer and the horizontal bonds directed out towards the viewer. In the stereochemical drawing, the solid tapered bonds project out from the plane of the paper, whereas those that are open project below. The example of Equation 1.6 is D-ribose. The configuration of each chiral carbon in D-ribose is D because the nonhydrogen substituent is drawn to the right in the Fischer projection. For sugars, the enantiomer is defined by the bottom chiral carbon when the carbon chain is oriented vertically with the carbonyl carbon at the top; thus the ribose depicted in Equation 1.6 is D.

The Fischer nomenclature leads, however, to ambiguities. A more rigorous and unambiguous method of identifying configuration for specifying absolute configuration of chiral tetrahedral centers as either R or S (from the Latin 'rectus' and 'sinister', respectively) is generally accepted. The procedure requires the assignment of priority to the four substituents that generate a **chiral center**, followed by a procedure to identify the arrangement as either R or S. There are four rules for assigning the priorities of substituents bonded to the same atom.

(1) The priority is assigned in order of decreasing atomic number of the four atoms directly bonded to the chiral center.

(2) When two or more atoms cannot be distinguished by step 1, the atoms bonded to each of the atoms with equal priority are given their own priorities. If the atoms of greatest priority do not differentiate the two groups, the second atoms are compared, then the third.

(3) Heavier isotopes are given priority over lighter isotopes; for example ^{2}H is given precedence over ^{1}H and ^{14}C over ^{12}C.

(4) Double bonds count as two bonds to the same atom.

The relative priorities of the four substituents are assigned in step 1 and labeled 1 to 4:

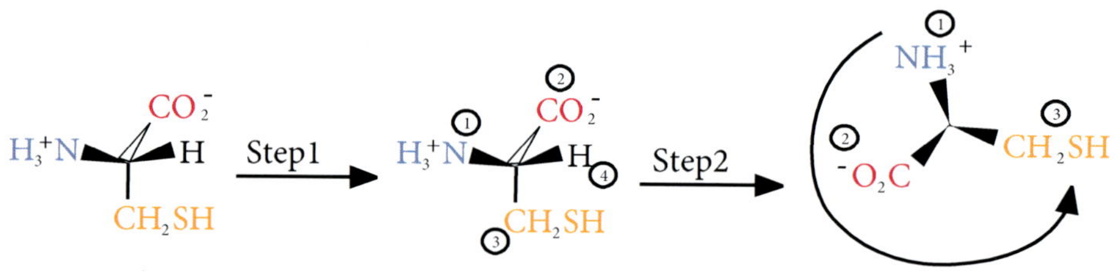

L-Cysteine (S)-Cysteine (1.7)

The carboxylate carbon here takes precedence over that with the thiol group because it has three bonds to oxygen. In step 2, the bond to the lowest priority group is oriented directly away from the observer, and the remaining three groups are viewed from above. If the arrangement of highest to lowest priority of the three groups is clockwise, the chiral center is assigned the R configuration; if counterclockwise it has the S configuration. The assignment of the S configuration in Equation 1.7 results from the counterclockwise arc that connects the substituents in order of precedence.

Determination of the interconversion energy barrier of enantiomers by separation methods. J. Krupcik *et al.* (2003) *J. Chromatogr. A* **1000**, 779–800.

A simple method to determine concentration of enantiomers in enzyme-catalyzed kinetic resolution. R. C. Zheng *et al.* (2007) *Biotechnol. Lett.* **29**, 1087–1091.

Molecular quantum similarity and chirality: enantiomers with two asymmetric centra. S. Janssens *et al.* (2007) *J. Phys. Chem.* **111**, 3143–3151.

---

2. Racemic mixtures

A mixture with equal amounts of two enantiomers of a compound is known as a **racemic mixture** or **racemate**. Two enantiomers have identical free energies when in a homogeneous environment, such as in solution, so when a chiral compound is formed by the chemical reaction of two nonchiral reactants, the product will be a racemic mixture. The exception is when the reaction involves an asymmetric catalyst, such as an enzyme. Any process that catalyzes the interconversion of enantiomers (i.e. a **racemization**) will necessarily result in a racemic mixture being formed. In the absence of any other chiral compound, the free energies of formation of two enantiomers must be identical.

Racemic macromolecules for use in X-ray crystallography. J. M. Berg & L. E. Zawadzke (1994) *Curr. Opinion Biotechnol.* **5**, 343–345.

Enantiomer-selective activation of racemic catalysts. K. Mikami *et al.* (2000) *Acc. Chem. Res.* **33**, 391–401.

Advances in chiral separation using capillary electromigration techniques. G. Gubitz & M. G. Schmid (2007) *Electrophoresis* **28**, 114–126.

---

3. Diastereomers

Diastereomers are stereoisomers that are not enantiomers. There are several common types. Molecules containing carbon–carbon and carbon–nitrogen double bonds will exist as two different geometric diastereomers if both of the double-bonded atoms have two different substituents (Equation 1.3). These diastereomers are differentiated with the designations *cis* and *trans* (Section 1.1.A.5).

The other common form of diastereomer occurs when a molecule contains more than one chiral center, with one having the same configuration in the two molecules but not the other:

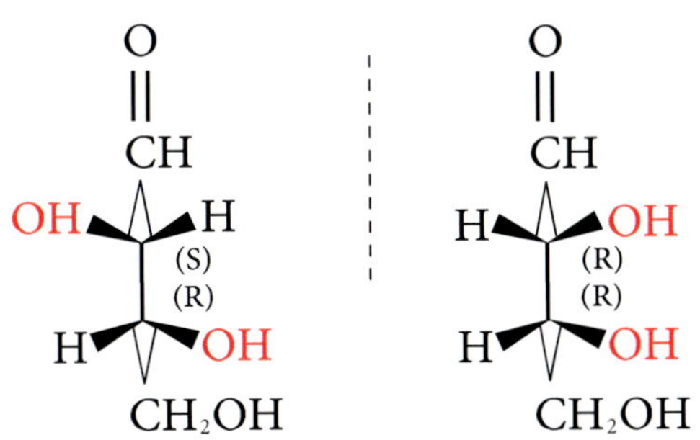

D-threose D-erythrose (1.8)

The two diasteromers are not mirror images. **Diastereomers have different chemical properties**.

Two atoms are considered to be **diastereotopic** if they would generate different diastereomers upon substitution with a different isotope (Section 1.1.B). Diastereotopic atoms are chemically different and can be distinguished by nuclear magnetic resonance (NMR).

Diastereoisomerism, contact points, and chiral selectivity: a four-site saga. R. Bentley (2003) *Arch. Biochem. Biophys.* **414**, 1–12.

Evaluation of experimental strategies for the development of chiral chromatographic methods based on diastereomer formation. N. R. Srinivas (2004) *Biomed. Chromatogr.* **18**, 207–233.

Total chemical synthesis and X-ray crystal structure of a protein diastereomer: [D-Gln 35]ubiquitin. D. Bang *et al.* (2005) *Angew. Chem. Int. Ed. Engl.* **44**, 3852–3856.

Crystallization-induced diastereomer transformations. K. M. Brands & A. J. Davies (2006) *Chem. Rev.* **106**, 2711–2733.

4. Epimers and Epimerization

Diastereomers related by the inversion of configuration at a *single* chiral center are known as epimers. This definition excludes enantiomers such as D- and L-alanine (Equation 1.4) because they are not diastereomers. It also excludes diastereomers that are related by inversion of more than a single chiral center. For example, D-glucose and D-galactose are epimers, as are D-glucose and D-mannose, but galactose and mannose are not:

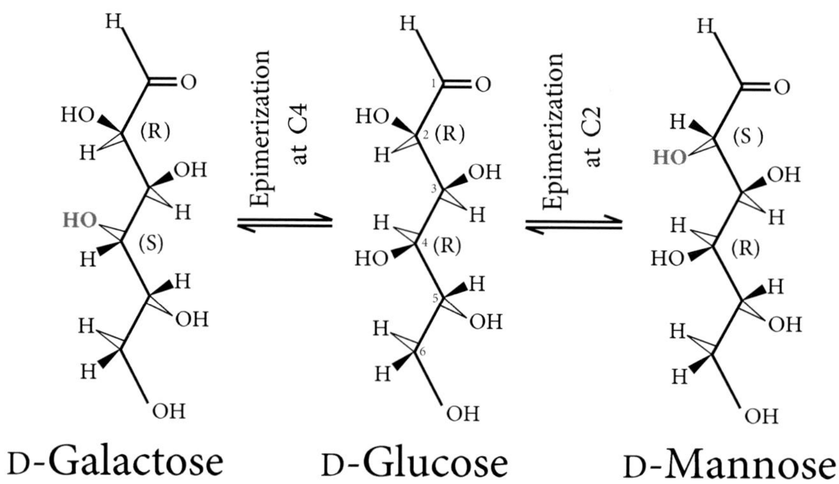

(1.9)

The configurations at C2 and C4 are labeled and distinguish these three sugars. Glucose is an epimer of both mannose and galactose because it differs from each by the configuration of a single chiral center. Mannose and galactose have different configurations at both C2 and C4 and therefore are not epimers.

The chemical conversion of one epimer to another is called **epimerization**; it contrasts with the racemization of enantiomers.

Mechanistic aspects of enzymatic carbohydrate epimerization. J. Samuel & M. E. Tanner (2002) *Nat. Prod. Rep.* **19**, 261–277.

Understanding nature's strategies for enzyme-catalyzed racemization and epimerization. M. E. Tanner (2002) *Acc. Chem. Res.* **35**, 237–246.

5. Cis *and* Trans *Isomers*

Molecules containing carbon–carbon and carbon–nitrogen double bonds can exist as two different geometric diastereomers if both of the double-bonded atoms have two different substituents. These diastereomers are differentiated with the designation *cis* and *trans*, or more formally as *Z* and *E* (from the German 'zusammen' for together and 'entgegen' for opposite). When similar substituents, or those with the highest priority, are on the same side of the ring or double bond, the configuration is referred to as *cis* or *Z*; when they are on opposite sides, they are referred to as *trans* or *E*. For example, maleate is the *cis* form of fumarate (Equation 1.3). In potentially ambiguous cases, the priority of the substituents of each double-bonded atom is determined by the rules for deciding priorities (Section 1.1.A.1).

The conformation about a single bond may also be noted as *cis* or *trans*, particularly when all of the substituents lie in a plane. For example, the highlighted C2 and C3 hydroxyl groups of the furanose ring form of ribose are *cis:*

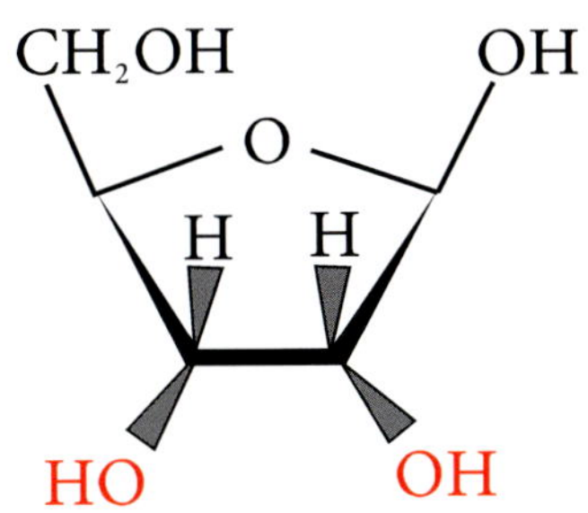

(1.10)

The *cis* conformation of a peptide bond has both C^{α} on the same side of the C–N amide bond:

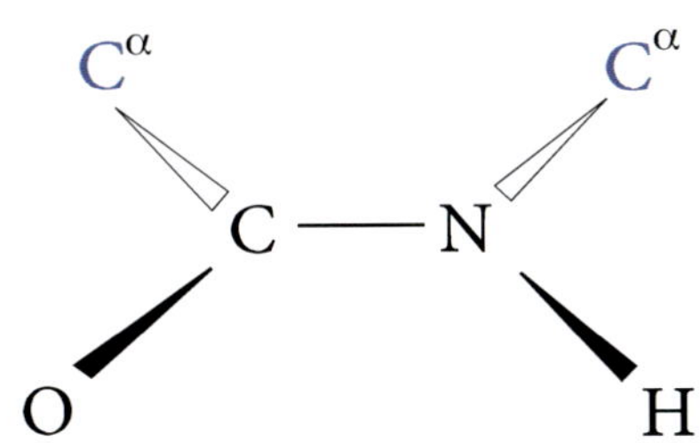

(1.11)

The peptide bond has partial double-bond character (Section 8.1), which tends to keep the four atoms planar. The notation *s-cis* is used to emphasize that the conformation about a single bond is being described, as in a *s*-1-*cis* long-chain aldehyde:

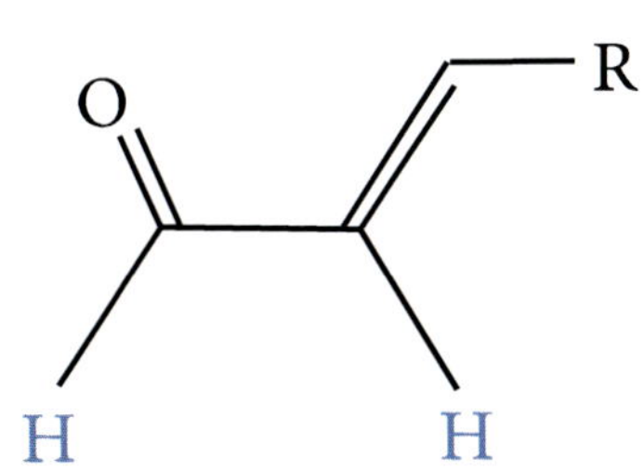

(1.12)

An example would be retinal.

Cis and *trans* isomers can often be interconverted by rotation about the central linkage, but at widely varying rates. The bond order of the central linkage correlates with the magnitude of the energy barrier to rotation. It ranges from high values (slow rotations) for double bonds, via an intermediate range for linkages having a partial double-bond character, down to low values (rapid rotation) for C–C single bonds. The magnitude of the rotational barrier determines whether or not the individual isomers are readily interconverted; if readily interconverted, they are considered different conformations; otherwise, they are different configurations. There is the potential for confusion in such instances unless the time scale is specified.

In molecular biology, the terms *cis* and *trans* are also often used to refer to genetic elements on the same or different nucleic acid molecules.

1.1.B. Prochiral

The designation prochiral indicates that substitution of an atom with a different isotope will alter the chirality of the molecule. A **prochiral center** is an atom with two identical substituents where substitution of either one with a different isotope would make that atom a chiral center. If this would generate enantiomers, the groups are **enantiotopic**, while if the substitution would produce diastereomers, the groups are **diastereotopic** (Section 1.1.A.3).

For example, the C^{α} atom in the amino acid glycine:

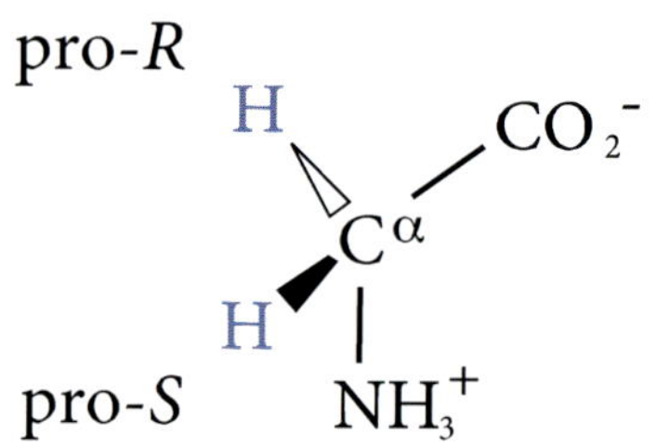

(1.13)

is a prochiral center because substitution of either of the two nearly identical H^{α} atoms with a heavier isotope would generate a chiral molecule. The two H^{α} atoms are enantiotopic because the replacement of one would generate different enantiomers of glycine. Enantiotopic groups are distinguished by using the prefix 'pro-' with the specification of the configuration generated. In chemical structures, the prochirality of an atom can be identified by R or S. The forward H atom of glycine is pro-S because substitution with ^{2}H there would generate (S)-[2-^{2}H]glycine, as indicated by the counterclockwise arc connecting the atoms in order of priority. In the peptide glycyl-L-serine, where the serine residue has a chiral center, the glycine H^{α} atoms are diastereotopic because substitution of either glycine H would generate diastereomers.

In the case of citrate:

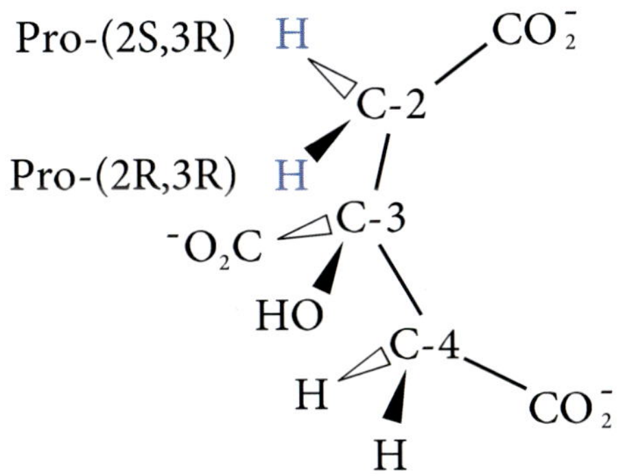

(1.14)

the H_2C groups 2 and 4 are enantiotopic; C3 is a prochiral center because substitution of any of the H atoms or carboxyl group attached to either C2 or C4 will generate chirality there. The upper (C2) of $-CH_2-COO^-$ is the pro-R group because substituting an isotope anywhere in the functional group would generate (3R)-citrate. C2 and C4 are also prochiral centers, because substituting either methylene H atom would generate a chiral center. The two H atoms on C2 are diastereotopic, because substituting deuterium at either of these positions would generate diastereomers; two new chiral centers would be generated at C2 and C3. The chirality at C3 would be the same for both H atoms, but different at C2.

Double-bonded C and N atoms that are planar have three substituents. They will be prochiral if the three substituents are different, because the two faces of the planar molecule are not equivalent. The faces of the molecule are differentiated by the designation of Re or si based on the priorities of the three substituents according to the rules of priority for substituents (Section 1.1.A.1). The two faces of pyruvate are not equivalent:

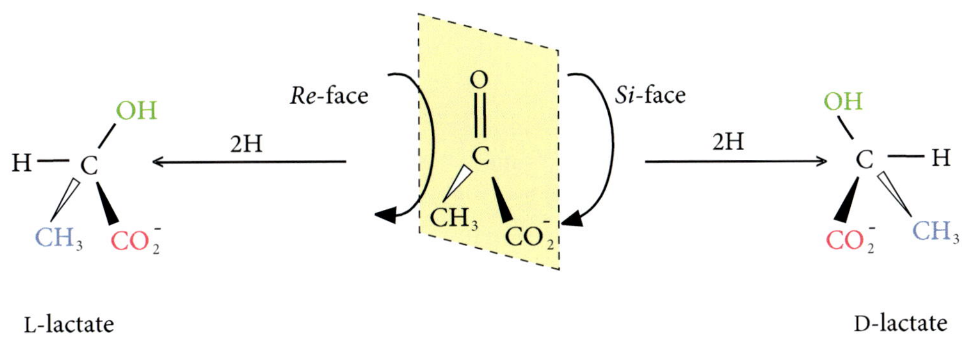

(1.15)

The circle connecting the three different substituents in their priority order describes a clockwise motion when viewed from the left, but a counterclockwise motion when viewed from the right. The two faces of pyruvate are enantiotopic because addition of chemical groups to the opposite sides of the central C atom would generate enantiomers. As shown in Equation 1.15, addition of 2 H atoms to the *Re* face would generate L-lactate, whereas addition to the opposite face would generate D-lactate. The faces would be diastereotopic if addition to the opposite faces would generate diastereomers.

The concept of enantiotopic and diastereotopic groups is important in molecular biology because these groups interact differently with chiral molecules, such as proteins.

Prochirality revisited. An approach for restructuring stereochemistry by novel terminology. S. Fujita (2002) *J. Org. Chem.* **67**, 6055–6063.

Analysis of the ^{13}C NMR spectra of molecules, chiral by isotopic substitution, dissolved in a chiral oriented environment: towards the absolute assignment of the pro-R/pro-S character of enantiotopic ligands in prochiral molecules. P. Lesot *et al.* (2004) *Chemistry* **10**, 3741–3746.

1.1.C. Tautomers

Tautomers are structural isomers that interconvert rapidly, so all the isomers will exist in a solution of the compound at equilibrium. In most cases, tautomers are generated by structural and electronic rearrangements caused by moving a single proton. This can occur via the solvent, with one water molecule or hydroxide ion removing the proton and another water molecule donating a proton. For example, glyceraldehyde-3-phosphate and dihydroxyacetone-phosphate are both tautomers of the enediol intermediate in their interconversion:

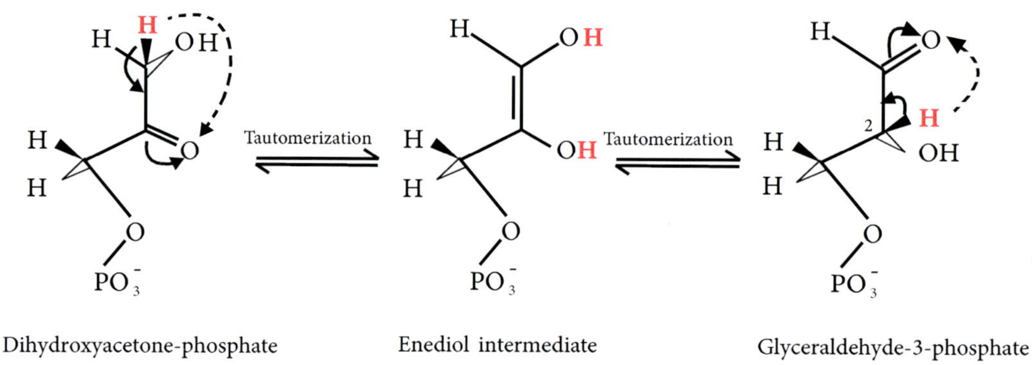

(1.16)

Glyceraldehyde-3-phosphate and the enediol are structural isomers that result when a proton is transferred from C2 to the aldehyde oxygen, as indicated by the dashed arrow; transfer of the proton is accompanied by a rearrangement of the electrons, as indicated by the solid curved arrows. Similarly, dihydroxyacetone-phosphate and the enediol intermediate are tautomers, as a single proton transfer can also cause their interconversion. The two H atoms that move in each tautomerization of Equation 1.16 are indicated in red.

The phrase 'rapidly' is not precise, so two molecules that are each tautomers of a third molecule may not be tautomers. For example, in Equation 1.16 glyceraldehyde-3-phosphate and dihydroxyacetone-phosphate are not interconverted rapidly and are not considered to be tautomers, even though both are considered tautomers of the enediol intermediate.

Another form of tautomerism that is important in molecular biology is the existence of both open-chain and ring forms of monosaccharides. Thus the three common forms of glucose

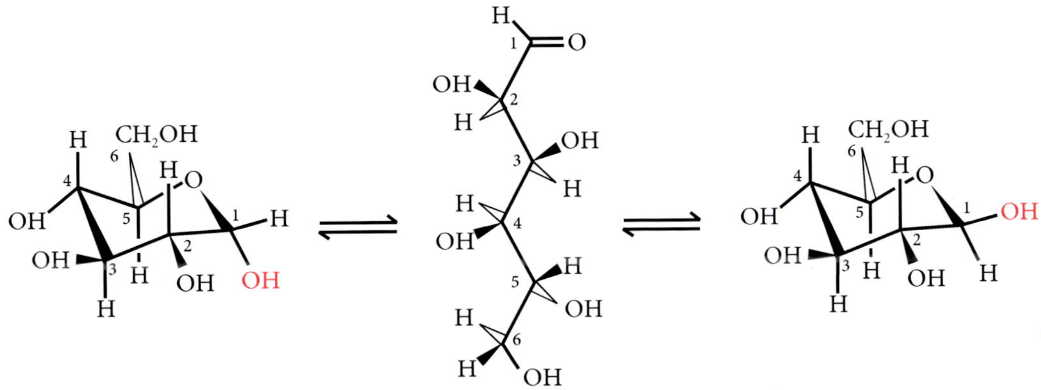

(1.17)

α–gluco-pyranose Aldose β-gluco-pyranose

are all tautomers because they interconvert rapidly in aqueous solution. The two cyclic pyranose forms are also epimers and diastereomers (Section 1.1.A).

The existence of tautomers can be detected spectroscopically if each tautomeric form gives rise to different spectral features. For example, resonances from all three tautomeric forms of glucose (Equation 1.17) are apparent simultaneously by proton NMR (^{1}H-NMR). One tautomer can predominate if it is stabilized in some way, such as by incorporating it into a crystal lattice or into a larger molecule. The equilibration of the tautomers in Equation 1.17 in aqueous solution can be monitored using the change in optical rotation after dissolving a crystal of either of the pure pyranose forms in water.

Often one tautomeric form is more stable than the other in solution, but the less-favored tautomer might be the biologically active form. Identifying the correct tautomers of the nucleobases of nucleic acids was a crucial step in discovering the double-helical structure of DNA, as it governs the pairing of bases by hydrogen bonding (Section 2.2.A). Keto-enol tautomerism, comparable to that at C1 in Equation 1.17, can change a carbonyl group, which is a hydrogen-bond acceptor, to a hydroxyl, which can be a hydrogen-bond donor. Tautomeric forms of all five nucleic acid bases (Figure 2-9) exist in solution but the rare tautomer occurs $<10^{-4}$ of the time.

Tautomerism of sterically hindered Schiff bases. Deuterium isotope effects on ^{13}C chemical shifts. A. Filarowski *et al.* (2005) *J. Phys. Chem. A* **109**, 4464–4473.

Differential solvation and tautomer stability of a model base pair within the minor and major grooves of DNA. F. Y. Dupradeau *et al.* (2005) *J. Am. Chem. Soc.* **127**, 15612–15617.

Hydrogen-bonded nucleic acid base pairs containing unusual base tautomers: complete basis set calculations at the MP2 and CCSD(T) levels. J. Rejnek & P. Hobza (2007) *J. Phys. Chem. B* **111**, 641–645.

1.2. CONFORMATIONS

Different conformations are nonidentical spatial arrangements of the atoms of a molecule achieved solely by rotations about single covalent bonds. Molecules with identical covalent structures but different conformations are known as **conformers**. The ability of two conformers to assume the same geometry spontaneously and become identical differentiates them from stereoisomers. The terms 'conformers' and 'stereoisomers' should not be confused, nor should 'configuration' and 'conformation'.

Conformers are usually interconverted rapidly because single bonds can usually rotate rapidly. Unless certain conformers are stabilized specifically, it is difficult, if not impossible, to isolate individual conformers. Any slow bond rotations will provide exceptions to this rule; one is rotation about the C–N peptide bond of polypeptides (Equation 1.11). It has partial double-bond character, so it is planar and the two possible *cis* and *trans* conformations are only interconverted slowly.

Any conformation of a molecule of known covalent structure can be specified by the rotations about its single bonds, generally measured by either the **torsion angle** (Section 1.2.A) or the **dihedral angle** (Section 1.2.B). In general, interactions between neighboring atoms, usually steric repulsions, mean that not all bond rotations have the same free energy and are equally probable. Ethylene glycol can serve as a simple model:

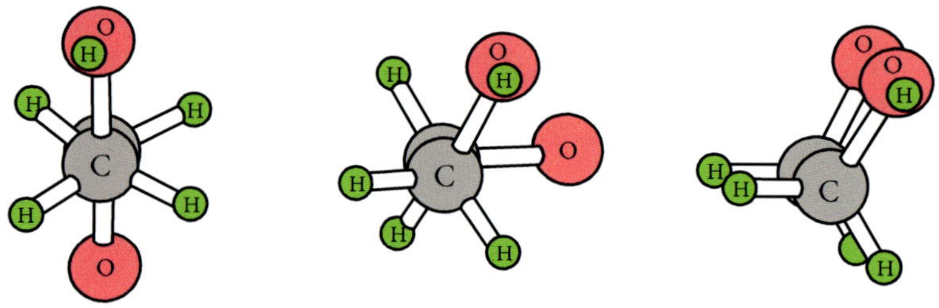

anti or *anti*periplanar — *gauche* or *syn* clinal — *eclipsed* or *syn* periplanar (1.18)

Individual torsion angles may be described qualitatively as ***anti***, ***gauche*** or ***eclipsed*** (Section 1.2.A.1), which are more formally named ***anti*periplanar**, ***syn*clinal** and ***syn*periplanar**. The term *gauche* generally means not lying in a plane, while *anti* is used to describe the relative orientation of two substituents on adjacent atoms of a molecule when their torsion angle is about ±180°. When the large substituents (hydroxyl groups in this case) are superimposed, the conformation is described as *eclipsed*; steric clashes between the two hydroxyl groups mean that this conformation is the least stable. When they are staggered by 60°, the conformation is *gauche*. When they are opposite, staggered by 180°, it is *trans* or *anti*; this conformation would have the fewest clashes between the hydroxyl groups and be the most stable. With constrained systems, such as cyclic molecules, certain common combinations of torsion angles can lead to descriptive names for groups of atoms; for example, the 2′-*endo* conformation of ribose describes all five of the torsion angles of the furanose ring (Figure 2-5).

A large biological macromolecule can have a stable, fixed 3-D structure, referred to as its **native conformation**, if it has sufficient stabilizing interactions between its various atoms. Although most of the conformation is fixed and it is considered a single conformation, parts of the molecule may still be flexible and able to undergo rotations about certain bonds. The average overall conformation can be considered a **macro-conformation**, whereas the variations resulting from flexibility define various **micro-conformations**. Interconverting different macro-conformations requires a cooperative change of a number of bond rotations simultaneously, whereas micro-conformations are interconverted by changes of just one or a few bond rotations. A cooperative change of a number of bond rotation angles, such as occurs in protein unfolding or the melting of double-stranded DNA (Chapter 5), will usually occur only slowly or infrequently under physiological conditions, although it can be speeded up dramatically under denaturing conditions. This relatively slow interconversion of the two macro-conformations is described as a **conformational change**, in which the macromolecule has been converted from one family of micro-conformations to an experimentally distinguishable family of other micro-conformations.

Conformational analysis. E. L. Eliel *et al.* (1967) Wiley-Interscience, NY.

1.2.A. Torsion Angle

Torsion angles within a molecule refer to the rotations about individual covalent bonds linking a pair of atoms; they are defined using two further atoms bonded to the first pair of atoms. Therefore, four atoms connected by three consecutive covalent bonds are used to define the torsion angle of the middle bond:

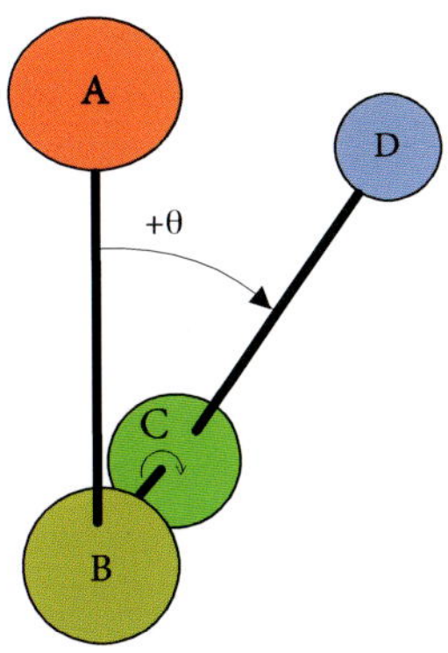

(1.19)

The torsion angle of the bond connecting atoms B and C is determined by looking down this bond and measuring the angle that the bond between atoms A and B must be rotated through to eclipse the bond between atoms C and D. Torsion angles are usually expressed as having values between −180° and +180°; the value is zero when the flanking bonds are eclipsed, and positive if the front bond is rotated in a clockwise direction. For biological macromolecules, the backbone torsion angles are defined by four contiguous backbone atoms.

Torsion angles about single bonds may also be described qualitatively. They are distinguished as being (a) positive or negative, (b) syn or anti and (c) periplanar or clinal:

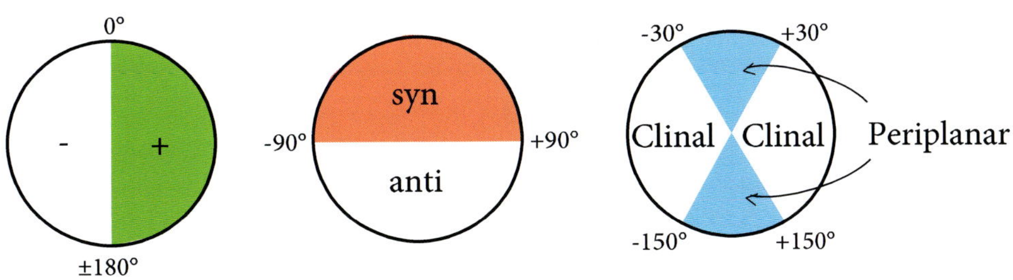

(1.20)

These define eight different rotamers that are indicated with their two-letter abbreviations:

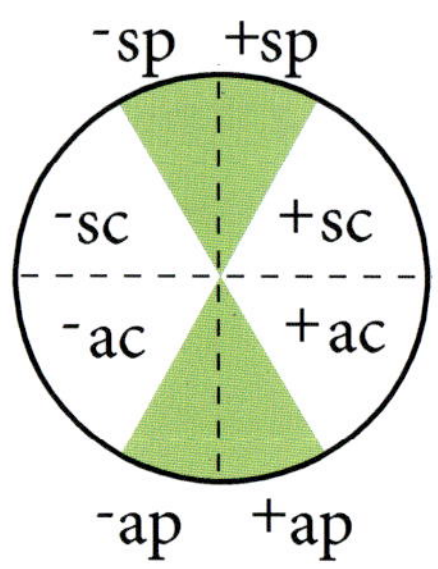

(1.21)

Torsion angles are important because they are commonly used to define the conformations of molecules and any differences between them.

Application of torsion angle molecular dynamics for efficient sampling of protein conformations. J. Chen *et al.* (2005) *J. Comput. Chem.* **26**, 1565–1578.

On updating torsion angles of molecular conformations. V. Choi (2006) *J. Chem. Inf. Model.* **46**, 438–444.

1.2.B. Dihedral Angle

A dihedral angle is the angle formed by intersecting lines normal to two planes:

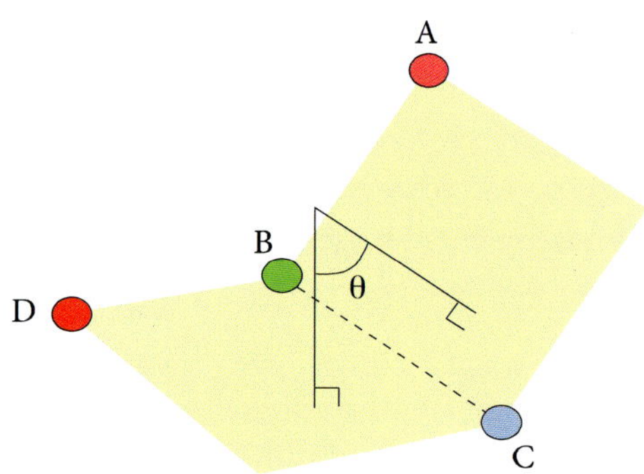

(1.22)

In this example, one plane is defined by atoms A, B and C, the other by atoms B, C and D. The dihedral angle θ is defined as the angle between the two lines perpendicular to each plane. The preferred nomenclature for rotations about single bonds in a molecule uses the conventions described for torsion angles. Unlike torsion angles, however, the dihedral angle can be measured between the two planes defined by any group of three atoms, whether or not they are bonded together. When four atoms are specified in order, such as atoms A, B, C and D, the dihedral angle is measured between the plane containing atoms A, B and C and the plane containing atoms B, C and D. When the four atoms are not linked by consecutive covalent bonds, and a torsion angle can not be defined, this dihedral angle is known as an **improper dihedral angle** or **improper torsion angle**.

Straightening out the dihedral angles. J. Clauwaert & J. Z. Xia (1993) *Trends Biochem. Sci.* **18**, 317–318.

Pairwise NMR experiments for the determination of protein backbone dihedral angle ϕ based on cross-correlated spin relaxation. H. Takahashi & I. Shimada (2007) *J. Biomol. NMR* **37**, 179–185.

1.3. CONFORMATIONS OF IDEALIZED POLYMERS

The conformations adopted by a polymer will depend on the structures and conformational properties of the monomers, the way that they are covalently linked together, and the environment, especially the relative interactions of the polymer with the solvent and with itself. In the simplest case, **where each monomeric unit is the same and each adopts only a single conformation, the linear polymer will adopt a helical conformation** (Figure 1-1). A helix is defined as a point that rotates at a given distance around an axis *z* while moving parallel to that axis. In a helical macromolecule, the helix will be characterized by the angle α and the translation *p* along axis *z* between adjacent monomers. It can be specified by the **helical repeat**, the number of monomers per turn (which need not be an integer), and the **pitch** of the helix, the vertical distance between adjacent turns. Whether the helix is left- or right-handed is determined by the sense of the rotation needed to advance along *z*: the helix should be considered a screw that needs to be advanced by turning it with a screwdriver. If the direction of rotation is that indicated by the fingers of the right hand, the helix is right-handed. Otherwise it is left-handed. Note that a helical repeat of exactly 2 does not produce a helical molecule but one that takes a zig-zag path (Figure 1-1C).

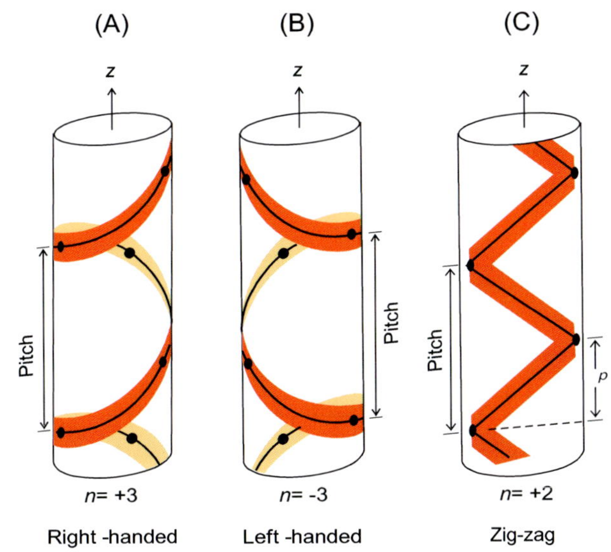

Figure 1-1. Description of helical conformations generated when each monomer of a polymer adopts the same conformation. The two helices in (A) and (B) are identical except for their handedness. Both have three monomers per turn, *n*, so α = 120° and pitch = 3 *p*. If *n* = 2, as in (C), the polymer does not adopt a helical path but is straight and zig-zag, as will be found in the β-strands of proteins (Figure 8-9); α = 180° and pitch = 2 *p*. In each case, the *z*-axis is vertical.

If, on the other hand, each monomer can adopt more than one conformation, the polymeric chain is likely to adopt a wide variety of micro-conformations, unless it is subject to interactions that favor one or more individual conformations over all the others. The 3-D structure will be specified by the torsion angles (Section 1.2.A) adopted by each monomer.

The ability to adopt a number of conformations is an entropic factor that stabilizes the flexible state, known as the **conformational entropy**. A single conformation will be adopted only if the interactions stabilizing that particular conformation are sufficiently strong to overcome the conformational entropy tending to keep the polymer unfolded. The number of micro-conformations and the conformational entropies of polymer chains can be very large. For example, if each residue can adopt an average of j conformations, and there are N residues in the polymer chain, the total number of conformations possible will be approximately j^N. If the reasonable assumption is made that j is 8 and N is 500, there will be 10^{452} (= 8^{500}) conformations possible, a truly astronomical number. Of course, some of these conformations will not be feasible because they would have atoms of the polymer overlapping in space, the **excluded volume effect**. There is still, however, much scope to be conservative and predict an astronomical number of conformations. For example, even a short polymer of 100 residues in which each residue could adopt only two different conformations could adopt more than 10^{30} different polymer conformations. If all these conformations have similar free energies, each would have only a very small probability of occurring in a molecule. At 25°C, the free energy contribution of the conformational entropy for a polymer in which each residue can adopt 10 conformations will be 1.36 N kcal/mol. Consequently, for any one conformation to predominate will require, on average, stabilizing interactions >1.36 kcal/mol/residue. In the absence of such stabilizing interactions, a polymer will tend to exist in many different conformations. Yet proteins and nucleic acids, as will be shown in Chapters 2 and 4, are able to adopt single folded conformations that predominate. Such stable conformations can be considered macro-conformations, in contrast to the micro-conformations that are adopted only transiently (Section 1.2).

A disordered polymer will usually have so many possible micro-conformations that not all can possibly be present within a population of molecules at any instant of time. For example, a reasonable sample consisting of 1 μmol of a polymer will contain only 6×10^{17} molecules. Moreover, micro-conformations can be converted no more rapidly than rotations can occur within the polymer backbone, which requires at least 10^{-10} s, so a single molecule can sample no more than 10^{+10} conformations per second, which is likely to be only a small fraction of the conformations possible. Although it may not be possible for all conformations to be present or sampled, very many micro-conformations will be present within a sample of polymer molecules, and only statistical averages of the properties of the population of molecules can be given.

With so many conformations possible, the conformational properties of random polymers are best calculated statistically, using the mathematical procedures developed for synthetic polymers. Such calculations require detailed knowledge of the conformational properties of the monomeric unit of the polymer and the relative energies of all of its possible micro-conformations. Note that with N monomer units, there are only $N - 1$ linkages between them, and the torsion angles of only $N - 2$ such linkages specify the conformation (specifying the other linkage will merely fix the orientation of the molecule in space).

Conformations of Macromolecules. T. M. Birshtein & O. B. Ptitsyn (1964) Wiley-Interscience, NY.

Statistical Mechanics of Chain Molecules. P. J. Flory (1969) John Wiley, NY.

1.3.A. RANDOM COILS

Polymers in which the conformational properties of each residue are independent of the conformations of all other residues, except for those adjacent in the polymer chain, are known as **random coils.**

Frequently the statistical properties are calculated of an ideal **unperturbed random coil**, in which the 3-D covalent structure of the polymer is considered, along with the conformational properties of each monomer unit, but not any interactions between distant parts of the polymer. In the 'unperturbed' state, no account is taken of the excluded volume effect, so impossible conformations in which nonbonded atoms occupy the same space are included. This is unrealistic but makes the calculations more feasible.

The average properties of such polymers are often compared with those of the hypothetical **random-flight chain** or **freely jointed chain**. This is not a realistic model either, but simply a mathematical string of vectors of fixed lengths representing the bonds between adjacent atoms; the atoms are not included, the chain has no volume, all bond angles have equal probability, and all rotations about the bonds are equally likely. A somewhat more realistic model is the **freely rotating chain**, in which a constraint of fixed-bond angles between monomers is introduced. When the actual conformational preferences of the monomer unit are taken into account, by permitting only the most favorable possible torsion angles, the model is known as the **rotational isomeric state** model. In this case, however, the rotations about adjacent bonds are not independent, as the corresponding atoms would interact, so the allowed torsion angles must be those for pairs of neighboring bonds rather than single bonds. Calculations on polymers where each monomer unit contributes more than one covalent bond to the polymer backbone often simplify the architecture by using a single **virtual bond** for each monomer, a vector joining the comparable atoms of adjacent monomeric units. For example, the torsion angles of the three covalent bonds that make up the backbone of one amino acid residue can be replaced by one rotation about a virtual bond linking adjacent C^{α} atoms. In spite of its simplification, the rotational isomeric state model can simulate experimental data reasonably well. Such computations are complex, however, and outside the scope of this volume.

Random-coil behavior and the dimensions of chemically unfolded proteins. J. E. Kohn *et al.* (2004) *Proc. Natl. Acad. Sci. USA* **101**, 12491–12496.

Secondary structures in long compact polymers. R. Oberdorf *et al.* (2006) *Phys. Rev. E* **74**, 051801.

1. End-to-end Distances

Of greatest interest with random polymers are the averages and the variation of their physical dimensions. The **root mean square** (r.m.s.) value of the distance, *r*, between two atoms of a hypothetical random-flight chain is given by:

$$< r^2 >_0^{1/2} = n^{1/2}\, l \tag{1.23}$$

where *n* is the number of linkages between monomers (= *N* – 1, where *N* is the number of monomers in the chain) and *l* is the distance between monomers in the polymer backbone. The angle brackets in Equation 1.23 indicate that it is the average over all conformations, and the subscript zero refers to the unperturbed state. Note that the dimensions of such a random coil increase only with the square root of the number of residues in the polymer chain.

The calculated distribution of end-to-end distances is usually expressed as either the **Gaussian distribution function** or the **radial distribution function**, which are illustrated in Figure 1-2 for a hypothetical random-flight chain. The Gaussian distribution function, W*(x, y, z)* d*x* d*y* d*z*, gives the probability that the end of the polymer chain is within the volume d*x* d*y* d*z* at coordinates (*x*, *y*, *z*); the origin is taken as the other end of the chain. This distribution is spherically symmetrical, so it is usually expressed as the radial distribution function, W(*r*) d*r*, which is the probability that the two ends of the chain are within a distance *r* and *r* + d*r* of each other. For unperturbed random-coil chains, the scale for *r* in random-flight chains is simply increased by the factor $C_n^{1/2}$ (Section 1.3.A.3).

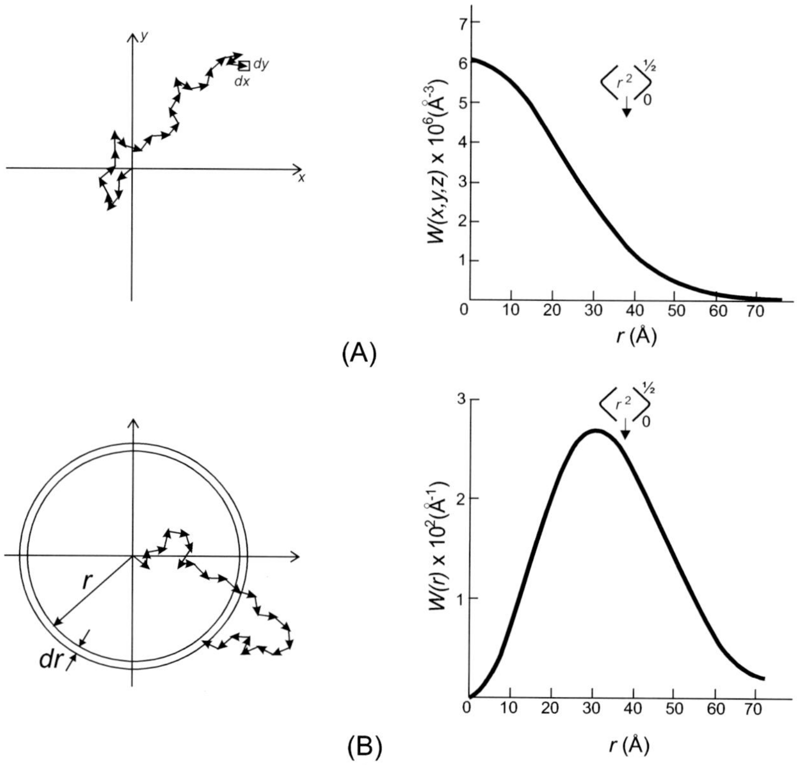

Figure 1-2. Illustration of Gaussian (A) and radial (B) distribution functions for the end-to-end distance of a freely jointed chain. On the *left* of each is a two-dimensional representation of how each distribution function is defined, giving the probability that the other end of the chain will lie within the enclosed area. The distribution functions are given by:

$$W(x,y,z)\,dx\,dy\,dz = \left(\frac{\beta}{\sqrt{\pi}}\right)^3 \exp(-\beta^2 r^2)\,dx\,dy\,dz$$

$$W(r)\,dr = \left(\frac{\beta}{\sqrt{\pi}}\right)^3 \exp(-\beta^2 r^2)\,4\pi r^2\,dr$$

where $r^2 = x^2 + y^2 + z^2$, $\beta = (3/2nl^2)^{1/2}$ and *n* is the number of freely jointed bonds of length *l*. On the *right* of each is the calculated distribution for a freely jointed polypeptide chain of 100 residues and a virtual bond length of 3.8 Å. The r.m.s. distance, $<r^2>_0^{1/2} = 38$ Å is indicated. The probability of the Gaussian distribution function reaches a maximum near the origin, whereas that of the radial distribution function approaches zero. The latter is simply a mathematical consequence of the decreasing volume of the spherical shell between *r* and *r* + d*r* as *r* decreases. From T. E. Creighton (1993) *Proteins: structures and molecular properties*, 2nd edn, W. H. Freeman, NY, p. 178.

One of the most direct methods of measuring distances between residues in a random coil-like polymer is to attach a fluorescent donor group to one and an acceptor to the other and measure the efficiency of the fluorescence energy transfer between them. It ideally varies inversely with the sixth power of the distance between them, but only if the donor and acceptor groups introduced have random orientations and do not interact with each other or modify the properties of the polymer; unfortunately fluorescent groups are usually large hydrophobic moieties that almost certainly attract each other in aqueous solution.

End-to-end distribution function of stiff polymers for all persistence lengths. B. Hamprecht & H. Kleinert (2005) *Phys. Rev. E* **71**, 031803.

Scaling exponents and probability distributions of DNA end-to-end distance. F. Valle *et al.* (2005) *Phys. Rev. Lett.* **95**, 158105.

End-to-end distance distributions and intrachain diffusion constants in unfolded polypeptide chains indicate intramolecular hydrogen bond formation. A. Moglich *et al.* (2006) *Proc. Natl. Acad. Sci. USA* **103**, 12394–12399.

2. Radius of Gyration

Another statistical measure often used with random coils is the average **radius of gyration (R_g)**, which is defined as the r.m.s. distance of the collection of atoms from their common center of gravity. For the random-flight chain:

$$< R_g >_0^2 = \frac{nl^2}{6}\frac{n+2}{n+1} \tag{1.24}$$

For large values of *n*, this becomes:

$$< R_g >_0^2 = \frac{nl^2}{6} = \frac{< r^2 >_0}{6} \tag{1.25}$$

This relationship holds for the unperturbed states of all very long polymers, so the radius of gyration is simply 0.408 ($= 6^{-1/2}$) times the average end-to-end distance.

3. Characteristic Ratio

The flexibility of an actual random coil is more limited than that of the hypothetical random-flight chain, and its dimensions tend to be considerably greater than those predicted by Equations 1.24 and 1.25. This increased stiffness is usually expressed as the **characteristic ratio,** C_n:

$$C_n = \frac{< r^2 >_0}{nl^2} \tag{1.26}$$

where $<r^2>_0$ is the square of the observed average end-to-end distance for the actual random coil, and nl^2 is this value for the random-flight chain (Equation 1.23). The value of C_n increases with increasing length of the polymer, *n*, reflecting the stiffness of the polymer backbone, so that segments distant in the polymer still do not have random orientations. In very long chains, however, C_n approaches a limit designated as C_∞; in this case, the distant segments are behaving as a truly random-flight chain.

Polypeptide chains of L-amino acids have values of C_∞ of approximately 9, while single-stranded polynucleotides generally have values of 10–15. Calculations in which all sterically allowed conformations are given equal weight predict values of C_∞ of 2.0–4.5, so energy differences between sterically allowed conformations must bias the chain towards greater extension.

A real polymer chain may be approximated as a polymer of freely jointed **statistical segments**, chosen to be long enough so that each is randomly oriented with respect to all other such segments. The stiffer the polymer chain, the longer the statistical segment. The statistical segment of a polypeptide chain is about 10 residues or 36 Å.

Another useful parameter is the **persistence length**, *a*. This is defined as the average projection of the end-to-end distance vector on the first bond of the chain, in the limit of infinite chain length. It can be considered a measure of the length over which the chain persists in the same direction as the first bond. For polymer chains composed of identical bonds of length *l*, the persistence length *a* is closely related to the limiting characteristic ratio, C_∞:

$$a = (C_\infty + 1)\, l/2 \qquad (1.27)$$

For random polypeptide chains, the value of the persistence length *a* is about 20 Å, or nearly six residues, while for double-stranded DNA it is approximately 500 Å, roughly 150 base pairs, depending upon the conditions. Double-stranded DNA is clearly much less flexible than an individual polypeptide chain.

Calculations of the ϕ–ψ conformational contour maps for *N*-acetyl alanine *N*′-methyl amide and of the characteristic ratios of poly-L-alanine using various molecular mechanics force fields. C. H. Lee & S. S. Zimmerman (1995) *J. Biomol. Struct. Dyn.* **13**, 201–218.

1.3.B. Excluded Volume Effects and Theta Solvents

Within an ensemble of many micro-conformations of an unfolded polymer, some will have all atoms in contact with the solvent, whereas others will have various parts of the polymer in contact with each other. Conformations where atoms would overlap in space are excluded, of course, which is the **excluded volume effect**. The relative energies of the individual micro-conformations will vary, depending upon the relative energetics of the different interactions. In a poor solvent, which does not interact favorably with a polymer, those conformations with contacts within the chain will be favored over those interacting only with the solvent. Consequently, the average dimensions of the population will be reduced. In contrast, a solvent that interacts favorably with the polymer will favor those conformations that are more extended, and the average dimensions of the population will be expanded. In a Θ **solvent**, known as **theta conditions**, the interactions of the polymer with the solvent are balanced to be the same energetically as those within the polymer. Polymers usually have relatively poor solubilities in Θ solvents.

In a Θ solvent, the solvent-induced preference of the chain for relatively compact conformations counterbalances exactly the effects of excluded volume. A real polymer then has dimensions like those calculated for the unperturbed state, in which excluded volume effects are ignored. Such ideal solvents are, however, generally not practical with biological polymers, such as polypeptide and polynucleotide chains, where there is a variety of side-chains on the polymer.

Contributions of short-range and excluded-volume interactions to unperturbed polymer chain dimensions. H. Yamakawa & T. Yoshizaki (2004) *J. Chem. Phys.* **121**, 3295–3298.

The optimized Rouse–Zimm theory of excluded volume effects on chain dynamics. J. H. Kim & S. Lee (2004) *J. Chem. Phys.* **121**, 12640–12649.

Corrections to scaling and crossover from good- to theta-solvent regimes of interacting polymers. A. Pelissetto & J. P. Hansen (2005) *J. Chem. Phys.* **122**, 134904.

1. Covalent Cross-links

The probability that the ends of a chain are spatially near each other in the polymer gives the probability that two functional groups on the polymer that are separated by the same number of residues will interact by reacting chemically with each other or forming a covalent cross-link. This would be an intramolecular reaction, and its probability can be expressed as the **effective concentration** of the two residues relative to each other within the polymer. **The ends of relatively flexible polymers with 30–100 residues are usually measured to have effective concentrations in the region of 10^{-3} M.** In contrast, the values measured are usually in the range of only 10^{-7} to 10^{-9} M for double-stranded DNA molecules with up to 10,000 base pairs, because DNA is relatively inflexible. The values can vary dramatically when residues are close in the polymer and will depend on the particular stereochemistry of the polymer. With longer chains, the effective concentration decreases with increasing length in proportion to $n^{-3/2}$.

Cross-linking two parts of a polymer chain with a covalent bond decreases the number of conformations that are possible and the conformational entropy of the random coil. The magnitude of this effect can be calculated by considering the probability, in the absence of the cross-link, that the ends of a random coil corresponding to the points of the cross-linkage would lie simultaneously within a small volume element, V (Figure 1-2). The smaller this probability, the greater the effect on the conformational entropy of constraining the ends to lie within this volume by a covalent cross-link. **The further apart in the covalent structure the residues that are cross-linked, the greater the decrease in conformational entropy**. Unfortunately, it is not certain what volume element V is appropriate for any particular example, and the energetic consequences of cross-links cannot be calculated with any degree of confidence.

Most importantly, **covalent cross-links stabilize any folded conformations with which they are compatible, because they destabilize the disordered state by decreasing its conformational entropy**.

Dissecting the roles of individual interactions in protein stability: lessons for a circularized protein. D. P. Goldenberg (1985) *J. Cell Biochem.* **29**, 321–335.

Loops, linkages, rings, catenanes, cages, and crowders: entropy-based strategies for stabilizing proteins. H. X. Zhou (2004) *Acc. Chem. Res.* **37**, 123–130.

Loop entropy and cytochrome c stability. L. Wang *et al.* (2005) *J. Mol. Biol.* **353**, 719–729.

1.4. STRUCTURE DATABASES: STRUCTURES ON THE WEB

Structure databases contain the 3-D structures of biological macromolecules determined by X-ray crystallography and NMR. The primary structure database is the Protein Data Bank (PDB), an archive of all publicly available 3-D structures of proteins, nucleic acids, carbohydrates, viruses and biomolecular complexes. The PDB is basically a collection of text files, each containing a structure entry deposited by those who determined the structure. An entry file consists of (1) text information of definition, source, references and comments, (2) the sequence information, (3) the secondary structure information and (4) the 3-D coordinates of all the atoms, as well as crystallographic structure factors and NMR experimental data. There is an efficient search engine for finding files of the desired structures. Other useful structure databases include the Nucleic acid Database (NDB) for nucleic acids and the Cambridge Structural Database (CSD) for organic and metal–organic compounds. The World Wide Web addresses for these databases are given in Table 1-1. The majority of databases in molecular biology are publicly available and freely accessible via the Web.

Table 1-1. Web addresses for the structure databases

Database	Address
PDB	www.rcsb.org
NDB	ndbserver.rutgers.edu
CSD	www.ccdc.cam.ac.uk

To view the known structures of macromolecules, it is recommended that you find the name of the data file in the PDB, then view the molecule using the program Jmol at http://molvis.sdsc.edu/fgij/index.htm.

Protein structure database search and evolutionary classification. J. M. Yang & C. H. Tung (2006) *Nucleic Acids Res.* **34**, 3646–3659.

Conformational specificity of non-canonical base pairs and higher order structures in nucleic acids: crystal structure database analysis. S. Mukherjee *et al.* (2006) *J. Comp. Aided Mol. Des.* **20**, 629–645.

The CATH domain structure database: new protocols and classification levels give a more comprehensive resource for exploring evolution. L. H. Greene *et al.* (2007) *Nucleic Acids Res.* **35**, D291–297.

~ CHAPTER 2 ~

DNA STRUCTURE

The discovery of the well-known double-helix structure of deoxyribonucleic acid (DNA) (Figure 2-1) was the seminal event in the origin of the field of molecular biology. This structure demonstrated graphically how just the three-dimensional (3-D) structures of biological macromolecules could be sufficient to illuminate the mechanisms of their biological functions. The DNA double-stranded helix held together by complementary base pairs immediately revealed how DNA could store and transmit all the genetic information in the genome of a living organism. It is now clear that the canonical Watson–Crick DNA double helix is not a rigid structure but, depending upon the conditions and the local nucleotide sequence, can adopt variant conformations that are both subtle and important. Even markedly different structural forms of DNA have been discovered. The structure of DNA is still central to molecular biology.

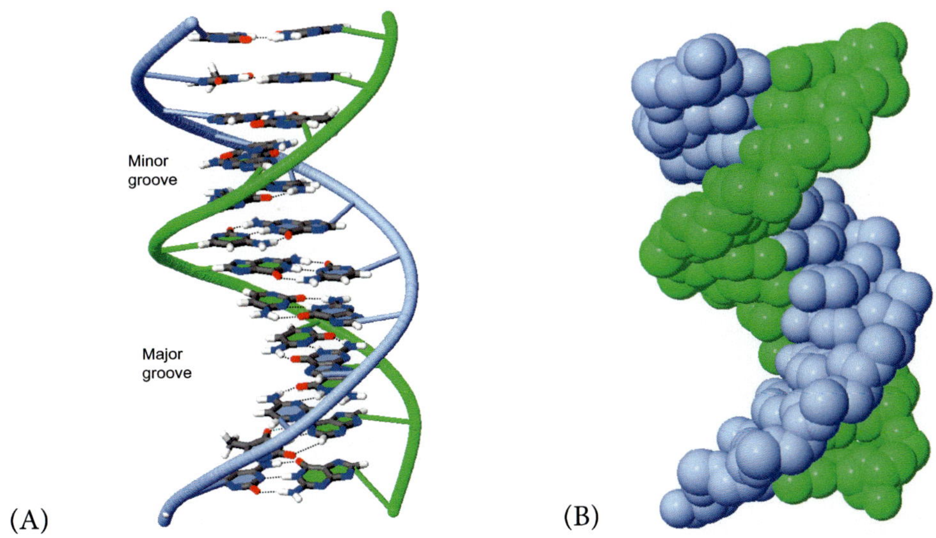

Figure 2-1. The double-stranded helical structure of B-DNA. This structure was determined by NMR with the dodecamer in which one chain has the sequence 5′–d(CTCGGCGCCATC)–3′ and the other is the complementary 5′–d(GATGGCGCCGAG)–3′. (A). The detailed structures of the nucleobases are presented, but the deoxyribose-phosphate backbone is depicted as a *simple cylinder*. The major and minor grooves are indicated, and some hydrogen bonds between bases are indicated as *dashed lines*. (B). Space-filling representation of the same structure in which all the atoms are depicted as spheres with their van der Waals radii. Figure generated from PDB file *2hkb* using the program Jmol.

DNA occurs naturally almost exclusively as the genetic material of organisms, the chromosomes containing the genetic information. Generally there is one DNA molecule per chromosome, and its detailed structure contains all the genetic information.

Principles of Nucleic Acid Structure. W. Saenger (1984) Springer-Verlag, NY.

Understanding DNA. C. R. Calladine & H. R. Drew (1992) Academic Press Inc., CA.

DNA Structure and Function. R. R. Sinden (1994) Academic Press Inc., CA.

Nucleic Acids in Chemistry and Biology. G. M. Blackburn & M. J. Gait (eds) (1996) Oxford University Press, NY.

Bioorganic Chemistry: nucleic acids. S. M. Hecht (ed.) (1996) Oxford University Press, NY.

Nucleic Acids: structures, properties and functions. V. A. Bloomfield *et al.* (2000) University Science Books, CA.

2.1. POLYNUCLEOTIDES

The nucleic acids DNA and ribonucleic acid (RNA) (Chapter 4) are polynucleotides that are produced by polymerizing nucleotides, either deoxynucleotides or ribonucleotides. Each **nucleotide** building block consists of three components: a sugar (**deoxyribose** or **ribose**), a phosphate group and a nucleobase. Two of the bases are derivatives of **purine** (Pu), **adenine** (A) and **guanine** (G), and two of **pyrimidine** (Py), **cytosine** (C) with **thymine** (T) in the case of DNA and **uracil** (U) in the case of RNA. When the bases are attached covalently to only the sugar moiety, they are known as the **nucleosides**: **deoxyadenosine**, **deoxyguanosine**, **deoxycytidine** and **thymidine** in the case of DNA (Table 2-1). In each nucleoside, the N9 atom of a purine or N1 of a pyrimidine is bonded in a **glycosidic bond** to the C1′ atom of the sugar (by convention, the numbers of atoms of the sugar are given primes). A **nucleotide** is a nucleoside joined to one or more phosphate groups by an ester linkage to the ribose or deoxyribose sugar (Figure 2-2 and Table 2-1).

Table 2-1. Names and abbreviations of the nucleotide constituents of nucleic acids and cyclic nucleotides

Name	Abbreviation
Ribonucleotides	
Uridine 5′-monophosphate, 5′-uridylic acid	UMP, pU
Guanosine 5′-monophosphate, 5′-guanylic acid	GMP, pG
Cytidine 5′-monophosphate, 5′-cytidylic acid	CMP, pC
Adenosine 5′-monophosphate, 5′-adenylic acid	AMP, pA
Deoxyribonucleotides	
Deoxythymidine 5′-monophosphate, 5′-deoxythymidylic, acid, thymidine 5′-monophosphate, 5′-thymidylic acid	dTMP, pdT
Deoxyguanosine 5′-monophosphate, 5′-deoxyguanylic acid	dGMP, pdG
Deoxycytidine 5′-monophosphate, 5′-deoxycytidylic acid	dCMP, pdC
Deoxyadenosine 5′-monophosphate, 5′-deoxyadenylic acid	dAMP, pdA

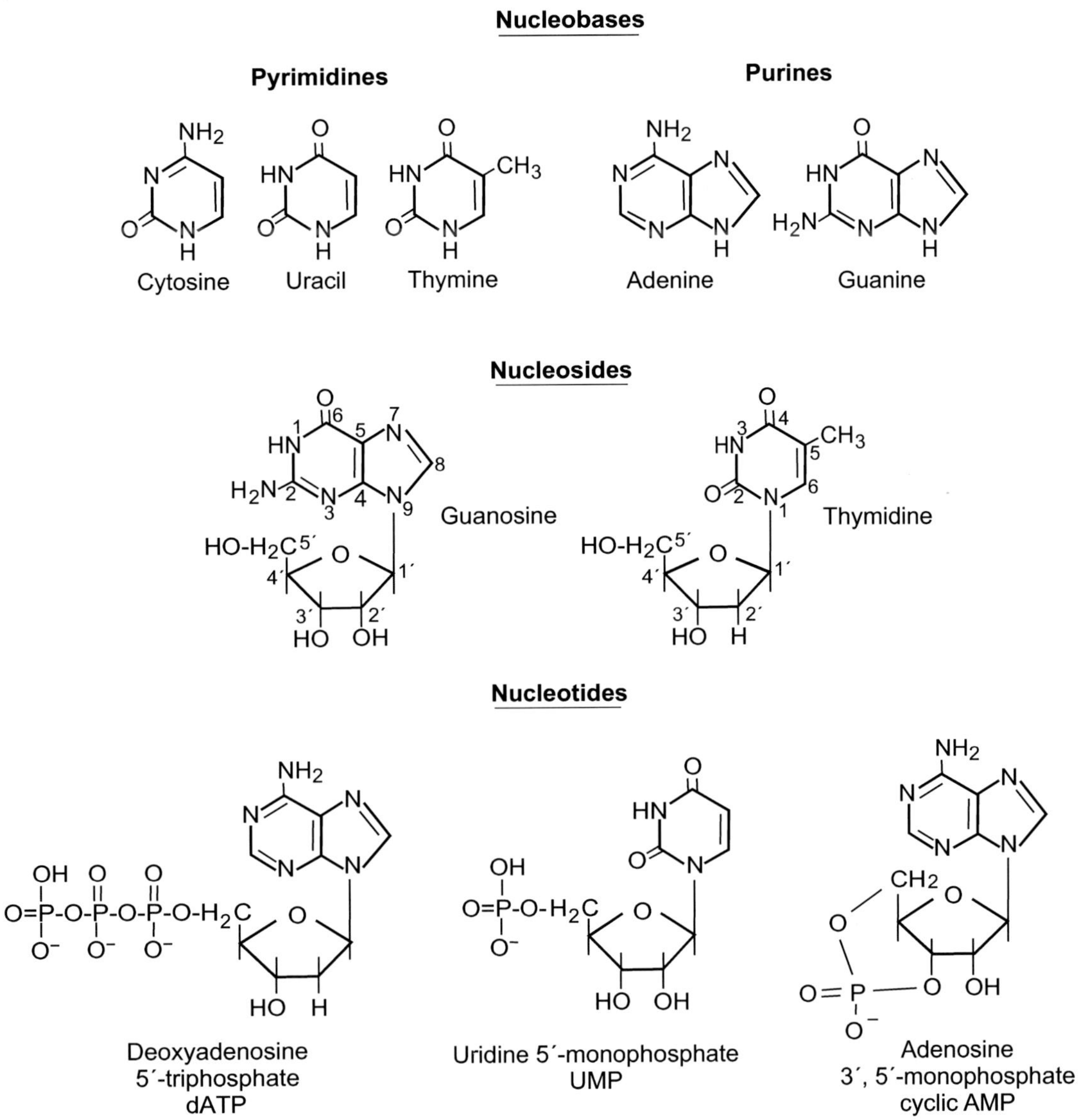

Figure 2-2. Structures of the five bases used in DNA and RNA and of representative nucleosides and nucleotides. The *syn* conformation (Equation 2.1) is shown for the nucleosides and nucleotides for reasons of space, but the *anti* conformation predominates in RNA and DNA.

Polynucleotides like DNA and RNA are produced by polymerizing the nucleotides, linking them by phosphodiester bridges, in which the 3′-hydroxyl group of the sugar moiety of one nucleotide is esterified to a phosphate group, which is, in turn, joined to the 5′-hydroxyl group of the adjacent sugar (Figure 2-3). **The chain of sugars linked by phosphodiester bridges comprises the backbone of the nucleic acid molecule**, which is invariant. The chemical structure of a short strand of DNA, containing one each of the four nucleotides, is illustrated in Figure 2-3. **The genetic information in the polynucleotide is made up of the sequence of the four bases that protrude from the backbone.**

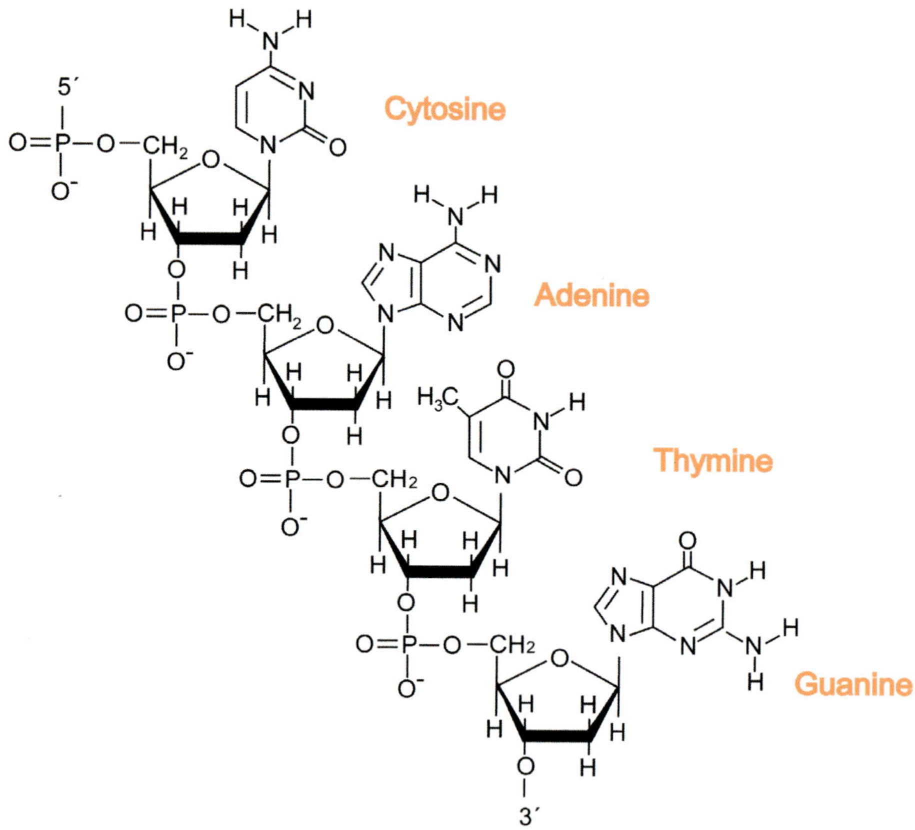

Figure 2-3. The chemical structure of a single strand of DNA.

In polynucleotides, the nucleotide building blocks are phosphorylated on both the C5′ and C3′ sugar atoms. Whether an isolated nucleotide is phosphorylated on the 5′- or 3′-sugar atom depends upon its origin. Most commonly, the phosphate group is attached to the C5′ atom, to produce a 5′-nucleotide. In certain nucleotides used in signal transduction in cells, such as cyclic AMP and cyclic GMP, a single phosphate group is esterified simultaneously to C3′ and C5′.

A linear polynucleotide has a 5′-end and a 3′-end, either or both of which can be phosphorylated. By convention, the sequence of a polynucleotide chain is described in the 5′- to 3′-direction. The general abbreviation for 5′-nucleotides is **pN**; for 3′-nucleotides it is **Np**. For convenience, DNA oligonucleotides are denoted with their sequence in a form like d(CGCGAATTCGCG). For short sequences, a more specific designation would indicate the phosphate groups ('p'), such as pCGCGAATTCGCG or pCpGpCpGpApApTpTpCpGpCpG.

The phosphate groups cause nucleotides and polynucleotides to be highly acidic and to exist as anions at neutral pH. The primary and secondary pK_a values for the 5′-phosphates of nucleotides are about 1 and 6, respectively, so they have two negative charges at neutral pH. Each phosphodiester group of DNA has an intrinsic pK_a of about 1 and will bear one negative charge at all except extremely acidic pH values. Its ionization is complicated in DNA, however, because DNA has many such groups in close proximity that affect each other and bind counterions tightly, so such polyelectrolytes have complicated ionization behavior (Section 2.3). Ionization of the nucleotides and polynucleotides prevents them from crossing cell membranes.

RNA (Chapter 4) differs from DNA in having ribose as the sugar and the base uracil instead of thymine. Ribose differs from deoxyribose only in having an additional O atom, while uracil differs from thymine only in lacking the methyl group. Yet these two small differences cause DNA and RNA to have different structures, functions and properties. The absence of the 2′-hydroxyl group on the ribose ring causes DNA to be about 100-fold less susceptible to alkaline hydrolysis than RNA (Figure 4-2). Thymine is believed to be used in DNA instead of uracil because it can be distinguished from the uracil that arises spontaneously in DNA by any deamination of cytosine nucleotides (Figure 2-33); if the cell detects uracil in DNA, an enzyme converts it back to cytosine. Both of these changes seem appropriate in view of the need for DNA to be extremely stable as the carrier of genetic information. In contrast, RNA is a much more dynamic molecule that need not be long-lived in most cases, but when necessary can be protected by binding proteins (Section 7.4). It is important to use the 'deoxy' prefix when there might be confusion with the ribonucleosides of RNA. It is usually omitted for thymidine, however, because this base is not normally encountered in RNA molecules; it is, however, found in some transfer RNA (Section 4.2.B), when it is referred to as ribosylthymidine or 5-methyluridine.

A single-stranded polynucleotide is very flexible, because rotations can take place about six bonds of each nucleotide of the backbone (Figure 2-4). There are very significant steric restrictions on what combinations of torsion angles can occur, but to depict this as with a Ramachandran plot for proteins (Figure 8-3) would be impractical, as it would require six dimensions. The number of distinct conformations for each of the bond rotations is believed to be about one for torsion angle α, two for δ, ζ and χ, and three for β, γ and ε.

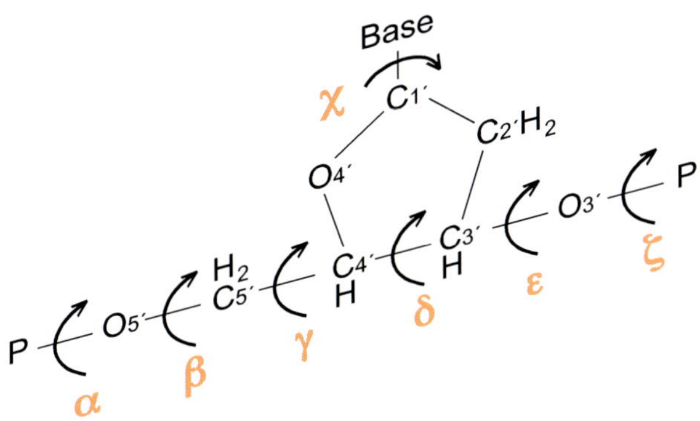

Figure 2-4. Definition of the torsion angles α, β, γ, δ, ε and ζ of a polynucleotide sugar-phosphate backbone and of the glycosyl angle χ.

2.1.A. The Deoxyribose Group

Sugars such as ribose and deoxyribose normally exist in solution as three different tautomeric forms in equilibrium, two cyclic forms and one linear (Equation 1.17). Replacing the H atom normally on C1′ by a base, to generate a nucleoside, however, locks the sugar into the cyclic form. In DNA and RNA, the base lies on the same side of the plane of the sugar as the 5′-carbon, and the configuration of the *N*-glycosidic link is β (Figure 2-2). This stereospecificity is possible because the nucleosides and nucleotides are synthesized by asymmetric enzymes that are usually stereospecific.

The deoxyribose ring does not have all five atoms lying in the same plane. The five torsion angles of the ribose ring are designated υ_0 to υ_4, starting with the bond between the O atom and C1′ and ending with that from C4′ to the O. All five torsion angles would be zero in the planar conformation, but this would be very strained. Such strain is relieved by having at least one of the five atoms lie out of the plane, by up to 0.5 Å, introducing a pucker to the ring. The carbons most likely to be out of the plane are the 2′ and 3′. If the nonplanar atom is on the same side as the base and the 5′-carbon, the conformation is said to be *endo*; if on the other side, the conformation is *exo*. These conformations can interchange rapidly, and they do so on the 10^{-9} s time scale in single-stranded nucleic acids, where one conformation is not greatly favored over another. The ribose ring conformations observed most frequently in nucleic acids are illustrated in Figure 2-5. The ribose pucker is important for nucleic acids because it governs the relative orientations of the phosphate substituents (Figure 2-6). In the most common helical structures (Section 2.2.B), the C2′-*endo* conformation occurs in B-DNA, C3′-*endo* in A-DNA. The purine nucleotides are C3′-*endo* in Z-DNA and the pyrimidine residues C2′-*endo*.

Nucleotides and nucleosides can exist with either of two orientations about the glycosidic bond (as defined by the torsion angle χ about the C1′–N glycosyl bond; Figure 2-4), commonly designated as *syn* and *anti*:

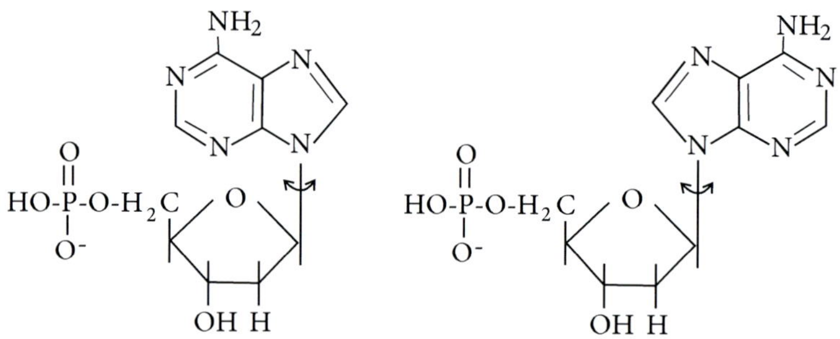

syn *anti* (2.1)

Rotation about this bond is hindered, so these two most stable conformations are not readily interconverted. For pyrimidines, only the *anti* conformation is stable, because otherwise the sugar residue would sterically interfere with the pyrimidine's C2 O atom. The *anti* conformation is found in all DNA structures, except for Z-DNA, where the purine residues are *syn*. In spite of that, most of the structures shown here are represented as *syn* because it is more compact to draw.

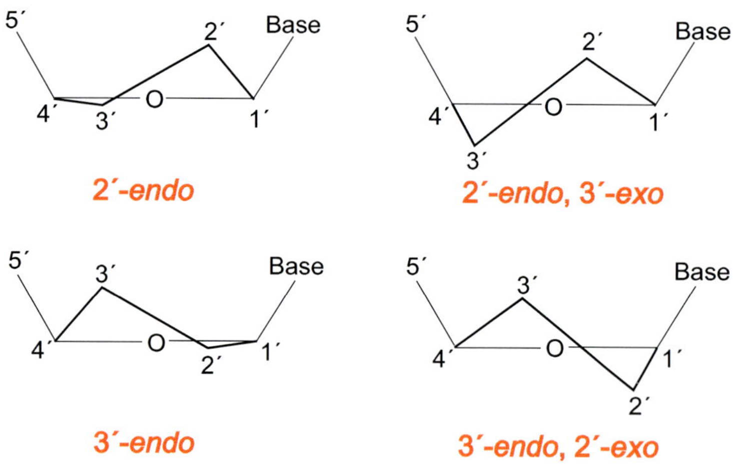

Figure 2-5. The conformations of the ribose and deoxyribose rings observed most frequently in nucleic acids. The plane of the ring is observed edge on, with the bond between C2′ and C3′ in front and the O atom at the rear.

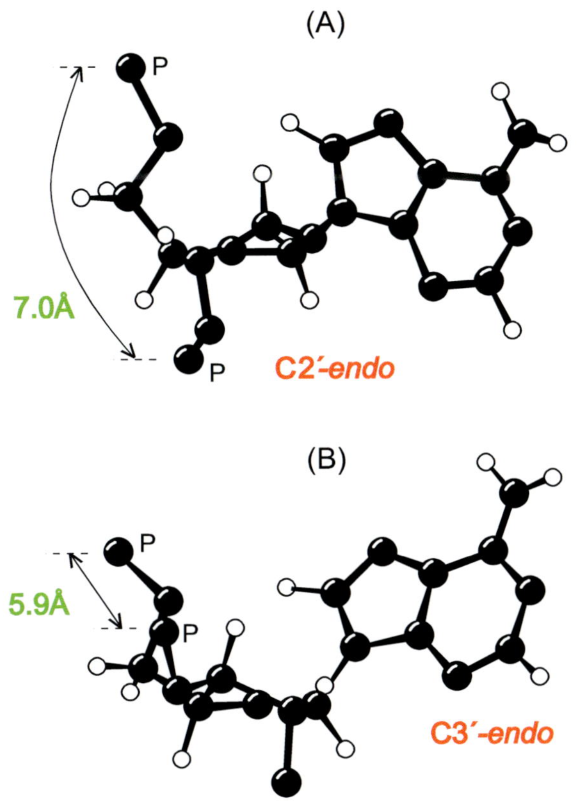

Figure 2-6. Effects on the backbone of nucleotides in the (A) C2′-*endo* conformation and (B) the C3′-*endo* conformation. These conformations occur in A-DNA and B-DNA, respectively. The distances between adjacent P atoms in the sugar-phosphate backbone are indicated.

2.1.B. Properties of the Bases

Each base is essentially planar and rigid, although its bonds are subject to the usual bending and stretching vibrations. The methyl group of thymine in isolation rotates on the 10^{-10} s time scale. The peripheral amino groups rotate more slowly, probably due to hydrogen-bond interactions with the aqueous solvent; that of cytosine rotates at an exceptionally slow rate of 35 revolutions per second (35 s^{-1}) at 0°C.

The bases are not ionized at neutral pH, but they can lose and accept protons at extremes of pH (Figure 2-7). The adenine, guanine and cytosine bases have amino groups that are very weak bases, with pK_a values of 2–4.5, so they are ionized only at acidic pH values. The bases guanine, uracil and thymine can lose a proton above pH 9. The presence of the ionized phosphate group on mononucleotides raises these pK_a values by 0.2–0.6 pH units (Table 2-2), and those of deoxyribonucleotides are 0.1–0.3 pH units greater than those of ribonucleosides. The variations in the pK_a values, plus the absence of an amino group on uracil and thymine, can be used to adjust the pH so that the various nucleotides have different net negative charges and can be separated readily by ion-exchange chromatography or electrophoresis.

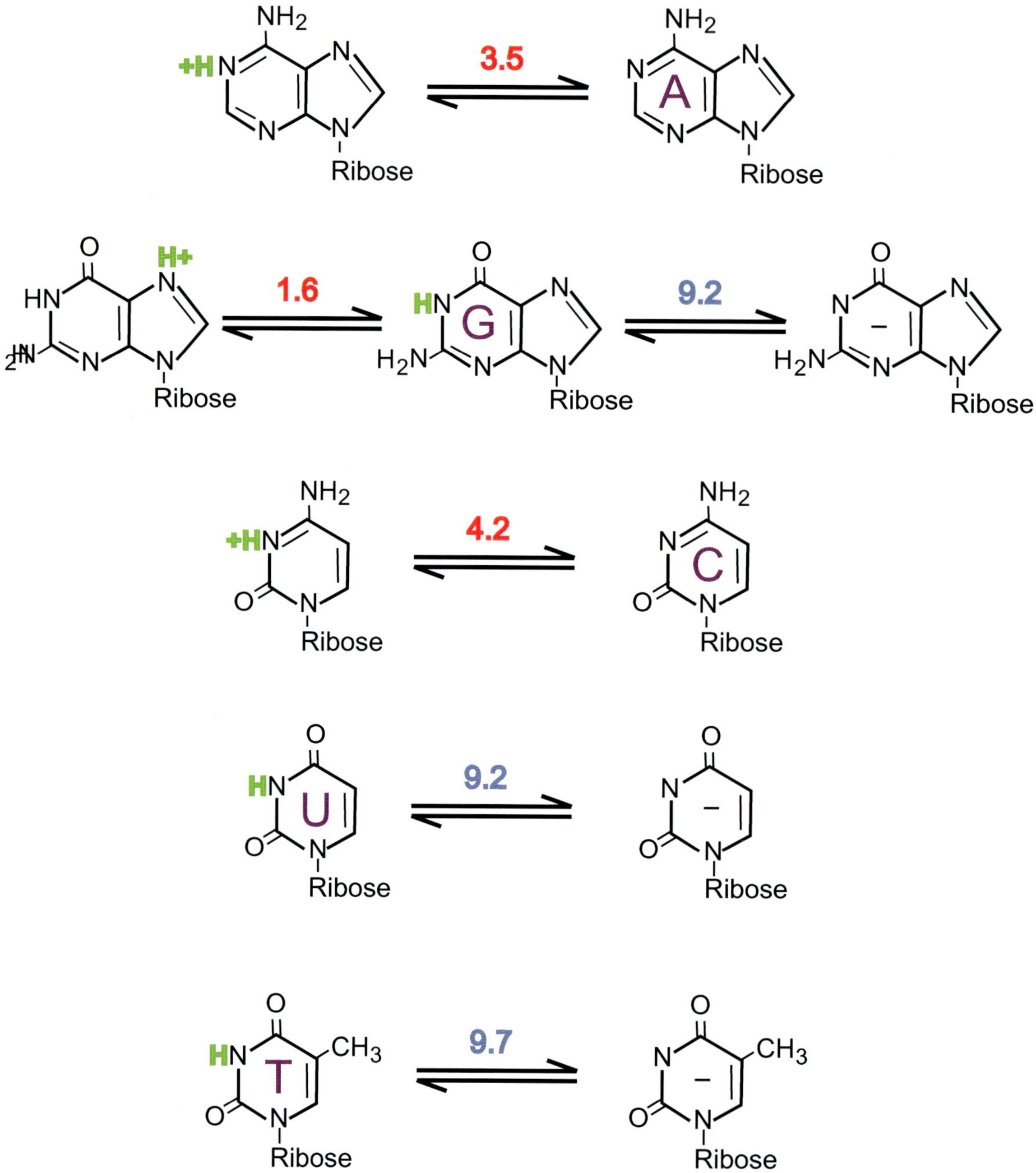

Figure 2-7. Ionization of the bases in nucleosides. The pK_a value is given above the *arrows*. The proton removed or added is indicated in *green*.

Table 2-2. Ionization and UV Absorbance Properties of Nucleoside Monophosphates.

Nucleotide	pK_a of ring $-NH_2$	UV absorbance[a]	
		λ_{max} (nm)	Molar absorbance (10^3 M^{-1} cm^{-1})
AMP	3.8	257	15.0
CMP	4.5	280	13.2
GMP	2.4	256	12.2
UMP		262	10.0
dAMP	4.4	258	14.3
dCMP	4.6	280	13.5
dGMP	2.9	255	11.8
dTMP		267	10.2
cAMP		256	14.5
cGMP		256.5	11.4

[a] Measured at pH 1 to 2.
Data from C.K. Mathews (1999) In *Encyclopedia of Molecular Biology* (T.E. Creighton, ed.) Wiley-Interscience, New York, p. 1676.

The peripheral groups on the bases are extremely important, as they can serve as donors or acceptors in hydrogen bonds (Figure 2-8). Each base has both donor and acceptor sites, so each has the potential to hydrogen bond with any of the bases. These determine their crucial tendency to pair with each other (Section 2.2.A). The peripheral groups are not fixed, however, as each base can occur in several tautomers (Section 1.1.C), which tends to reverse their hydrogen-bonding potentials (Figure 2-9). These other tautomeric forms are rare, however, and those described in Figure 2-8 predominate under normal conditions.

The aromatic nature of the bases causes them to absorb ultraviolet (UV) light with characteristic absorbance spectra (Table 2-2), which allows them to be detected, identified and quantified. Excitations of the electrons in the bases are classified as being either perpendicular to the plane of the base or in the plane. The in-plane transitions involve primarily the π electrons of the base (those on the planar surface of the base) and thus are designated as $\pi \rightarrow \pi^*$ transitions. The out-of-plane transitions involve excitation of nonbonding electrons of the N or O atoms into the π electron system and are designated as $n \rightarrow \pi^*$. The $\pi \rightarrow \pi^*$ transitions predominate in the absorbance of light, but the $n \rightarrow \pi^*$ transitions can be important for circular dichroism. The spectral properties of adenine, guanine and cytosine are dependent on the pH, due to the ionization of their amino groups, which also helps in their identification. The close proximity of bases in helical polynucleotides alters their absorbance properties, which is very useful for monitoring the breakdown of the helical structure (Section 5.1).

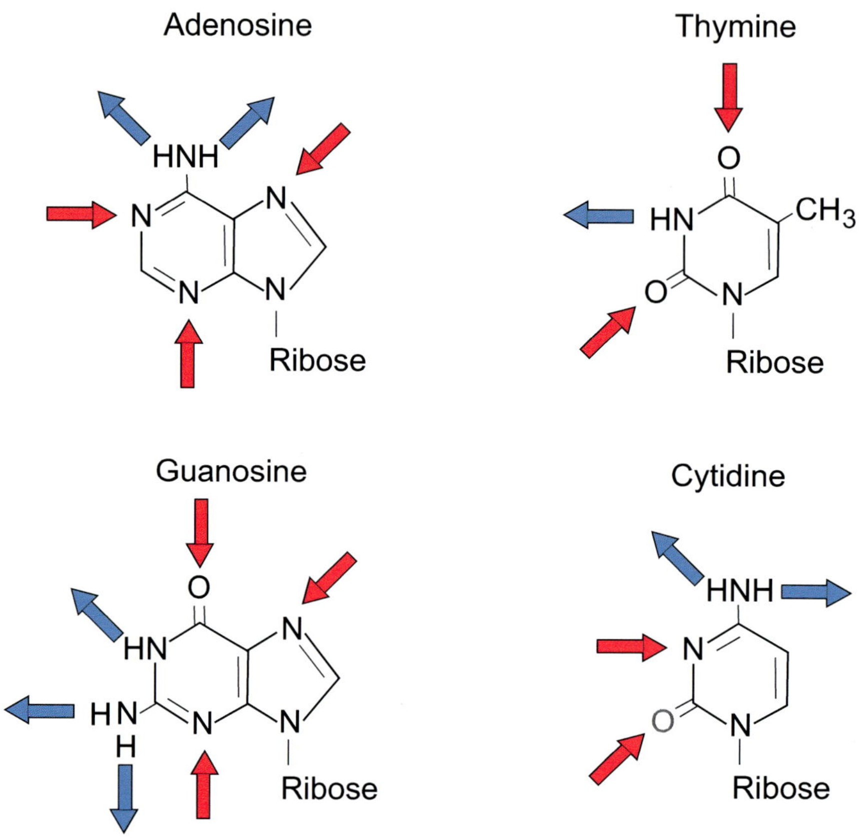

Figure 2-8. Hydrogen-bonding capabilities of the bases of DNA. H atoms that may be donated in hydrogen bonds are indicated by *blue arrows*, whereas potential acceptors of H atoms are indicated by *red arrows*.

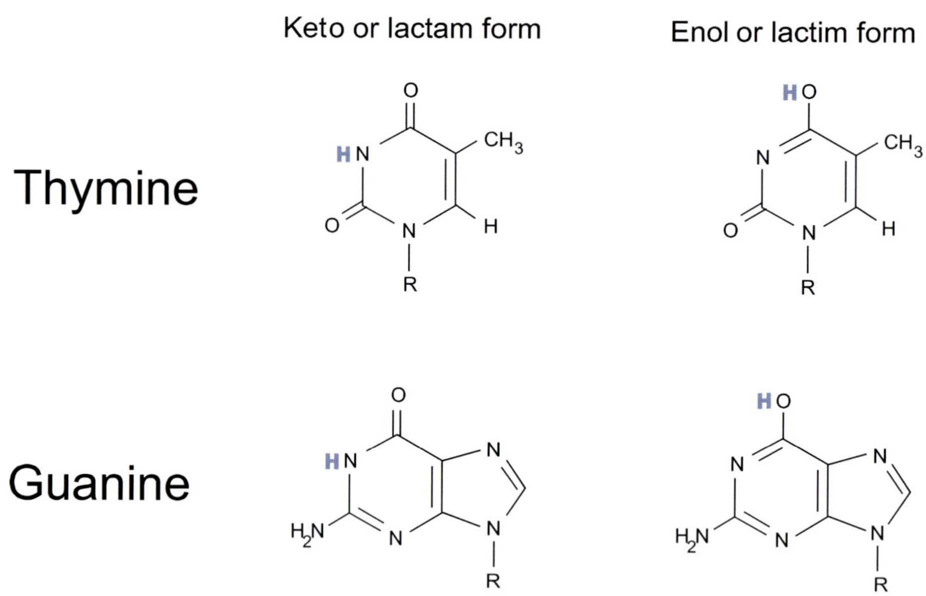

Figure 2-9. Possible tautomeric conversions for thymine and guanine bases. The hydrogen that changes position is indicated in *blue*. Cytosine and adenine residues can undergo similar proton shifts, in some cases involving the amino group attached to the ring.

The nucleobases are only very slightly fluorescent at room temperature in aqueous solution. Their quantum yields vary between 3×10^{-5} and 1.2×10^{-4}, while their fluorescent lifetimes are in the picosecond range. These weak fluorescent properties are not very useful, as they can be overwhelmed by impurities.

The hydrophobicities of the bases are important for their interactions; they were measured by partitioning the bases between water and cyclohexane, the organic solvent considered most pertinent to hydrophobicity, as it contains a minimum of water. The ribose moiety is so polar that nucleosides are almost totally excluded from cyclohexane. Consequently, the ribose moiety was substituted by methyl, butyl and tetrahydrofuryl moieties:

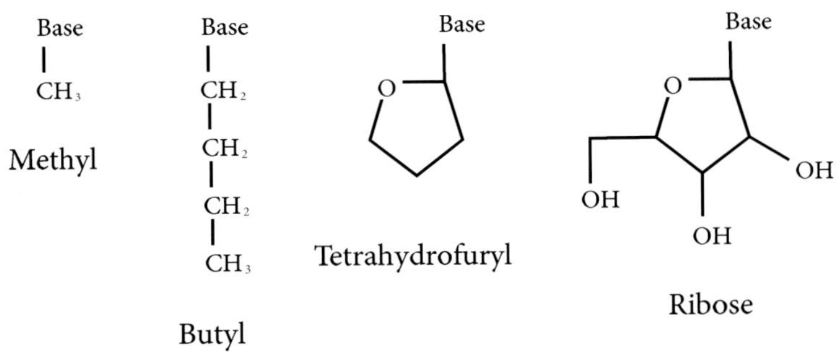

(2.2)

The free energies of transfer measured in these systems are tabulated in Table 2-3. The relative hydrophobicities of the bases are compared in Figure 2-10.

Table 2-3. Relative hydrophobicities of the nucleic acid bases as measured by the free energy of transfer from water to cyclohexane of their butyl and tetrahydrofuryl derivatives

	Free energy of transfer (kcal/mol)	
Base	**Butyl derivative**	**Tetrahydrofuryl derivative**
Purine	1.1	2.6
Thymine	2.2	4.1
Adenine	2.8	4.3
Uracil	3.7	5.0
Pyrimidin-2-one	4.1	5.6
Hypoxanthine	5.7	7.1
Cytosine	6.0	7.3
Guanine	6.7	7.5

Data from P. Shih *et al.* (1998) *J. Mol. Biol.* **280**, 421–430.

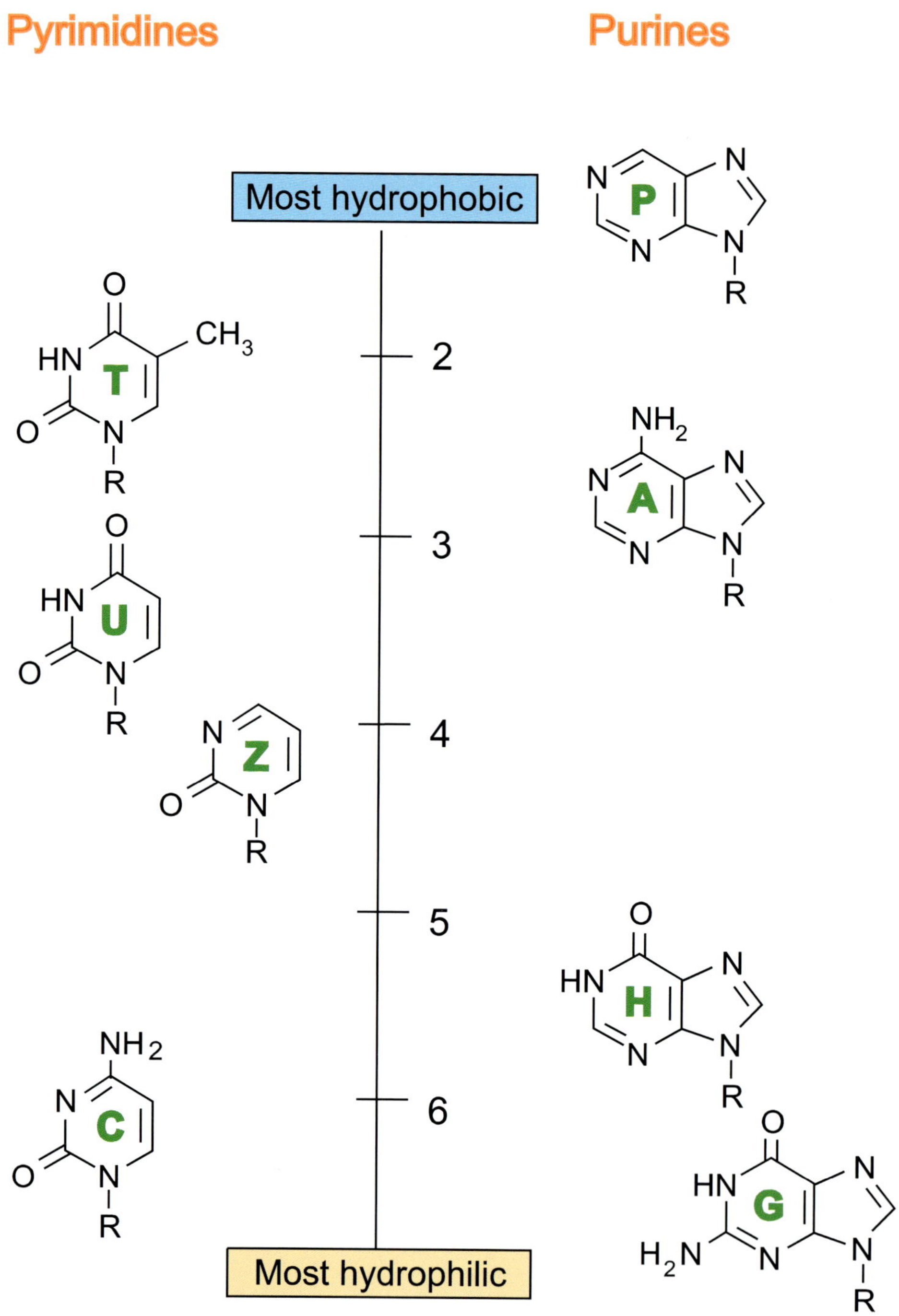

Figure 2-10. The relative hydrophobicities of the nucleic acid bases. The hydrophobicity scale is in units of kcal/mol. The structures of the bases are given: P, purine; T, thymine; A, adenine; U, uracil; Z, pyrimidin-2-one; H, hypoxanthine; C, cytosine; G, guanine. Data from P. Shih *et al.* (1998) *J. Mol. Biol.* **280**, 421–430.

The heat capacities of the bases are significantly more positive than would be predicted on the basis of just their surface areas accessible to water. N, O and P atoms of small molecules and amino acids make negative contributions to ΔC_p, probably because they increase the distortion of water hydrogen bonds. The anomalous properties of the nucleobases may indicate that many of their polar atoms are hydrated by water molecules whose hydrogen bonds are less distorted than in bulk water, which would produce a more positive ΔC_p.

2.1.C. Modifications of the Bases

The nucleotides of DNA can be modified by a number of chemical processes (Section 2.6). Some of these are very important *in vivo*, as any change of the base of a nucleotide is likely to produce a mutation in the organism, a change in the genetic information that is passed on to the next generation. The bases in DNA are sensitive to the action of numerous chemicals, such as nitrous acid (HNO_2) and hydroxylamine (NH_2-OH), and various alkylating reagents, such as dimethyl sulfate (Section 2.6). Alkylating agents are strong electrophiles that can attack chemically the approximately 20 nucleophilic sites in DNA. Bases that are alkylated become positively charged and have altered chemical properties, including their hydrogen-bonding patterns. DNA also can undergo numerous spontaneous changes, most commonly deamination, especially of cytosine to uracil, and the loss of purine bases (Figure 2-33-C) by direct attack of oxygen radicals. Oxidation and radiation can cause many types of damage to DNA; UV light damages DNA bases by many different mechanisms, including the production of oxygen radicals that lead to oxidative damage and the formation of dimers between adjacent pyrimidine bases by addition across their C5–C6 double bonds:

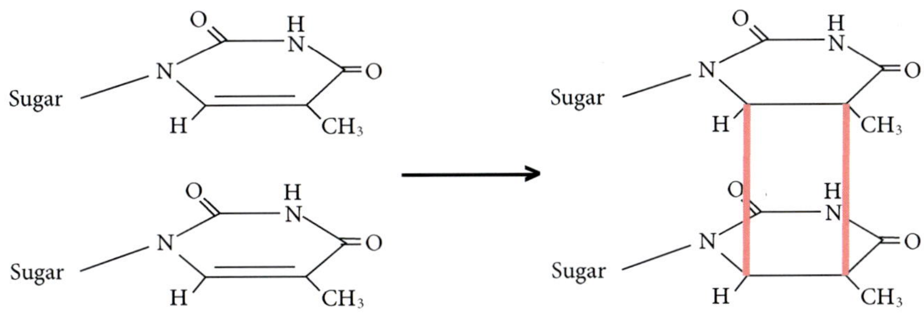

(2.3)

Thymine residues are particularly susceptible to this reaction, although cytosine dimers and thymine–cytosine pairs are also produced.

Each human genome of 3 billion base pairs is believed to suffer between 100 and 5000 of these modifications each day! Fortunately, enzymes are present to reverse most of these unavoidable modifications.

Most natural nucleic acids contain some bases that have been modified specifically; the most common are illustrated in Figure 2-11. Most widespread is **5-methylcytosine**, which is present in up to 5% of the cytosine residues in most eukaryotic DNAs. These cytosine bases are modified by **DNA methyltransferases**, enzymes that catalyze the transfer and covalent attachment of a methyl group from the cofactor S-adenosylmethionine to DNA, with the release of S-adenosylhomocysteine. Methylation occurs at cytosine bases to form C^5-methylcytosine or N^4-methylcytosine, or at adenine bases to generate N^6-methyladenine (Figure 2-11). Such DNA methylation is involved in biological phenomena ranging from replication of prokaryotic DNA and protection of host DNA to gene regulation in eukaryotes and embryonic development. Other methylated bases are found in RNA (Chapter 4).

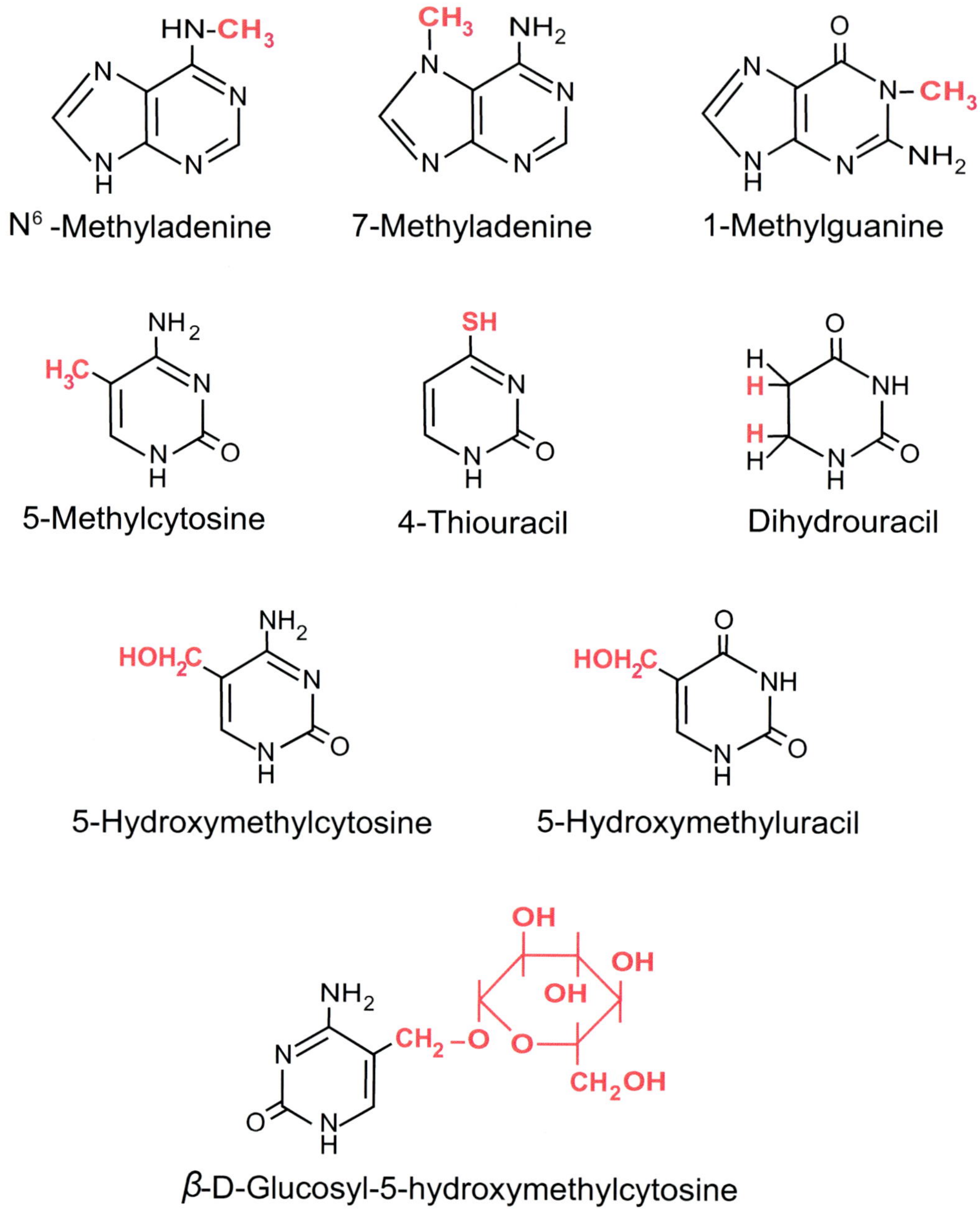

Figure 2-11. Structures of the most common modified nucleobases found in natural nucleic acids. The modifications are in *red*.

Certain organisms, such as the T-even (T2, T4 and T6) bacteriophages, contain four forms of 5-methylcytosine in place of all the cytosine nucleotides: the unmodified 5-methylcytosine, when glycosylated at the hydroxymethyl position with glucose in either the *alpha* or *beta* configuration, or when glycosylated with cellobiose (glucosyl-1,4-β-D-glucose). Other bacteriophages contain 5-hydroxymethyluracil in place of all thymine bases. This modification can also occur by oxidation of thymine residues.

A large number of nucleoside and nucleobase analogs (Figure 2-12) have been synthesized and developed as drugs useful in treating cancer and infections caused by viruses, bacteria and other parasites. These analogs usually act after being converted enzymatically to analogs of the deoxyribonucleotides. This must be done within the cell, because nucleotides generally cannot cross cell membranes, due to their ionization. The modified nucleotides usually function by inhibiting DNA replication in cells, because deoxyribonucleotides normally function only as building blocks for DNA synthesis.

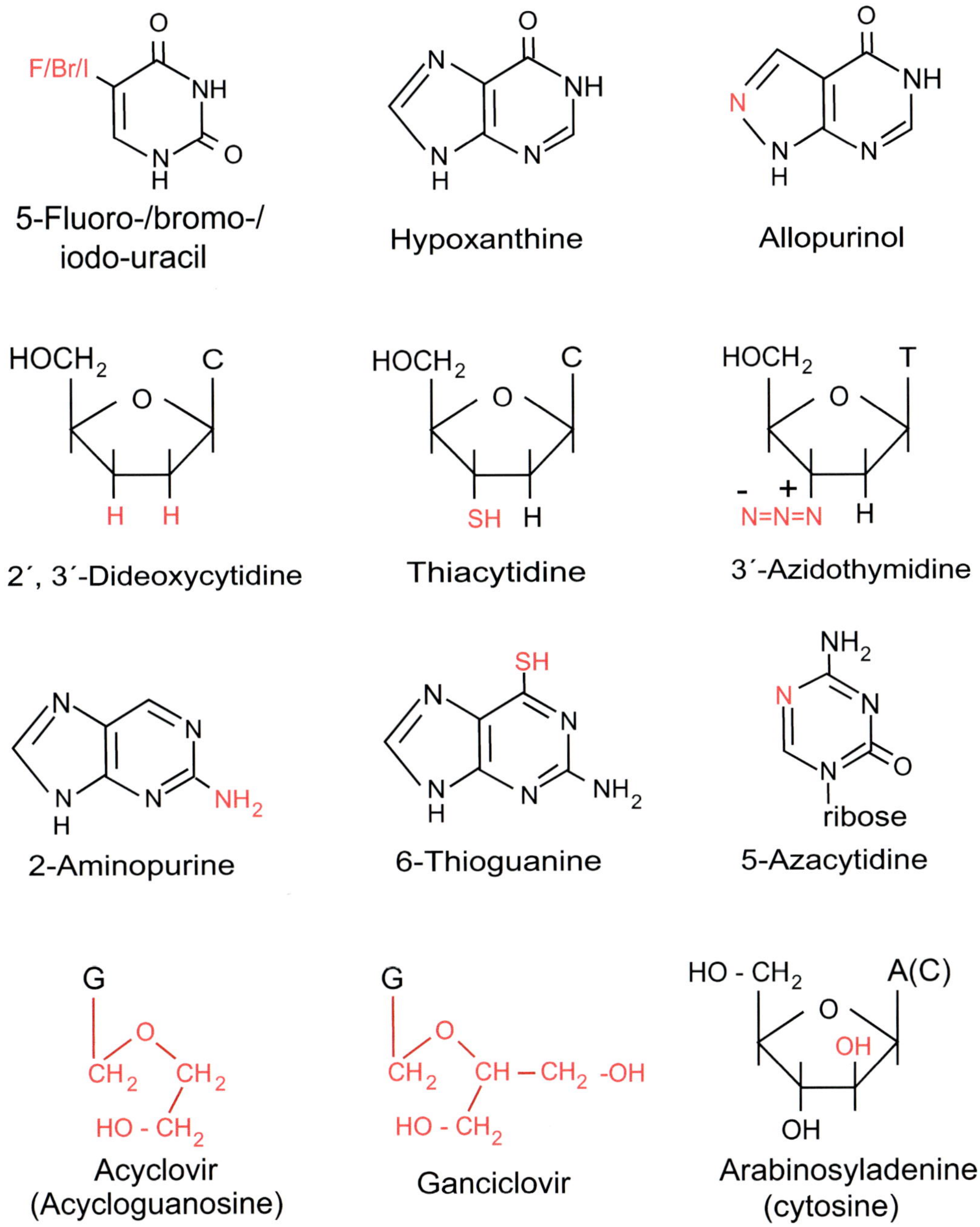

Figure 2-12. Structures of some important and useful analogs of nucleobases and nucleosides.

The fluorinated derivatives 5-fluorouracil and 5-fluorodeoxyuridine, after conversion to the nucleotides, inhibit the synthesis of dTTP and consequently the replication of DNA. The pyrimidines modified by bromine are useful experimentally because of the greater density imparted by the Br atom, making it possible to separate brominated DNA from normal DNA by density-gradient centrifugation. The Br atom is similar in size to a methyl group, so 5-bromouracil can be mistaken as thymine by the cell. Other analogs have the ribose moiety replaced by arabinose, which is the 2-epimer of ribose (Section 1.1.A.4).

Nucleobases in molecular recognition: molecular adducts of adenine and cytosine with COOH functional groups. S. R. Perumalla *et al.* (2005) *Angew. Chem. Int. Ed.* **44**, 7752–7757.

Efforts towards expansion of the genetic alphabet: pyridone and methyl pyridone nucleobases. A. M. Leconte *et al.* (2006) *Angew. Chem. Int. Ed.* **45**, 4326–4329.

Hydration of mononucleotides. D. Liu *et al.* (2006) *J. Am. Chem. Soc.* **128**, 15155–15163.

Syn- and *anti*-conformations of 5′-deoxy- and 5′-O-methyl-uridine 2′,3′-cyclic monophosphate. T. Grabarkiewicz & M. Hoffmann (2006) *J. Mol. Model.* **12**, 205–212.

Thermodynamic properties of enzyme-catalyzed reactions involving cytosine, uracil, thymine, and their nucleosides and nucleotides. R. A. Alberty (2007) *Biophys. Chem.* **127**, 91–96.

2.2. DNA THREE-DIMENSIONAL STRUCTURES

DNA molecules normally exist as two base-paired strands, associated in opposite orientations. An adenine residue in one strand is paired with a thymine in the other strand (A•T or T•A) **and each guanine residue is paired with a cytosine** (G•C or C•G; Figure 2-13). Consequently, the nucleotide sequence of each strand is sufficient to dictate the sequence of the other strand. This is vital to the replication of DNA molecules, in that the two strands are separated and the sequence of each is used as a template to synthesize the other strand with the complementary sequence (Section 6.1). Single-stranded DNA molecules occur only rarely in nature, primarily in a few bacterial viruses where the single-stranded DNA is packaged and protected within the virus. Under most circumstances *in vivo*, a single-stranded DNA molecule will be used as a template to synthesize the second strand, catalyzed by a DNA polymerase enzyme (Section 6.1). Even though the two strands are not linked by any covalent bonds and can be dissociated in various ways (Section 5.1), together they are commonly referred to as a single molecule.

How sequence defines structure: a crystallographic map of DNA structure and conformation. F. A. Hays *et al.* (2005) *Proc. Natl. Acad. Sci. USA* **102**, 7157–7162.

NMR structures of damaged DNA. M. Lukin & C. de Los Santos (2006) *Chem. Rev.* **106**, 607–686.

Exploring DNA structure with Cn3D. S. G. Porter *et al.* (2007) *CBE Life Sci. Educ.* **6**, 65–73.

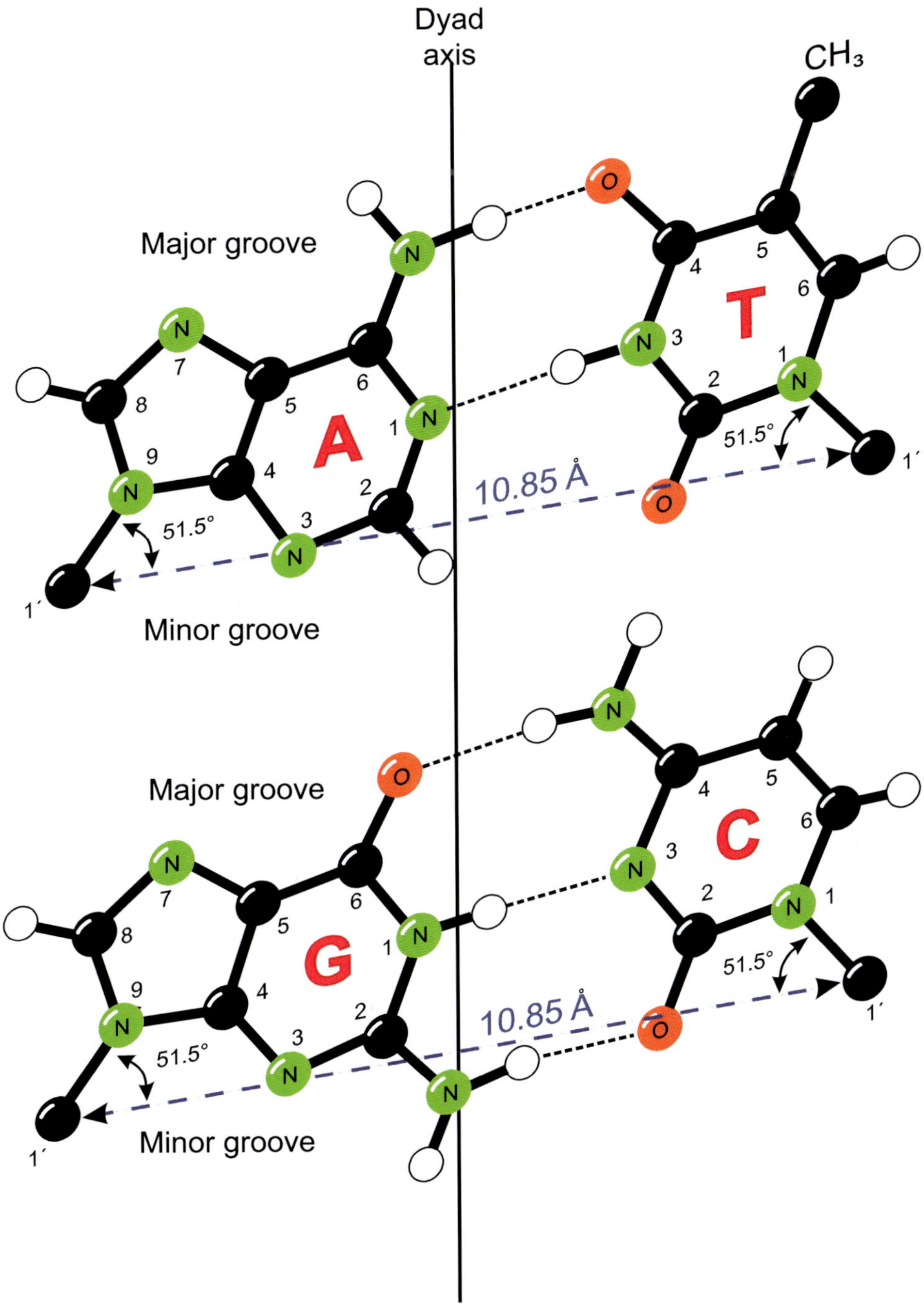

Figure 2-13. The A•T and G•C Watson–Crick base-pairs. H atoms are *white*, and hydrogen bonds are shown as *dashed lines*. The *black spheres* are C atoms, the *green spheres* are N atoms, and the *red spheres* are O atoms. The distance between the C1′ atoms of the two ribose moieties is the same in both base pairs and makes equal angles to the glycosidic bonds to the bases. This gives DNA a series of pseudo-two-fold symmetry axes (often referred to as dyad axes) that pass through the center of each base pair; this symmetry axis is perpendicular to the axis of the double helix. The edges of the base pair that contact the major and minor grooves of the double helix are indicated.

2.2.A. Base Pairing and Stacking

The base pairs normally found in DNA are known as the **Watson–Crick base pairs**, after the discoverers of the DNA double helix. **The specificity of base pairing is provided by both hydrogen bonds and the shapes of the pairs.** The most obvious determinant of base pairing is the hydrogen bonds between the two pairs. An equally important consideration, however, is their identical geometries: the line joining the C1′ atoms of the complementary ribose moieties has the same length in both base pairs and makes equal angles to the glycosidic bonds to the bases; consequently, **both base pairs are perfectly compatible with the same structure of the backbone.** This gives DNA a series of pseudo-two-fold symmetry axes (often referred to as **dyad axes**) that pass through the center of each base pair perpendicular to the axis of the double helix. Note that there are three hydrogen bonds between cytosine and guanine, but only two between adenine and thymine, and C•G base pairs correspondingly tend to be the more stable (Table 5-1). It should also be noted that these hydrogen-bonding patterns require the correct tautomeric state (Section 1.1.C) of each base. Other tautomers are possible (Figure 2-9 and Equation 2.8) and many other types of base pairs are possible and known to occur (Figure 2-14) in unusual DNA structures (Section 2.2.C). The Watson–Crick base pairings are among those that are intrinsically most stable, even in the absence of the double helix, but the double helix further minimizes the other possibilities.

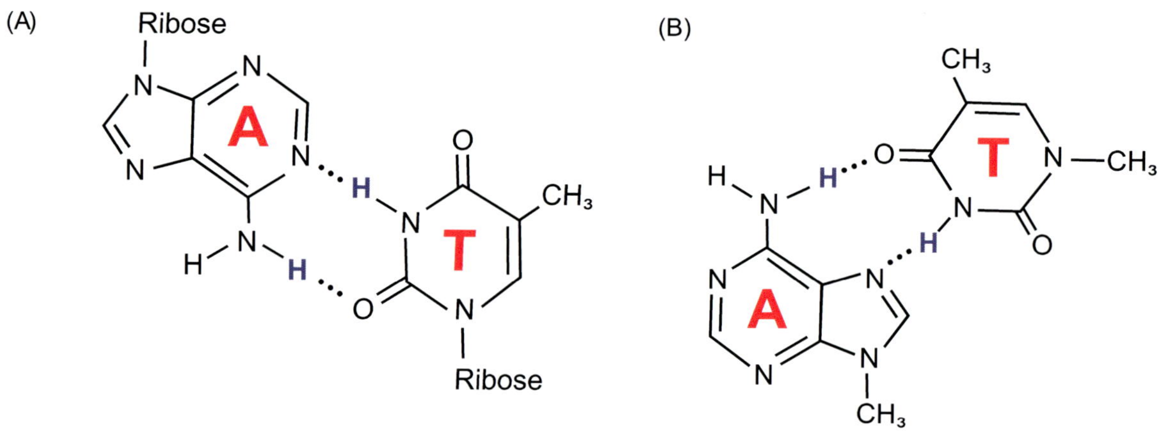

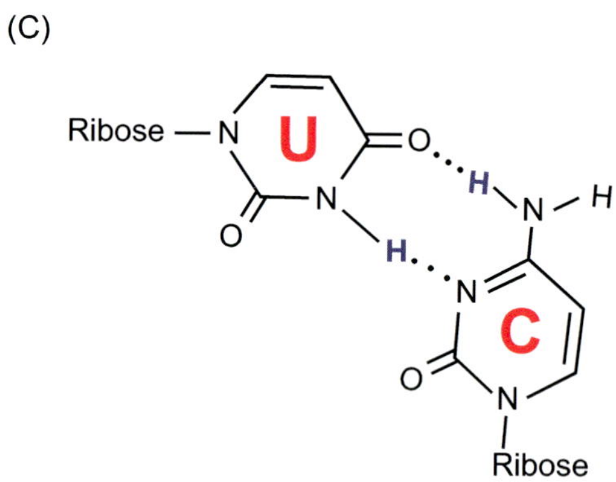

Figure 2-14. Examples of nonWatson–Crick base pairs. The hydrogen bonds are *dotted*. (A) ‘Reverse Watson–Crick’ between adenine and thymine. (B) Hoogsteen pairing between adenine and thymine observed in the crystal structure of 9-methyladenine plus 1-methylthymine. (C) A hypothetical pairing between cytosine and uracil.

Bases in a DNA double helix are stacked on top of each other and in close contact, although they usually do not maximize their contact area but are offset somewhat (Figure 2-17); the resultant structure is probably a compromise between the optimum geometry of base pairing and that of the deoxyribose-phosphate backbone. The base pairs within a DNA duplex often have significant departures from a strict coplanar geometry. The conformational parameters associated with the DNA base pairs are defined in Figure 2-15.

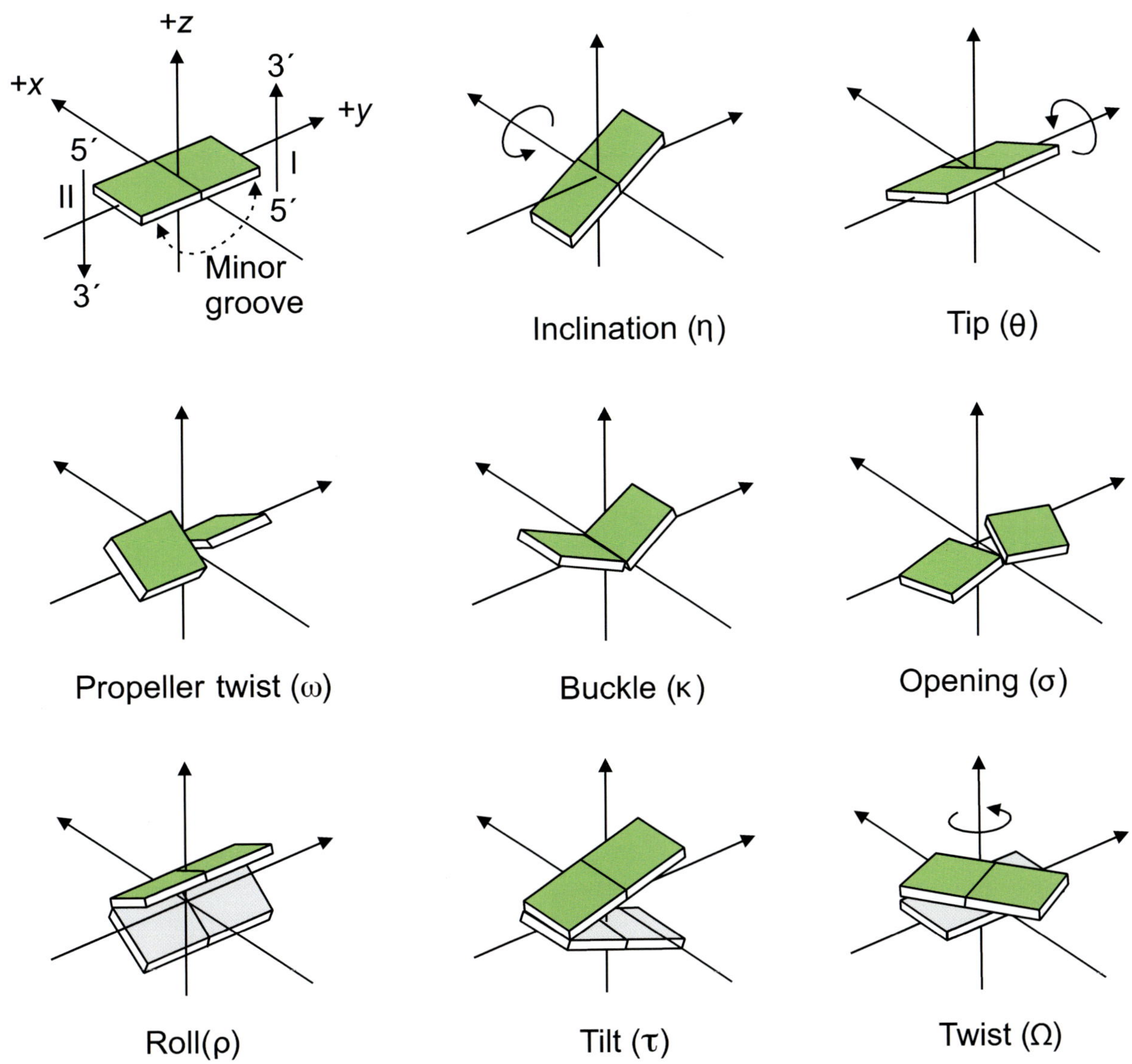

Figure 2-15. Definitions of various base-pair conformations (*upper two rows*) and two successive base pairs (*bottom row*). The motions of the bases in the *top row* are coordinated, whereas they are opposed in the *middle row*.

The stacking interaction is believed to be a complex phenomenon that is stabilized by charge transfer and van der Waals interactions between the π electrons of the aromatic rings; it also minimizes the exposure of the bases to the aqueous solvent. The bases are relatively polarizable and not remarkably hydrophobic (Figure 2-10), due to their polar N and O atoms, but they have large interfaces.

Stacking is an intrinsically favorable interaction for the bases in water, in that individual purines and pyrimidines free in aqueous solution tend to form aggregates of at least five bases, depending upon the concentration. The equilibrium constant for stacking one base on another is in the range of 1–10 M^{-1}. Base stacking is energetically more favorable than pairing because the individual hydrogen bonds formed between bases are not much more stable than those that the bases form with water, so there is little to be gained energetically by them forming hydrogen bonds with each other. On the other hand, base stacking is favored in water because it minimizes the contact of nonpolar surfaces with water. In the absence of water, the bases do interact by hydrogen bonding rather than base stacking.

The stacking interaction has been measured in several model systems (Tables 2-4 and 2-5). One is the stacking interaction within a short DNA duplex having a C•G base pair at the end (Figure 2-16). These data illustrate the predominance of van der Waals interactions and the hydrophobic interaction, in that large nonpolar aromatic groups stack much more strongly than do natural DNA bases. Nevertheless, there are large differences in the strengths of the stacking interaction between adjacent base pairs (Table 2-5), which are believed to be the result of electrostatic interactions between permanent dipoles of the bases. Consequently, the stacking patterns vary with the helix type and the particular base–base step. Purines stack more avidly than pyrimidines, probably because of their larger surface areas. In B-DNA, the 5′-purine-p-pyrimidine-3′ (Pu–Py) base-pair step has a good stacking overlap and is energetically favorable, whereas the 5′-pyrimidine-p-purine-3′ (Py–Pu) base-pair step has a poor stacking overlap and is thus less stable (Figure 2-17).

Table 2-4. Thermodynamic parameters at 25°C for stacking of two individual nucleic acid bases, as measured for the reaction: unstacked dinucleoside phosphate ↔ stacked dinucleoside phosphate

	kcal/mol		
Dinucleotide phosphate	ΔG	ΔH	T ΔS
ApA	+0.64	−5.31	+5.95
ApU	1.15	−8.39	9.54
GpC	0.55	−7.79	8.34
CpG	0.27	−4.80	5.07
UpU	0.86	−7.79	8.65

Data from R. C. Davis & I. Tinoco, Jr (1968) *Biopolymers* **6**, 230.

Table 2-5. Stacking energies of base pairs

Dinucleotide base pairs	Stacking energies (kcal/mol per stacked pair)
(GC)•(GC)	−14.59
(AC)•(GT)	−10.51
(TC)•(GA)	−9.81
(CG)•(CG)	−9.69
(GG)•(CC)	−8.26
(AG)•(CT)	−6.78
(AT)•(AT)	−6.57
(TG)•(CA)	−6.57
(AA)•(TT)	−5.37
(TA)•(TA)	−3.82

Data from R. L. Ornstein *et al.* (1978) *Biopolymers* **17**, 2341.

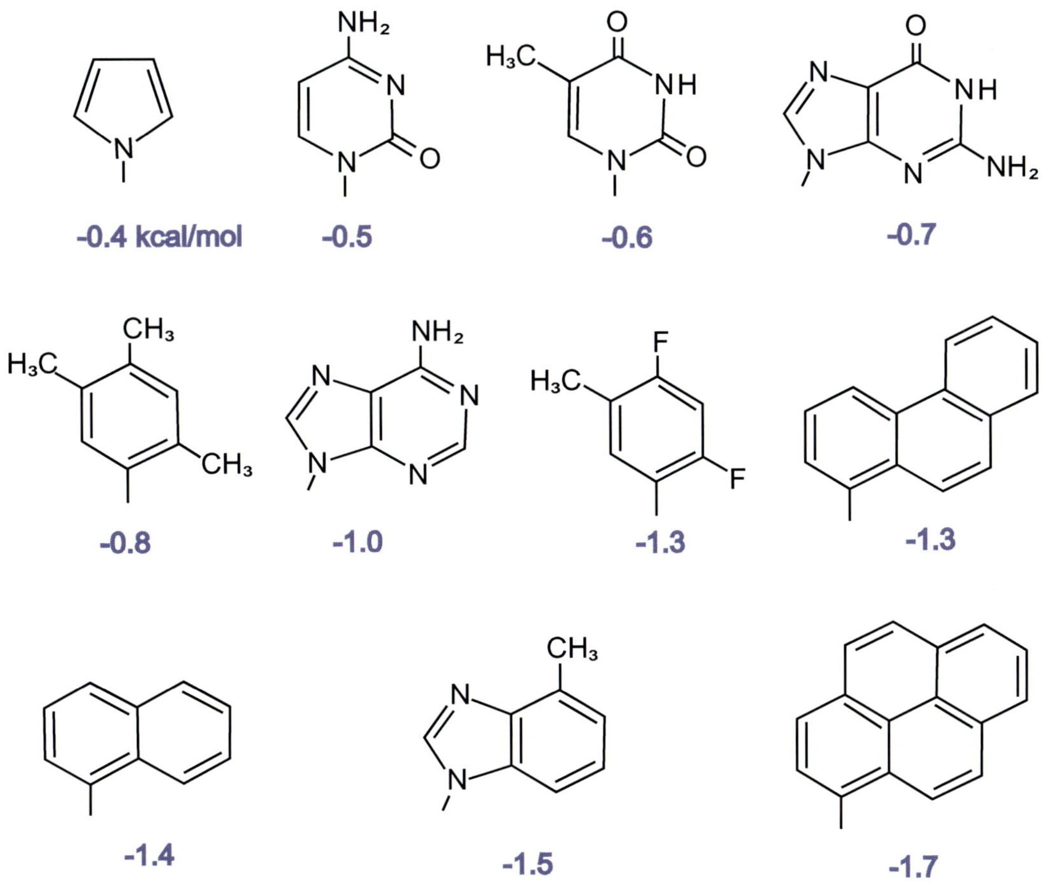

Figure 2-16. Free energies (in kcal/mol) for the stacking of the indicated natural bases and of nonnatural aromatic compounds at the end of a short DNA duplex with a C•G base pair. Data from F. M. Guckian *et al.* (2000) *J. Am. Chem. Soc.* **122**, 2213–2227.

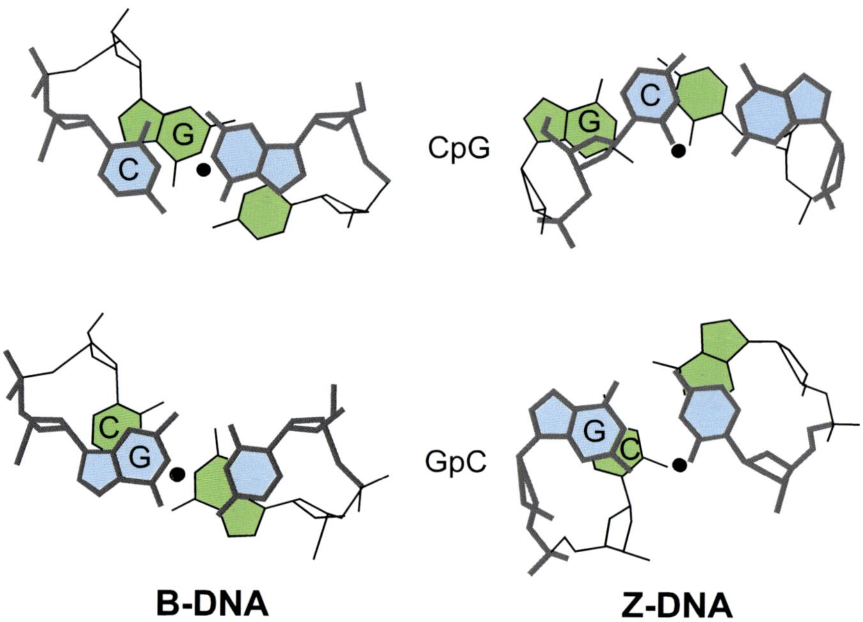

Figure 2-17. Base stacking interactions in B-DNA (*left*) and Z-DNA (*right*) showing pyrimidine–purine step stacking (*top*) and purine–pyrimidine step stacking (*bottom*). The upper base is indicated *blue* with a *gray outline*. The *circle* indicates the helix axis. The pyrimidine–purine steps have very small stacking interactions in B-DNA so these steps are deformed relatively easily.

Nature of base stacking: reference quantum-chemical stacking energies in ten unique B-DNA base-pair steps. J. Sponer *et al.* (2006) *Chemistry* **12**, 2854–2865.

Ab initio determination of the ionization potentials of DNA and RNA nucleobases. D. Roca-Sanjuan *et al.* (2006) *J. Chem. Phys.* **125**, 084302.

Electronic structure of DNA nucleobases and their dinucleotides explored by soft X-ray spectroscopy. Y. Harada *et al.* (2006) *J. Phys. Chem. A* **110**, 13227–13231.

Sensitivity of hydrogen bonds of DNA and RNA to hydration, as gauged by $^{1}J_{NH}$ measurements in ethanol–water mixtures. M. N. Manalo *et al.* (2007) *J. Biomol. NMR* **37**, 257–263.

Hydration and stability of nucleic acid bases and base pairs. M. Kabelac & P. Hobza (2007) *Phys. Chem. Chem. Phys.* **9**, 903–917.

2.2.B. Double Helices

Double-stranded DNA almost always adopts a helical structure (Figure 2-1) **that is much more rigid and extended than a single strand.** The antiparallel double-stranded structure of DNA was originally inferred by Watson and Crick from diffraction patterns of fibers of natural DNA that were produced by causing it to precipitate. Those studies uncovered two distinct forms of DNA, designated A and B, depending upon whether the fibers were dry (A form) or wet (B form), below or above about 85% relative humidity. Watson and Crick were able to infer from the more simple diffraction patterns of the B form that the DNA has a double-helical structure with planar base pairs perpendicular to the helix axis, and 10 base pairs per turn. Model building produced a structure very close to that accepted today (Figure 2-1) and most DNA *in vivo* appears to be in the B form. On the other hand, further studies of DNA fibers with different nucleotide sequences indicated a variety of double-helical structures (Table 2-6), confirming that even double-stranded DNA does not have a closely defined structure.

Short oligonucleotides can be crystallized, and their detailed structures have been elucidated using X-ray crystallography. **The detailed structures obtained are remarkably variable**, however, indicating that the canonical double helix is not a rigid structure but relatively flexible. The DNA structures in crystals appear to be very subject to crystal-packing forces. Structure determination of DNA molecules by nuclear magnetic resonance (NMR) does not suffer from these crystal constraints, but the traditional NMR approach is inherently more difficult with DNA than with proteins because DNA molecules are long and cylindrical and usually lack the elaborate tertiary structures found in proteins that bring atoms distant in the covalent structure into close proximity. In addition, DNA molecules have a lower density of H atoms. Therefore, the number of short distances of less than 5 Å between H atoms that can be observed by NMR (known as NOEs) is comparatively small in oligonucleotides. NOEs are typically found only between adjacent base pairs, so there are usually few long-range NOEs to define the overall structure. As a result, the structure tends to be defined only locally by NMR, and not very accurately. Fortunately, the orientational information contained in residual dipolar couplings has been very useful in determining accurate DNA structures.

The three most common types of DNA double helices are B-, A- and Z-DNA. Their structures are illustrated in Figure 2-18 and compared in Table 2-7. Details of virtually all the known structures of DNA can be obtained from the **Protein Data Bank (PDB; www.rcsb.org)** and from **http://ndbserver.rutgers.edu**. They can be viewed most readily online using the program **Jmol** at **http://firstglance.jmol.org**.

Table 2-6. Helical forms of DNA and RNA characterized by fiber diffraction

Polynucleotide	Relative humidity (%)	Counterion	Structure[a]	Helix symmetry	Helical rise (Å)	Turn angle (°)	Tilt (°)
Native DNA	75	Na	A	11/1	2.56	32.7	20.3
	92	Na	B	10/1	3.38	36.0	–5.9/ –2.1
	57–66	Li	C	10/1	3.32	38.6	–8.0/ –1.0
Poly (dA–dT)	75	Na	D	8/1	3.04	45.0	–16.4
	<98	Na	A	11/1			
	66	Li	B	10/1			
	56	Li	W	8/2	6.1/2	43.5/ 46.5	15.0/ 16.2
	56	Li	P	24/4	18.55/ 4	36.5– 54.4	18.0– 23.3

Table 2-6. Helical forms of DNA and RNA characterized by fiber diffraction - continued

Polynucleotide	Relative humidity (%)	Counterion	Structure[a]	Helix symmetry	Helical rise (Å)	Turn angle (°)	Tilt (°)
Poly(dG–dC)	<92	Na	A	11/1			
	81	Li	B	10/1			
	56	Li	W	10/2	6.72/2		
	43	Na	Z	6/5	3.63	–30.0	
Poly(dA–dC)•poly(dG–dT)	66	Na	A	11/1			
	66–92	Na	B	10/1			
	66	Na	Z	6/5			
Poly(dA–dG)•poly(dC–dT)	66	Na	C	9/2	3.23	48.0	
	95	Na	B	10/1			
Poly(dA)•poly(dT)	70	Na	B′	10/1	3.29	36.0	7.9/ 1.0
	71		H(A/B)	10/1	3.23		26.5/ 18.0
Poly(dG)•poly(dC)	75	Na	A	11/1	2.56	32.7	
	92	Na	B	10/1			
Poly(dT)•poly(dA)•poly(dT)		Na	A		3.26		7.2/ 9.2
DNA•RNA hybrid	33–92	Na	A	11/1			
Poly(I)•poly(dC)	75	Na	A′	12/1			
Poly(dA)•poly(U)			H(B/A)	11/1	3.06	32.7	13.0
Poly(dI)•poly(C)			H(B/A)	10/1	3.06	32.7	15.0
Poly(A)•poly(U)	<92	Na	A	11/1	2.56	32.7	16.0
Poly(I)•poly(C)	<92	Na	A′	12/1	3.0	30.0	10.0
Poly(U)•poly(A)•poly(U)	72	Na	A	11/1	3.04	32.7	11.7/ 13.6

[a] H, heteromorphic helices, where (A/B) indicates that the first strand is in the A conformation and the second in the B conformation; W, wrinkled helices with dinucleotide repeat; Z, left-handed helix with dinucleotide repeat; P, pleiomorphic helix with hexanucleotide repeat. Data from W. Guschlbauer (1990) in *Concise Encyclopedia of Polymer Science and Engineering* (J. Kroschwitz, ed.), Wiley-Interscience, NY.

Table 2-7. Structural Parameters of A-, B- and Z-DNA in solution

Structural parameter	A-DNA	B-DNA	Z-DNA
Helical sense	Right-handed	Right-handed	Left-handed
Base pairs/turn	11	10.4	12
Rise per base pair (Å)	2.3	3.3	3.7
Helical pitch (Å)	25.4	34	45
Repeat unit (base pairs)	1	1	2
Tilt (°)	19	0	-9
Glycosyl angle	*anti*	*anti*	*anti* at C, *syn* at G
Sugar pucker	C3′-*endo*	C2′-*endo*	C2′-*endo* at C, C3′-*endo* at G
Phosphate conformation (α/ζ)(°)	−88/−44	−40/−98	−146/80 at C, 60/−58 at G
Helical diameter (Å)	25	20	18
Major groove	Narrow and deep	Wide and deep	Flattened
Minor groove	Wide and shallow	Narrow and deep	Narrow and deep

Data from A. H.-J. Wang & H. Robinson (1999) in *Encyclopedia of Molecular Biology* (T. E. Creighton, ed.), Wiley-Interscience, NY.

The specificity of DNA structure clearly arises from the hydrogen bonds between the base pairs, and they also stabilize the double-stranded structure, just as hydrogen bonds stabilize folded protein molecules. Even though they can be replaced by hydrogen bonds to water, they form cooperative networks in double-stranded DNA in which they stabilize each other. The net strengths of individual hydrogen bonds between the base pairs have been estimated by comparing the stabilities of analogs of the base pairs lacking the hydrogen-bonding groups: values of between −0.25 and −1.6 kcal/mol per hydrogen bond have been measured for the normal base pairs, so **hydrogen bonds undoubtedly contribute to the stabilities of the double-stranded structures of DNA**. The variations in the degree of stabilization are probably due to variation in the stacking interactions between adjacent base pairs. **Incorrect base pairing has a drastic affect on the stability of the double helix, each incorrect base pair within the interior of the double helix decreasing its stability by 1–4 kcal/mol**; the most important effect is believed to be the need to distort the mismatched pair into a nonstandard geometry, leading to unfavorable conformational interactions in neighboring pairs. Mismatches at the ends of oligonucleotides have little or no effect.

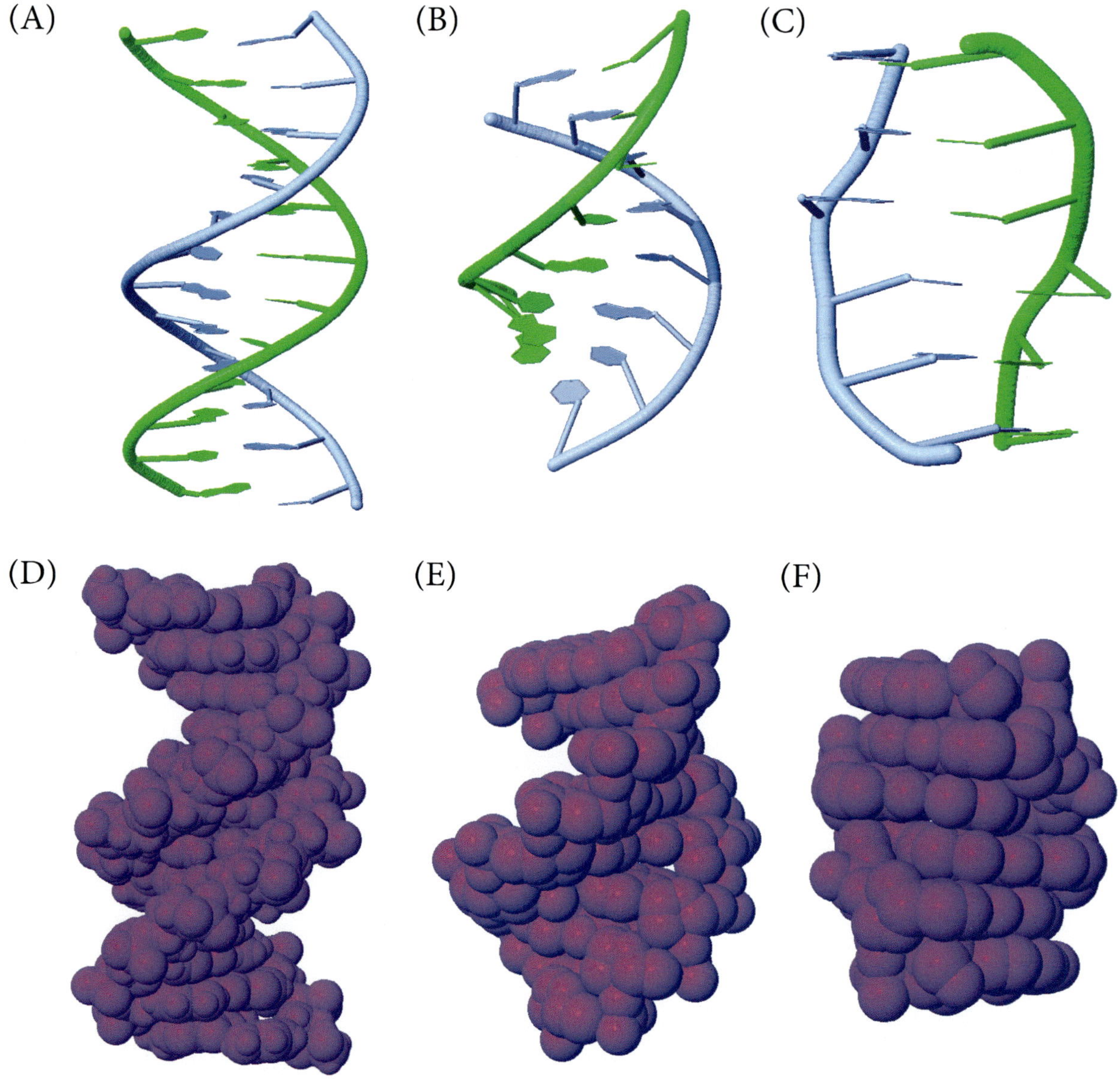

Figure 2-18. Side views of the double helices of B-DNA (A and D), A-DNA (B and E) and Z-DNA (C and F). (A–C) Skeletal models; (D–F) space-filling models. Figures generated from PDB files *1duf*, *115d* and *1dcg*, respectively, using the program Jmol.

A notable feature of all DNA double helices is the presence of a wide **major groove** and a narrow **minor groove** along the surface (Figure 2-1). Both grooves are important because they provide access to the base pairs within the interior of the helix, where proteins and other ligands can recognize the nucleotide sequence. Many ligands are observed to bind in these two grooves, as well as relatively fixed water molecules (Figure 2-19).

The DNA double helices keep the ionized phosphate groups on the outside, in contact with water and counterions, whereas the relatively nonpolar bases are in the interior. Consequently, nucleic acids are very soluble in water but can be precipitated most readily and reversibly by adding nonpolar alcohols, such as isopropanol.

The negative charge associated with each of the phosphate groups makes DNA an extreme polyelectrolyte. These negative charges are generally neutralized by the positive charges of metal ions, polyamines or proteins. Metal ions, such as sodium, potassium or magnesium ions, are used in the screening of the DNA negative charges by interactions with the phosphate O atoms and are crucial for the stability of the DNA structure. This negative charge also repels nucleophilic species, such as hydroxide ions, so phosphodiester bonds are much less susceptible to hydrolytic attack than other esters, such as carboxylic acid esters, which is crucial for maintaining the integrity of the DNA structure and the genetic information stored there. The absence of the 2′-hydroxyl group in DNA further increases its resistance to hydrolysis (Figure 4-2) and probably is the reason why DNA, rather than RNA, tends to be used for storing genetic information.

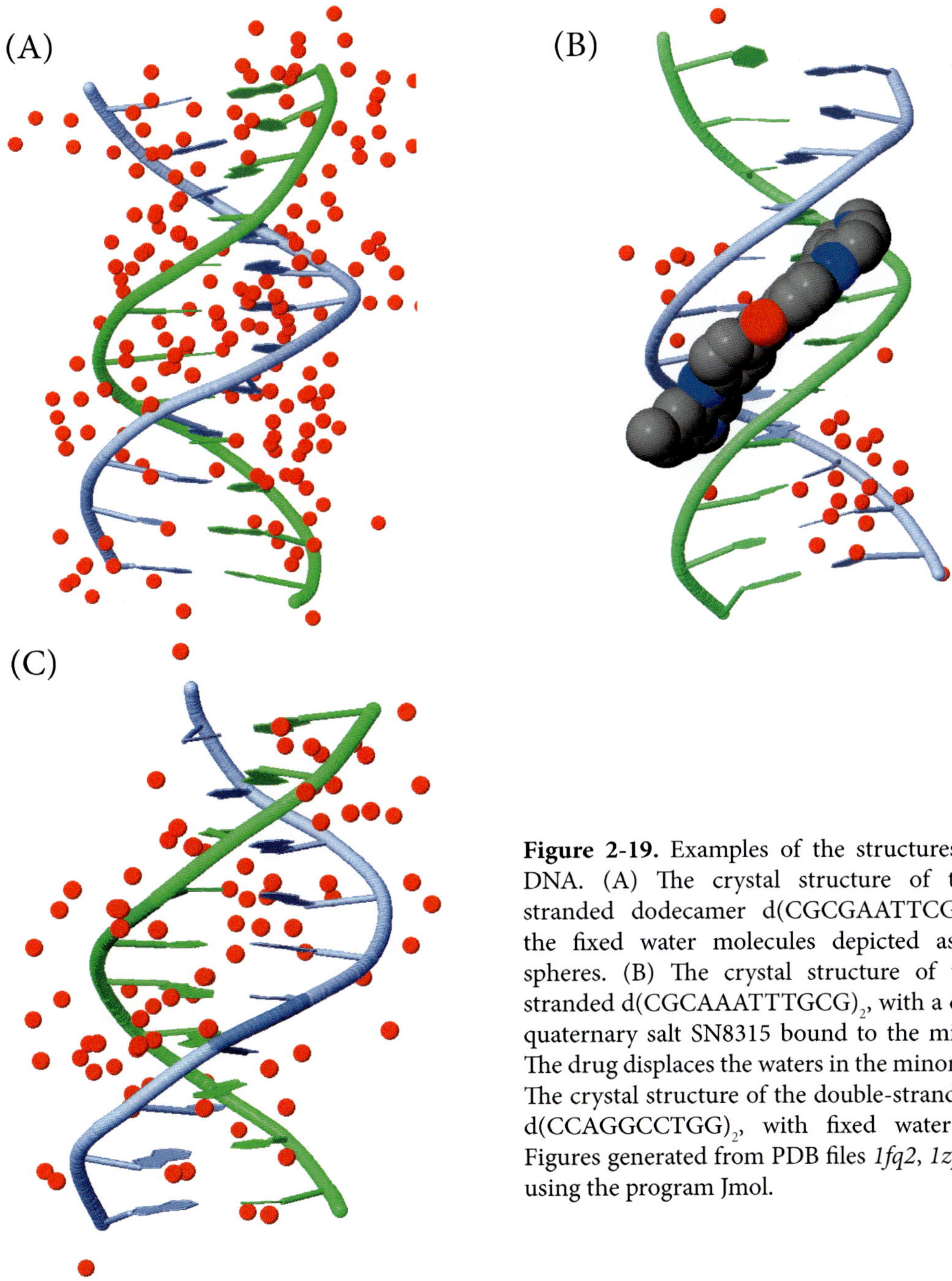

Figure 2-19. Examples of the structures of B-form DNA. (A) The crystal structure of the double-stranded dodecamer d(CGCGAATTCGCG)$_2$, with the fixed water molecules depicted as small red spheres. (B) The crystal structure of the double-stranded d(CGCAAATTTGCG)$_2$, with a quinolinium quaternary salt SN8315 bound to the minor groove. The drug displaces the waters in the minor groove. (C) The crystal structure of the double-stranded decamer d(CCAGGCCTGG)$_2$, with fixed water molecules. Figures generated from PDB files *1fq2*, *1zph* and *2d25* using the program Jmol.

The solvent-accessible surface areas of each nucleotide in seven different double helices are presented in Table 2-8, along with their intrinsic molecular volumes. The intrinsic volumes are similar within 3%, but the accessible surface areas vary by up to 19%. The ionized phosphate groups make up about 47% of the total accessible surface area, the polar groups another 14% and nonpolar atoms about 39%.

Table 2-8. The accessible surface areas of the charged, polar and nonpolar atoms, the overall surface areas, and the intrinsic volumes per base pair for seven nucleic acid duplexes

	Surface areas (Å^2)				
Duplex	**Charged**	**Polar**	**Nonpolar**	**Total**	**Intrinsic volume (Å^3)**
Poly(A)•poly(U)	75.9	36.6	56.2	168.7	272.9
Poly(A)•poly(dT)	72.4	27.6	64.8	164.7	282.3
Poly(dA)•poly(U)	81.0	26.2	66.7	173.8	273.1
Poly(dA)•poly(dT)	78.9	12.3	67.7	158.8	277.2
Poly(dAdT)•poly(dAdT)	74.2	16.4	59.7	150.3	284.1
Poly(dGdC)•poly(dGdC)	78.4	20.5	72.2	171.0	275.4
Poly(dIdC)•poly(dIdC)	71.9	19.1	51.7	142.6	284.0
Average	76 ± 5	23 ± 7	63 ± 8	162 ± 10	281 ± 8

A glossary of DNA structures from A to Z. A. Ghosh & M. Bansal (2003) *Acta Crystallogr. D* **59**, 620–626.

DNA coiled coils. J. L. Campos *et al.* (2005) *Proc. Natl. Acad. Sci. USA* **102**, 3663–3666.

Overview of the structure of all-AT oligonucleotides: organization in helices and packing interactions. L. Campos *et al.* (2006) *Biophys. J.* **91**, 892–903.

Estimation of strength in different extra Watson–Crick hydrogen bonds in DNA double helices through quantum chemical studies. D. Bandyopadhyay & D. Bhattacharyya (2006) *Biopolymers* **83**, 313–325.

Molecular design and synthesis of artificial double helices. Y. Furosho & E. Yashima (2007) *Chem. Rec.* **7**, 1–11.

1. B-DNA

B-DNA is the predominant DNA structure under normal physiological conditions. The B-DNA double helix is right-handed, with its base pairs perpendicular to the helix axis, which passes through the centers of the base pairs (Figure 2-1). The pitch of the helix is determined by the stacking between the adjacent base pairs. The major and minor grooves are roughly equivalent in depth, 8.5 Å and 7.5 Å, respectively, but the width of the major groove is about 12 Å, whereas that of the minor groove only about 6 Å. The bases that define the sequence are more accessible in the major groove, and this is where proteins tend to bind to specific nucleotide sequences (Section 7.2).

The first detailed structure of B-DNA determined by X-ray crystallography (Figure 2-19) was that of a dodecamer with the self-complementary sequence $d(CGCGAATTCGCG)_2$. It exhibited a number of interesting structural features that now appear to be general: the minor groove is narrower at the AATT region of the double helix than at the CGCG ends and is filled by a spine of water molecules that form hydrogen bonds to both the O2 atoms of thymines and the N3 of adenines; the base pairs in the central AATT region of the helix have high propeller twist angles, which enhances stacking of the bases along each strand of the double helix (A•T base pairs are generally more amenable to propeller twisting than G•C base pairs because of being paired by only two hydrogen bonds); finally, the sugar pucker of the deoxyribose ring favors the C2′-*endo* conformation, although conformations ranging from C1′-*exo* to O4′-*endo* are also present. This reflects the flexibilities of B-DNA structures, which depend upon the exact sequence and the environment. In general, runs of As lead to a very narrow minor groove and a high propeller twist, which introduce a bend in the helix axis. Several DNA decamer oligonucleotides with 'mixed' sequences, for example d(CCAGGCCTGG) (Figure 2-19-C), have slightly narrower or wider minor grooves, depending on the sequence. In general, the A•T regions have a narrower minor groove than the G•C regions.

Most DNA molecules in solution have the B-DNA structure, and this structure will be assumed here unless stated otherwise.

Accurate representation of B-DNA double helical structure with implicit solvent and counterions. L. Wang *et al.* (2002) *Biophys. J.* **83**, 382–406.

Sequence-dependent DNA structure: a database of octamer structural parameters. E. J. Gardiner *et al.* (2003) *J. Mol. Biol.* **332**, 1025–1035.

Molecular dynamics simulations of DNA with polarizable force fields: convergence of an ideal B-DNA structure to the crystallographic structure. V. Babin *et al.* (2006) *J. Phys. Chem. B* **110**, 11571–11581.

2. *A-DNA*

B-DNA can be converted to A-DNA under conditions of low hydration and by adding alcohols; the conversion is reversible and occurs on the microsecond time scale. A-DNA was first recognized from fiber X-ray diffraction analysis, where the hydration of the insoluble samples can be varied drastically. The double helix of A-DNA is short and fat, with the base pairs and backbone wrapped further away from the helix axis than in B-DNA (Figure 2-18). The base pairs are tilted significantly relative to the helix axis, by about 19°, and display a minor propeller twist. The equivalent of the major groove of B-DNA is deep and very narrow, with a width of only about 3 Å and a depth of about 13 Å. In contrast, the minor groove has a breadth of 11 Å and a shallow depth of 3 Å.

The A-form of DNA is often observed when short duplexes are characterized by X-ray crystallography. This is now thought to be due to the addition of alcohols to induce crystallization of the DNA. The propensity to form A-DNA is sequence-dependent: **guanine-rich regions readily form A-DNA, whereas stretches of adenine resist it**. This may be due to guanine-rich sequences having a characteristic intrastrand guanine–guanine stacking interaction in the A-DNA double helix. Also, the terminal base pairs from one helix abut the minor groove surface of the neighboring helix in the crystal lattice, minimizing the accessibility of solvent to the wide minor groove. That a low-humidity environment favors formation of the A-type helix indicates that the displacement of surface solvent molecules by hydrophobic base pairs provides a driving force in stabilizing the A-DNA conformation. Complex ions such as the physiological polyamines:

putrescine: $^{+}H_3N{-}(CH_2)_4{-}NH_3^{+}$

spermidine: $^{+}H_3N{-}(CH_2)_4{-}^{+}NH_2{-}(CH_2)_3{-}NH_3^{+}$

spermine: $^{+}H_3N{-}(CH_2)_3{-}^{+}NH_2{-}(CH_2)_4{-}^{+}NH_2{-}(CH_2)_3{-}NH_3^{+}$ (2.4)

Cobalt(III)hexamine and neomycin can also stabilize the A-DNA conformation in DNA containing stretches of G•C base pairs. Shorter helices with fewer base pairs convert to A-DNA more readily than do longer helices.

Double-stranded RNA and a number of DNA•RNA hybrids also adopt the A-DNA conformation. The ribose and 2′-deoxyribose sugars are all in the C3′-*endo* conformation, and the 2′-hydroxyl groups of the ribose form hydrogen bonds to adjacent nucleotides in the chain.

The dynamics of the B–A transition of natural DNA double helices. D. Jose & D. Porschke (2005) *J. Am. Chem. Soc.* **127**, 17120–16128.

B-DNA helix stability in a solvent-free environment. E. S. Baker & M. T. Bowers (2007) *J. Am. Mass Spectrom.* **18**, 1188–1195.

---

3. *Z-DNA*

The Z-DNA structure was unexpected and is unusual in being left-handed. Its name originates from the zig-zag path of the phosphate groups along the DNA backbone. Relative to B- and A-DNA, the Z-DNA double helix is tall and thin, with a diameter of about 18 Å and a helical pitch of about 45 Å (Figure 2-18 and Table 2-7). Z-DNA is favored by alternating Py–Pu sequences, especially (dC-dG)$_n$, and it was discovered in the structure of d(CGCGCG)$_2$. The repeat unit is a CpG dinucleotide, with the glycosyl conformation of cytosine being *anti* and that of guanine being *syn* (Equation 2.1). The sugar pucker for cytosine is C2′-*endo*, while that for guanine is predominantly C3′-*endo* (Figure 2-5). There are 12 base pairs per helical turn, the base pairs are slightly inclined by −9°, and the helical twist angle per dinucleotide repeat is −60°, about −8° for the CpG and −52° for the GpC. The rise per dinucleotide repeat is about 7.4 Å. The Z-DNA conformation has been observed in the crystal structures of several other oligonucleotides, including fragments containing nonalternating Pu–Py sequences and A•T and G•T base pairs.

The single groove of Z-DNA is narrow and deep, with an opening of only 6–7 Å. The equivalent of the major groove is almost nonexistent, being very shallow and exposed to the solvent. Spermine (Equation 2.4) and hydrated metal ions interact with the phosphates and the N7 atoms of guanine bases and neutralize the negative charge. Binding of spermine molecules may help to stabilize the Z-DNA helix, because some are found in the deep groove and bridge it by interacting with phosphate groups on both sides.

DNA is in a dynamic state and forms Z-DNA when circumstances stabilize it, then relaxing to right-handed B-DNA. DNA with alternating dCdG sequences, like poly(dCdG), can interconvert especially easily between the B- and Z-DNA conformations. This is somewhat surprising, because this interconversion requires a substantial change of the purines from *anti* to *syn* conformations (Equation 2.1 and Figure 2-20) plus a change in the hand of the helix. Each time a DNA segment turns into Z-DNA, two B–Z junctions form. There is continuous stacking of bases across this junction, except that one base pair at the junction is broken and the bases on each side are extruded. The tendency to form the Z-conformation is increased by negative supercoiling (Section 3.1), increasing the ionic strength, and by the presence of proteins that bind only to Z-DNA. The left-handedness of Z-DNA is especially adept at relieving the strain of negatively supercoiled DNA, for example during gene transcription.

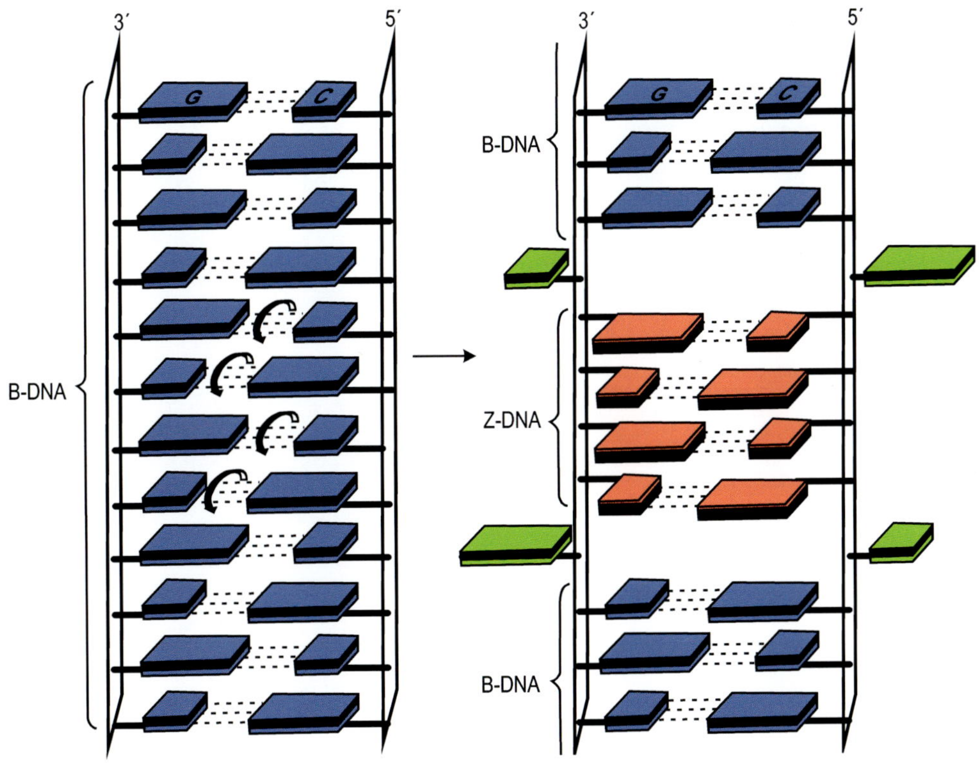

Figure 2-20. Interconversion of B- and Z-DNA requires disrupting the base pairing and flipping the planes of the nucleotide bases. Modified from A. Rich *et al.* (1984) *Ann. Rev. Biochem.* **53**, 799.

Crystal structure of a junction between B-DNA and Z-DNA reveals two extruded bases. S. C. Ha *et al.* (2005) *Nature* **437**, 1183–1186.

The hydration structure of a Z-DNA hexameric duplex determined by a neutron diffraction technique. T. Chatake *et al.* (2005) *Acta Crystallogr. D* **61**, 1088–1098.

Structure of d(TGCGCG)•d(CGCGCA) in two crystal forms: effect of sequence and crystal packing in Z-DNA. S. Thiyagarajan *et al.* (2005) *Acta Crystallogr. D* **61**, 1125–1131.

The rare crystallographic structure of $d(CGCGCG)_2$: the natural spermidine molecule bound to the minor groove of left-handed Z-DNA $d(CGCGCG)_2$ at 10°C. H. Ohishi *et al.* (2007) *Biochem. Biophys. Res. Commun.* **358**, 24–28.

2.2.C. Other DNA Structures

The double-helical structures of DNA described above are not so stable that they predominate under all conditions and with all nucleotide sequences. Certain sequences of DNA form 3-D structures that are not the common A-, B- or Z-DNA double helices.

1. Hoogsteen Base Pairing

Early analysis of nucleic acid base pairs observed that the individual nucleobases cytosine and guanine in water associate to form the Watson–Crick base pair, but that adenine and thymine in isolation preferred to pair differently from that observed in Watson–Crick base pairs. The N7 atom of adenine base paired with N3 of thymine instead of N1 (Figures 2-14 and 2-21). This type of A•T base pair was named the Hoogsteen base pair. The corresponding G•C Hoogsteen base pair is more stable if the cytosine is protonated at the N3 position (which is often depicted as C^+; Figure 2-21). An important difference between the two types of base pairs is that the C1′–C1′ distance is only 8.65 Å in the Hoogsteen base pair, significantly shorter than the 10.5 Å of the Watson–Crick base pair. Consequently, Hoogsteen base pairs are not compatible structurally with the B-DNA structure, but they do occur, usually in regions containing significant distortion or near sites of drug intercalation. They are physically possible, but they destabilize the B-DNA structure and are energetically unfavorable.

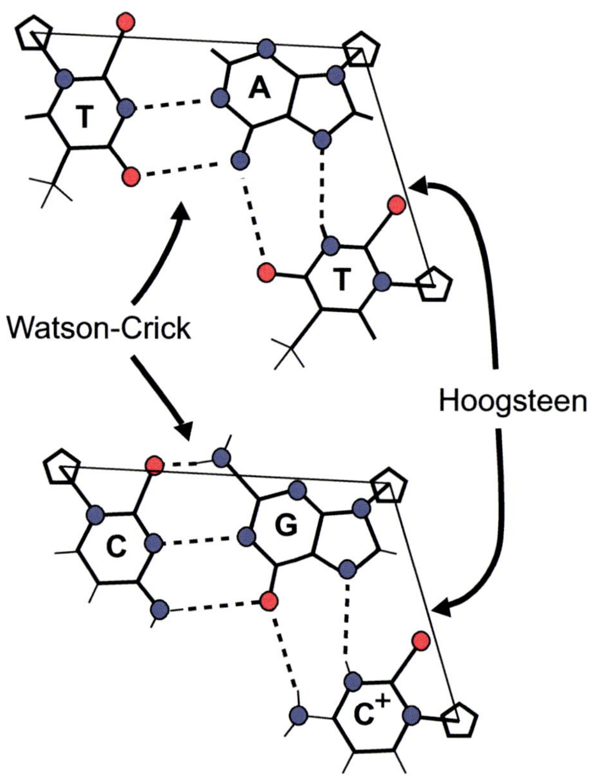

Figure 2-21. Comparison of the Watson–Crick and Hoogsteen base pairs for A•T and G•C. Hydrogen bonds are depicted as *dashed lines*; the two C1′ atoms are depicted as *open pentamers*, and the distance between them is indicated by a *thin line*. N atoms are *blue spheres* and O atoms are *red spheres*.

Certain chemical modifications increase the stability of the Hoogsteen base pair, and modified bases may be used when a Hoogsteen base pair is needed. For example, the modified nucleoside 3-isodeoxyadenosine (iA) forms a stable base pair with thymine using the Hoogsteen conformation: the amino group at the C8 position of 8-amino-adenine can form an additional hydrogen bond with the O2 of thymine. The stable d(CG[iA]TCG)$_2$ duplex illustrates that the iA•T base pair is fully compatible with the normal Watson–Crick base pair.

The Hoogsteen and other alternative base pairs can be generated when certain proteins or drugs bind to DNA (Section 2.5). The antibiotics triostine A and echinomycin bind to –CG– sequences by intercalating (Section 2.5.B) on both sides of a CG dinucleotide base pair. The A•T and G•C$^+$ base pairs adjacent to the antibiotic rings adopt Hoogsteen structures, because this improves the stacking interactions between them and the antibiotic.

The A•T and G•C$^+$ Hoogsteen base pairs have important roles in the structures of nucleic acid triple helices (Section 2.2.C.2). Hoogsteen base pairing is not restricted to DNA but also occurs in transfer RNA (Section 4.2.B), where the nucleic acid bases are often modified and involved in tertiary interactions.

X-ray and NMR studies of the DNA oligomer d(ATATAT): Hoogsteen base pairing in duplex DNA. N. G. Abrescia *et al.* (2004) *Biochemistry* **43**, 4092–4100.

Theoretical study of the Hoogsteen–Watson–Crick junctions in DNA. E. Cubero *et al.* (2006) *Biophys. J.* **90**, 1000–1008.

Hoogsteen base-pairing revisited: resolving a role in normal biological processes and human diseases. G. Ghosal & K. Muniyappa (2006) *Biochem. Biophys. Res. Commun.* **343**, 1–7.

2. Triple Helices

Adding increasing amounts of poly(U) to poly(A) first produces the expected 1:1 poly(A)•poly(U) duplex but then a 2:1 complex of three strands is also formed. In the triple-helical structure, the extra strand of poly(U) is bound to the poly(A) strand of the poly(A)•poly(U) duplex using Hoogsteen base pairs (Figure 2-21). The triple helix is important for its possible role in gene transcription and in the design of antisense oligonucleotides (Section 5.3.D).

Triplexes can be either parallel or antiparallel (Figure 2-22). In the parallel motif, a third strand (rich in thymine and cytosine) binds to a purine (adenine and guanine)-rich region within a standard double helix and lies parallel to the purine strand with which it makes contact. In the antiparallel motif, the third strand is rich in guanine and adenine (or thymine) and binds in antiparallel orientation to a purine region of double-helical DNA. In each case, **the third strand lies in the major groove, and the double helix of the other two strands remains largely undistorted**. One example of the 3-D structure of a triple-stranded DNA molecule is shown in Figure 2-23. The third strand forms Hoogsteen base pairs with the purine-rich strand of the duplex. The parallel motif is more stable if the N3 of cytosine is protonated, so it is more stable at low pH; the antiparallel motif is largely insensitive to pH. Triple helices have a greater concentration of ionized phosphate groups than do double helices, so they are more sensitive to the nature and concentration of counterions.

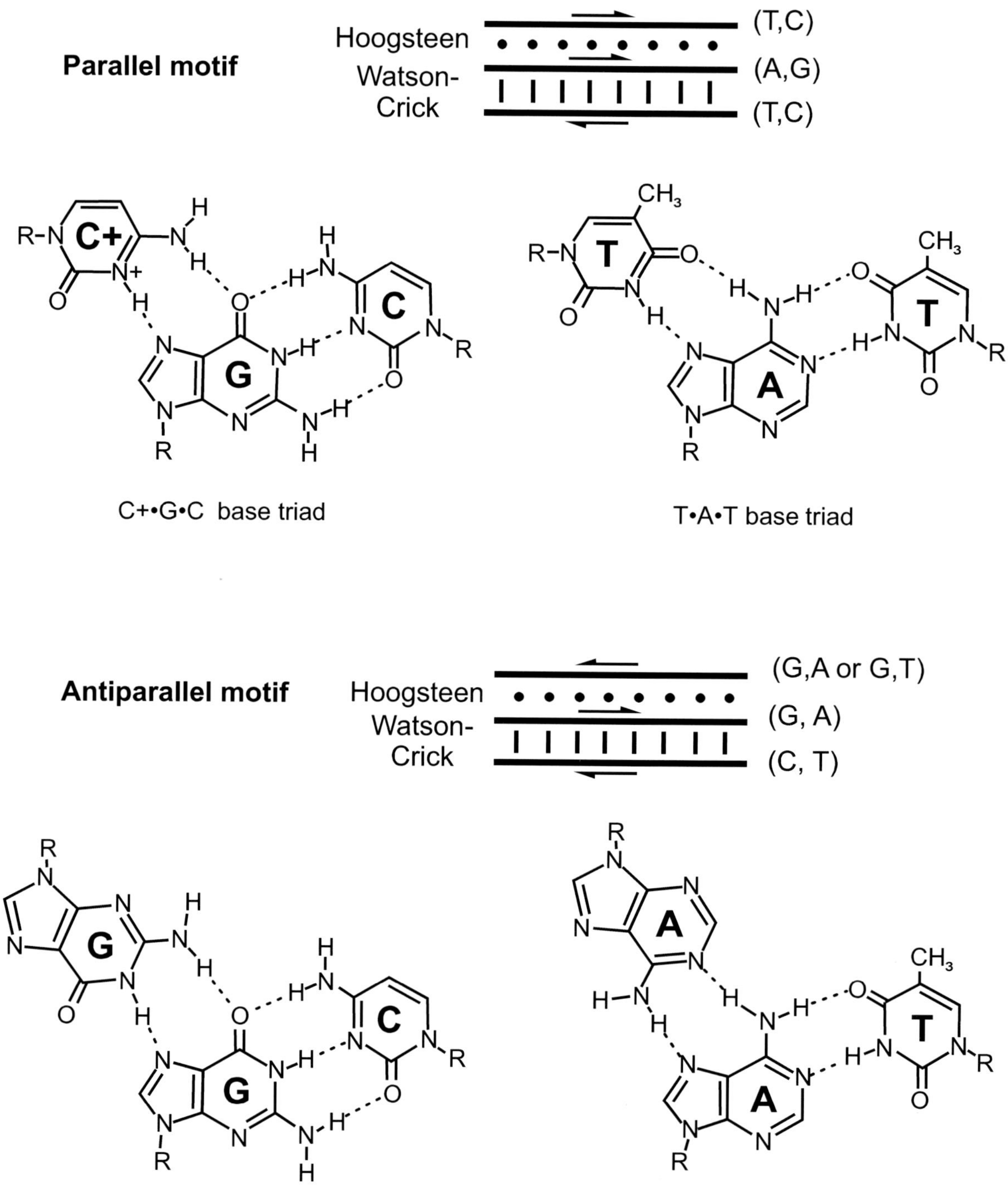

Figure 2-22. Base pairing in triple-helical DNA. The common base triads used in the two types of triplexes in the parallel (*top*) and antiparallel (*bottom*) motifs are described. The orientations of the three strands in the parallel and antiparallel motifs are shown above. In each case, the first and second strands use standard Watson–Crick base pairing, whereas the third strand at the *top* uses Hoogsteen base pairing to the second strand. The T•A•T triad is also found in the antiparallel motif. Adapted from E. T. Kool (1996) *Ann. Rev. Biophys. Biomol. Structure* **25**, 1–28.

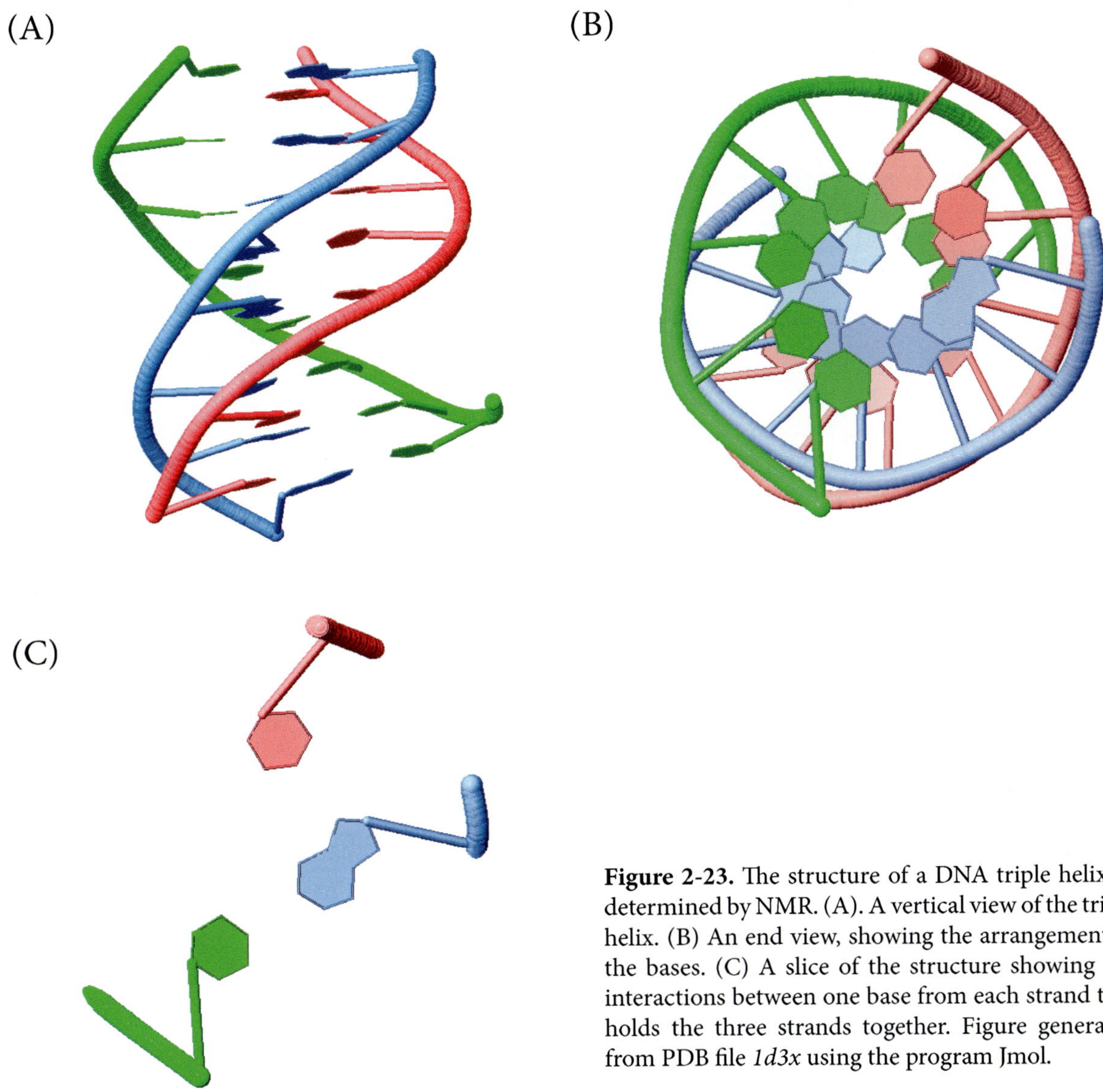

Figure 2-23. The structure of a DNA triple helix as determined by NMR. (A). A vertical view of the triple helix. (B) An end view, showing the arrangement of the bases. (C) A slice of the structure showing the interactions between one base from each strand that holds the three strands together. Figure generated from PDB file *1d3x* using the program Jmol.

Formation of stable triple helices of the (Py)•(Pu)•Py) type is relatively straightforward, but it is more difficult to design oligonucleotides that can bind to a specific DNA duplex using all four bases.

Triplex-forming oligonucleotides: principles and applications. K. M. Vasques & P. M. Glazer (2002) *Quart. Rev. Biophys.* **35**, 89–107.

Effect of divalent cations and cytosine protonation on thermodynamic properties of intermolecular DNA double and triple helices. P. Wu *et al.* (2002) *J. Inorg. Biochem.* **91**, 277–285.

Monitoring denaturation behaviour and comparative stability of DNA triple helices using oligonucleotide–gold nanoparticle conjugates. D. Murphy *et al.* (2004) *Nucleic Acids Res.* **32**, e65.

---

3. H-DNA: Intramolecular Triple Helices

Stretches of $(GA)_n$•$(TC)_n$ sequences in genomic DNA are hypersensitive to nucleases during active gene transcription. They undergo a notable change in DNA structure when in a plasmid that is either negatively supercoiled (Section 3.1) or at low pH. The structural change is due to the formation of an intramolecular triple helix for part of the $(GA)_n$•$(TC)_n$ sequence, with extrusion of a single-stranded loop (Figure 2-24). This structure is known as H-DNA. Other DNA segments with stretches of pyrimidines on one strand and purines on the other have also been shown to adopt the H-DNA structure. The triple helix is formed by the pyrimidine strand and half of the purine strand; the other half of the purine strand is single-stranded. The triple helix is stabilized by Hoogsteen base pairs (Figure 2-21), which accounts for the stabilizing effect of low pH. The single-stranded loop accounts for the susceptibility to nucleases. Negative super-coiling (Section 3.1) introduces the required topology of the chain for the H structure.

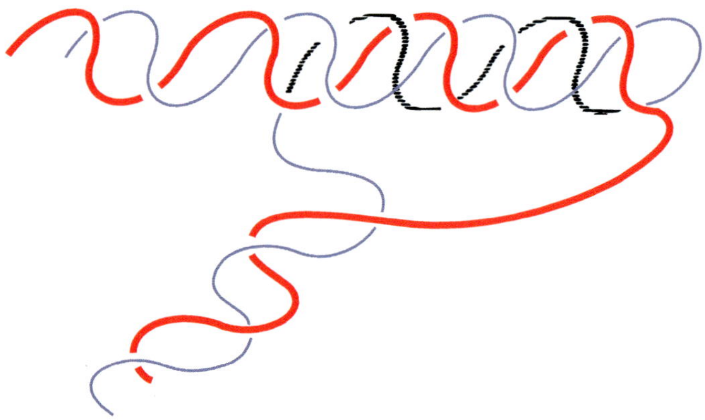

Figure 2-24. Schematic model of H-DNA, with an intramolecular triple helix. The homopurine strand is indicated with the *thick red line*, the homopyrimidine strand with a *thin blue line*. The *dashed line* is the half of the homopyrimidine stand incorporated into the triple helix. Adapted from S. M. Mirkin & M. D. Frank-Kamenetskii (1994) *Ann. Rev. Biophys. Biomol. Structure* **23**, 541–576.

When the pyrimidine-rich strand folds back to form a triplex, cytosines from one half of the sequence should interact with G•C but not A•T base pairs in its other half. Conversely, thymines in one half should interact with A•T but not G•C base pairs from the other half. Consequently, a homopurine–homopyrimidine that forms H-DNA should have a sequence that is a mirror repeat.

H-DNA and related structures. S. M. Mirkin & M. D. Frank-Kamenetskii (1994) *Ann. Rev. Biophys. Biomol. Structure* **223**, 541–576.

H-DNA:DNA triplex formation within topologically closed plasmids. S. L. Broitman (1995) *Prog. Biophys. Mol. Biol.* **63**, 119–129.

4. Four-stranded Structures: Guanine Quartet

The ends of eukaryotic chromosomes, known as **telomeres**, have DNA sequences that are rich in guanine and contain a short highly repetitive sequence, such as $(TTGGGG)_n$. The guanine-rich strand at the 3′-end exists as a single-strand and is capable of forming a higher ordered DNA structure in which four strands associate, known as the **guanine- or G-quartet**. One arrangement of the G-quartet is a four-fold symmetric motif in which the four guanine strands are in parallel orientations (Figure 2-25). The guanine residues are held together by a cyclic head-to-tail network of hydrogen bonds and a nonhydrated metal ion ($K^+ > Na^+ > Cs^+ > Li^+$, depending on its size) is located at the center. The four guanine residues have deoxyribose conformations that alternate between *anti* and *syn*.

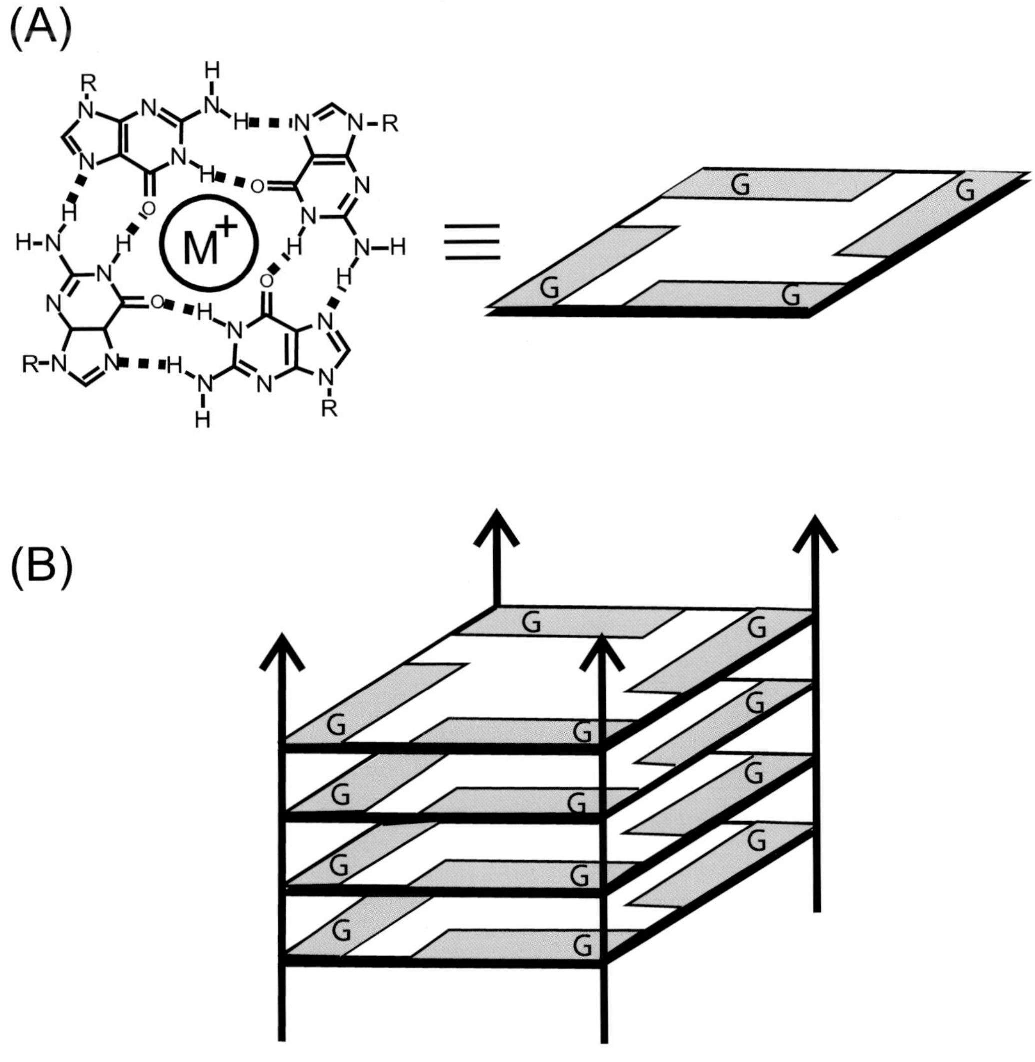

Figure 2-25. Models of the G-quartet. (A) The cyclic array of four guanine bases, each of which is a donor in two hydrogen bonds and an acceptor in another two. At the center is a pocket for interaction with a cation (M). (B) Schematic diagram of how stacks of G-quartets form quadruple helical structures. Adapted from J. R. Williamson (1994) *Ann. Rev. Biophys. Biomol. Structure* **23**, 703–730.

Other four-stranded DNA structures can be formed by tetrads resulting from the association of Watson–Crick base pairs, through either the minor or major groove side of the base pairs. For example, the d(GGGGTTTTGGGG) sequence forms a hairpin, and two such hairpins join to make an antiparallel hairpin dimer, thereby producing a four-stranded helical structure, with the thymines forming loops at either end (Figure 2-26). Structures stabilized by minor-groove tetrads differ from the canonical guanine quadruplex and represent a unique structural motif. For example, the cyclic oligonucleotide dpCCGTCCGT self-associates using intermolecular G•C base pairs and forms a symmetrical dimer stabilized by two G•C•G•C tetrads. The overall 3-D structure is similar to that found in other cyclic and linear oligonucleotides of related sequences, but the relative position of the two base pairs is slipped along the axis defined by the base pairs. This contrasts with other minor-groove G•C•G•C tetrads in which the G•C base pairs are aligned directly. An analogous arrangement of G•C base pairs occurs between the terminal residues of contiguous duplexes in some DNA crystals.

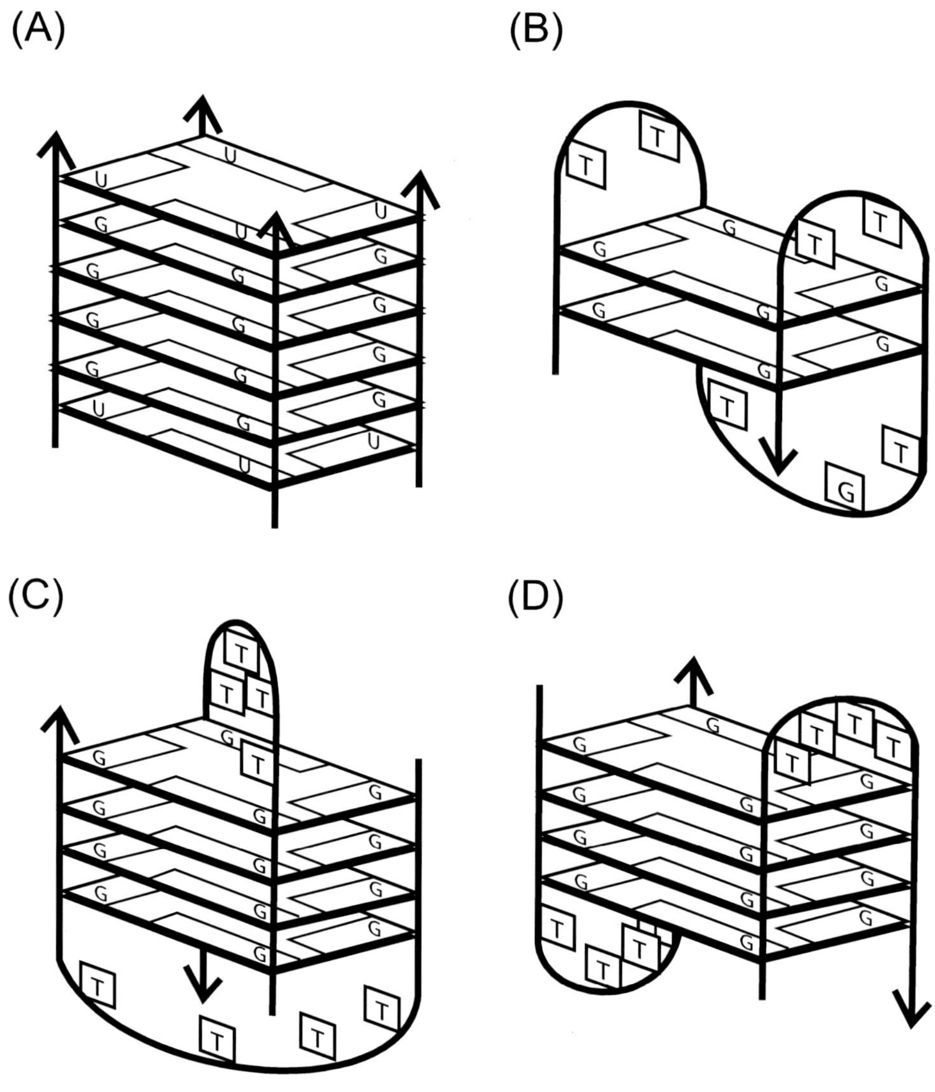

Figure 2-26. Schematic representations of G-quartet structures formed by the following oligonucleotides: (A) ribo$(UGGGGU)_4$; (B) d(GGTTGGTGTGGTTGG); (C, D) d(GGGGTTTTGGGG) with (C) Na^+ or (D) K^+. Adapted from J. R. Williamson (1994) *Ann. Rev. Biophys. Biomol. Structure* **23**, 703–730.

Telomere end-binding proteins control the formation of G-quadruplex DNA structures *in vivo*. K. Paeschke *et al.* (2005) *Nature Struct. Biol.* **12**, 847–854.

Quadruplex DNA: sequence, topology and structure. S. Burge *et al.* (2006) *Nucleic Acids Res.* **34**, 5402–5415.

Topology variation and loop structural homology in crystal and simulated structures of a bimolecular DNA quadruplex. P. Hazel *et al.* (2006) *J. Am. Chem. Soc.* **128**, 5480–5487.

Four-stranded DNA structures can be stabilized by two different types of minor groove G•C•G•C tetrads. N. Escaja *et al.* (2007) *J. Am. Chem. Soc.* **129**, 2004–2014.

5. *i-Motif*

The complementary DNA strand of the telomere TTGGGG sequence has the sequence CCCCAA. It adopts an unusual structure at low pH: a four-stranded structure, known as the **i-motif**, in which two parallel duplexes intercalate with one another in an antiparallel orientation, greatly extending each chain (Figure 2-27). Each parallel duplex consists of C•C$^+$ base pairs in which one cytosine is protonated.

(A) (B)

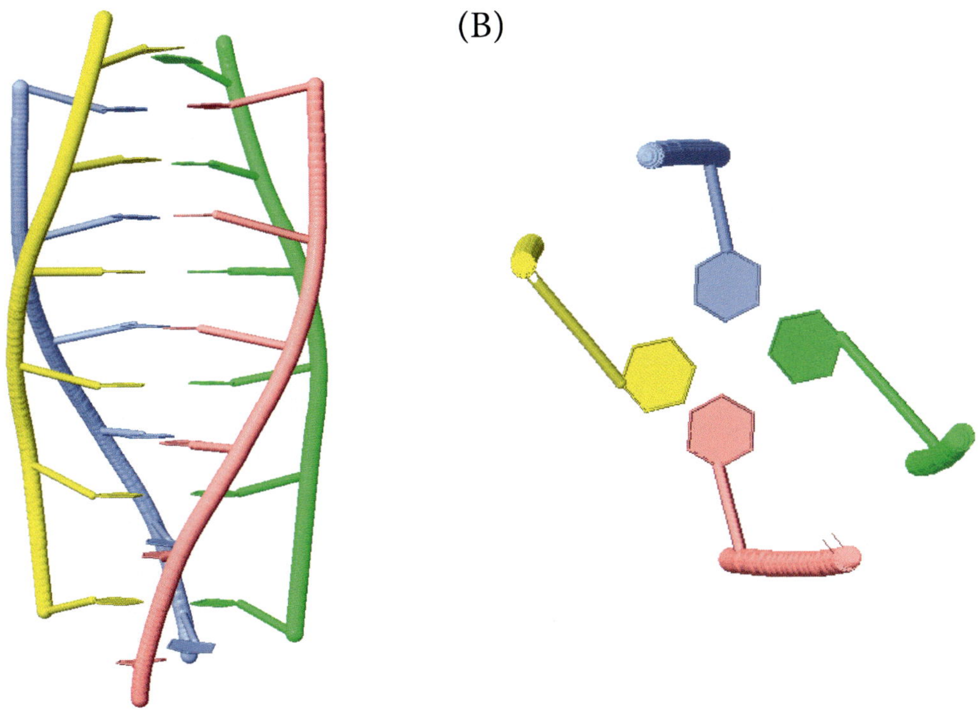

Figure 2-27. The four-stranded i-motif DNA structure determined by NMR. In this case, the sequence of each strand is 5′-d(TCCCCC)-3′, and the cytosine bases are protonated. (A) A vertical view, showing how the base pairs occur between only one pair of strands, but those of the other pair are intercalated between them. Consequently, each oligonucleotide chain is greatly elongated. (B) A slice of the structure showing two base pairs, one from each pair of oligonucleotide chains. The blue and rose bases are hydrogen-bonded, as are the yellow and green, which are above the blue and rose pair. Figure generated from PDB file *225d* using the program Jmol.

Crystal structures of a DNA octaplex with i-motif of G-quartets and its splitting into two quadruplexes suggest a folding mechanism of eight tandem repeats. J. Kondo *et al.* (2004) *Nucleic Acids Res.* **32**, 2541–2549.

DNA octaplex formation with an i-motif of water-mediated A-quartets: reinterpretation of the crystal structure of d(GCGAAAGC). Y. Sato *et al.* (2006) *J. Biochem.* **140**, 759–762.

6. *Inverted Repeat Sequences and Palindromes*

When the sequences of the two strands of DNA are identical, they are said to be 'palindromic'. A palindromic word or text reads the same in both directions. An impressive example is the Latin phrase *Sator arepo tenet opera rotas*, which translates roughly as 'Arepo the sower holds the wheels at work'. In the form:

S A T O R

A R E P O

T E N E T

O P E R A

R O T A S (2.5)

it reads the same in both directions, as well as from top to bottom, and bottom to top. In nucleic acid sequences, the term palindrome is traditionally (although not strictly correctly) applied to double-stranded DNA or RNA (which is two lines) when the nucleotide sequence is the same when read forwards from one strand, then backwards from the other, complementary, strand. In both cases, the sequence is read from 5′ to 3′, as the two strands are antiparallel. Such sequences also have complementary two-fold rotational symmetry:

5′-CCA•TGG-3′
3′-GGT•ACC-5′ (2.6)

They are more correctly referred to as **inverted repeats**. They occur naturally, especially in **introns**, which interrupt the coding sequence in eukaryotic genes.

The two halves of each strand of a sequence that is said to be palindromic are also complementary to each other, with the potential for the two halves of the same strand to fold back on itself, to form a hairpin structure (Figure 2-28). If it happens in both strands, a cruciform structure is formed. The loops of the cruciform could be any size and any sequence, not necessarily palindromic. Such inverted repeat sequences are very common, being present in sites recognized by restriction enzymes (Section 6.3.A) and genetic repressors and in sequences of transposable elements.

A palindromic repeat sequence adopts a stable fold back structure under supercoiling. A. K. Shukla & K. B. Roy (2006) *J. Biochem.* **139**, 35–39.

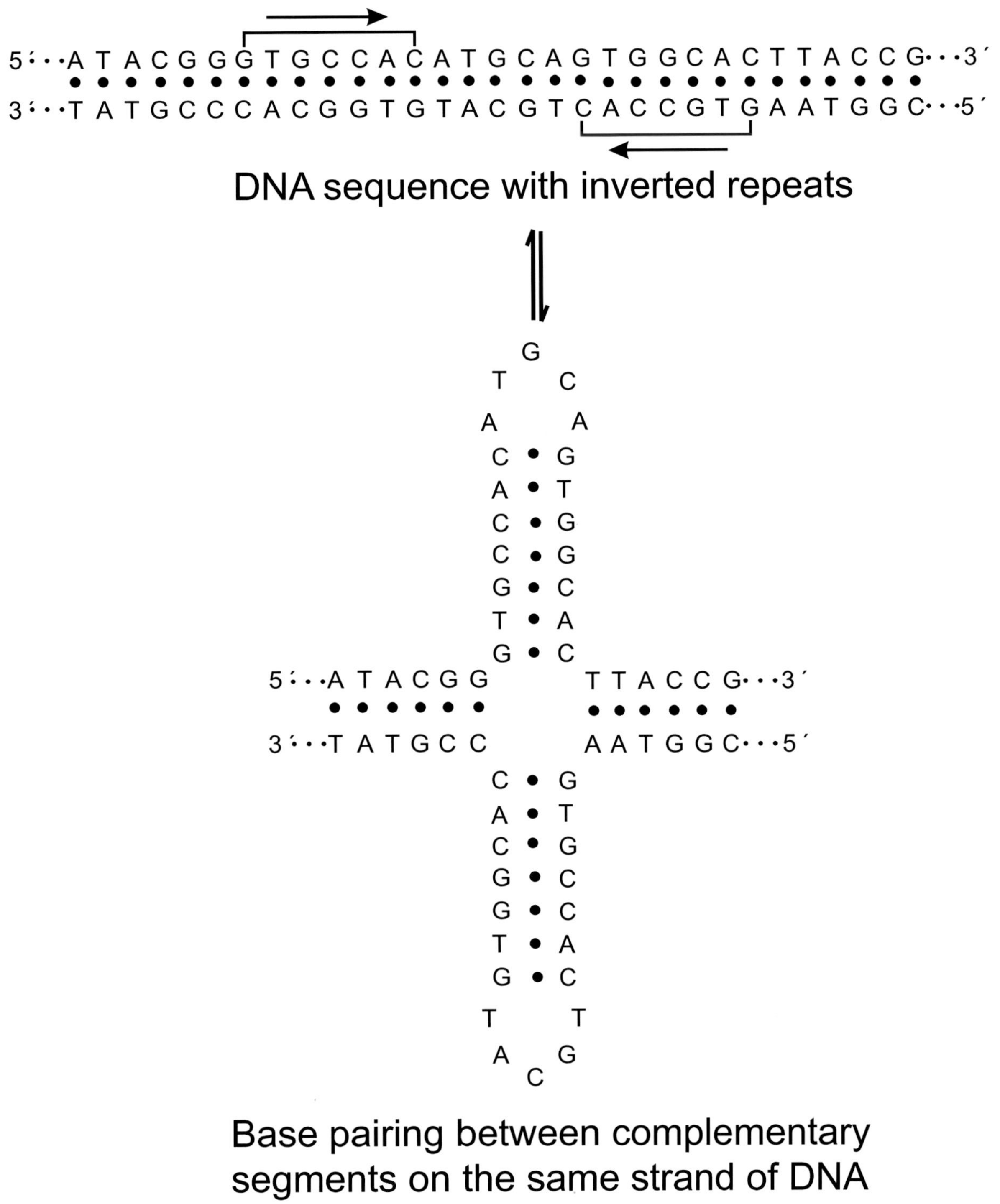

Figure 2-28. Cruciform structure generated from a double helix with inverted repeat sequences. The cruciform is a distorted form that is normally less stable than the linear double helix.

7. *Helical Junctions: Cruciforms, Holliday Junctions*

The **Holliday junction** is a key intermediate in DNA recombination. Four homologous DNA strands join into a structure that looks like a **cruciform** when observed by electron microscopy. A cruciform structure can also be generated in a double helix if the sequence has inverted repeats (Figure 2-28); it does not happen spontaneously in linear DNA molecules but can be induced by supercoiling (Section 3.1). The detailed structure of the cruciform DNA structure was determined when specific sequences were designed to form several types of so-called **immobile junctions** or **helical junctions**. An example of a four-way junction composed of four oligonucleotides is illustrated in Figure 2-29-A. Each of the four strands is paired with two other strands to form two arms of double helix. The junction is formed because there is no homologous two-fold sequence symmetry flanking the central branch point. Multi-valent metal ions, such as Mg^{2+} and $Co(NH_3)_6^{3+}$, or very high concentrations of monovalent cations, cause the structure to change to a stacked X-shape (Figure 2-29-B). The stacked X-structure is formed by the pairwise coaxial stacking of helices to adopt a right-handed, antiparallel cross with a cross-over angle of about 60° or 120° (Figure 2-29-C). Coaxial stacking can occur in two different ways and produce two different conformers that can interconvert, at rates that depend upon the salt concentration and type. The salt-dependence of these conformational transitions has been traced to the charge on one backbone phosphate group at the junction (Figure 2-29-B). Without the charge on this group, the stacked X-structure occurs in the absence of metal ions.

A number of B-DNA molecules in a crystal lattice have helix–helix packing contacts that are similar to the X-structure.

Happy Hollidays: 40th anniversary of the Holliday junction. Y. Liu & S. C. West (2004) *Nature Rev. Mol. Cell Biol.* **5**, 937–944.

Stereospecific effects determine the structure of a four-way DNA junction. J. Liu *et al.* (2005) *Chem. Biol.* **12**, 217–228.

Solution formation of Holliday junctions in inverted-repeat DNA sequences. F. A. Hays *et al.* (2006) *Biochemistry* **45**, 2467–2471.

The stacked-X DNA Holliday junction and protein recognition. P. A. Khuu *et al.* (2006) *J. Mol. Recognit.* **19**, 234–242.

8. *Parallel-stranded DNA Duplexes*

Double-stranded DNA structures in which the strands are parallel, rather than the normal antiparallel, do not occur naturally, although examples have been encountered in Sections 2.2.C.4 and 2.2.C.5. Other examples have been designed. In one series, the structures involve adenine- and thymine-containing DNA sequences that use nonWatson–Crick base pairs. The stabilities of these duplexes are modest; for example, the melting temperature (Section 5.1) of one comprising 21 nucleotides is 15°C lower than that of the corresponding antiparallel duplex. Another parallel motif using nonWatson–Crick base pairs is known as **Π-DNA**; the d(CGA) sequence has the greatest propensity to form this structure.

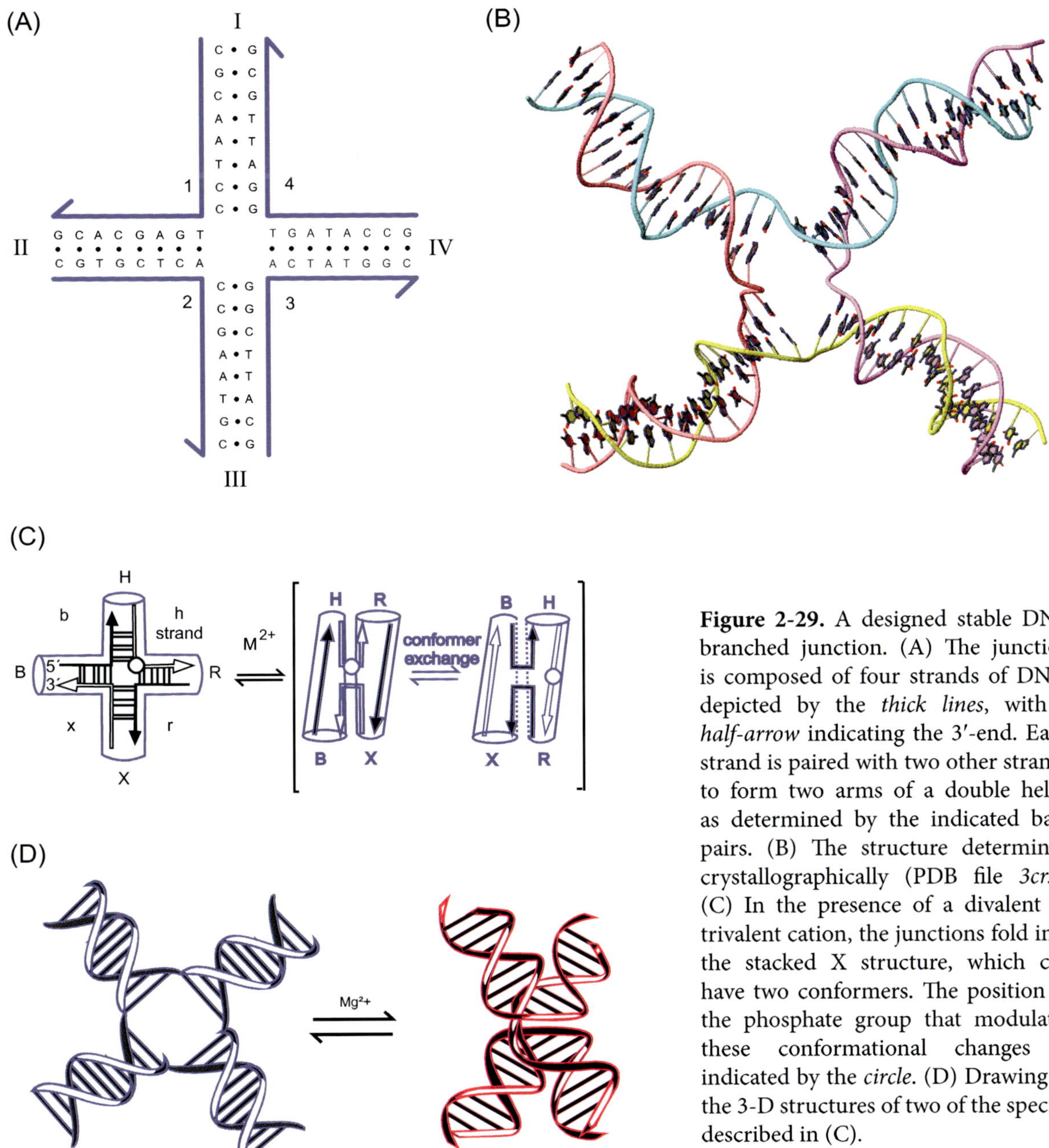

Figure 2-29. A designed stable DNA branched junction. (A) The junction is composed of four strands of DNA, depicted by the *thick lines*, with a *half-arrow* indicating the 3′-end. Each strand is paired with two other strands to form two arms of a double helix, as determined by the indicated base pairs. (B) The structure determined crystallographically (PDB file *3crx*). (C) In the presence of a divalent or trivalent cation, the junctions fold into the stacked X structure, which can have two conformers. The position of the phosphate group that modulates these conformational changes is indicated by the *circle*. (D) Drawing of the 3-D structures of two of the species described in (C).

An important requirement for a parallel duplex with diverse base pairs is that the two glycosyl bonds within a base pair must come from opposite directions, because of the identical chain polarity. For the normal nucleic acid bases, this can be accomplished using an alternative Watson–Crick base pair conformation, but the A•T and G•C base pairs are then not iso-structural, due to their hydrogen-bonding restrictions. Therefore, it has not been easy to design a stable parallel-strand duplex in which all four normal bases can be incorporated in random order.

Overcoming this difficulty required the use of alternative nucleosides, 2′-deoxyisoguanosine (iG) and 2′-deoxy-5-methylisocytosine (iC), which can form stable atypical Watson–Crick base pairs with the normal 2′-deoxycytosine (C) and 2′-deoxyguanosine (G), respectively. Indeed, oligodeoxynucleotides containing iG and iC can form remarkably stable parallel-stranded duplexes with the complementary (G,C)-containing DNA or RNA strands.

Stretches of parallel-stranded double-helical DNA can occur within antiparallel-stranded Watson–Crick DNA in looped structures or in the presence of sequence mismatches.

The biochemical significance of parallel DNA duplexes. H. M. Buck (2003) *Nucleosides Nucleotides Nucleic Acids* **22**, 1549–1552.

Parallel DNA double helices incorporating isoG or m^5isoC bases studied by FTIR, CD and molecular modeling. F. Geinguenaud *et al.* (2005) *Spectrochim. Acta A* **61**, 579–587.

Parallel and antiparallel DNA: fluorescence quenching of ethidium bromide by 7-deazapurines. H. Li & F. Seela (2005) *Nucleosides Nucleotides Nucleic Acids* **24**, 865–868.

2.3. DNA AS A POLYELECTROLYTE: HYDRATION AND COUNTERIONS

Nucleic acids are very soluble in aqueous solution because of their preponderance of ionized phosphate groups, which interact very favorably with water. **Nucleic acids bind twice as much water as do proteins**, some 0.6 g of water per gram of nucleic acid. Of the surface of DNA accessible to solvent, about 45% is made up of phosphate O atoms, 35% of sugar atoms and only 20% of atoms of the nucleobases. DNA is fully hydrated with about 20 water molecules per nucleotide when it adopts the B-structure. The first hydration shell of DNA consists of about 11–12 water molecules per nucleotide that contact the backbone as well as the functional groups of the bases. Water molecules are observed to be most fixed in a 'spine of hydration' in the minor groove that depends upon its width. A narrow minor groove has a single spine of water molecules along its floor, whereas a double ribbon of water molecules is associated with a wide minor groove, bridging the N or O atoms at the edges of the bases to the O4′ atoms of the sugar ring. It is now generally accepted that the hydration shell surrounding the DNA molecule plays an important role in DNA structure and its recognition by proteins and other ligands. Specific water molecules play critical roles in the sequence-specific recognition by proteins, and they modulate the binding of sequence-nonspecific DNA-binding proteins to DNA of random sequences (Section 7.2.A).

The degree of hydration of the double helices is not correlated with the types of atoms exposed to the solvent (Table 2-8), indicating more specific interactions between nucleic acid and water molecules. AT-rich sequences are observed to have a spine of water molecules in the minor groove. In contrast, GC-rich sequences appear to have patches of water molecules in the major groove that would be disrupted by the methyl group of the thymine nucleotide in AT-rich regions. Both types of hydration occur in both the minor and major grooves of poly(dIdC)•poly(dIdC). The minor groove hydration spine of each nucleotide is estimated to contain eight water molecules, the major groove hydration 13. These are a considerable fraction of the 24 water molecules that can fit within the first hydration layer of each nucleotide of a double helix, so some probably involve the second and third layers of water molecules.

A major factor in the solvation of DNA is the single charge on each ionized phosphate group of the backbone, which should be spread amongst the four O atoms. **The polyelectrolyte nature of nucleic acids results in them retaining a significant number of counterions, even when these ions are absent from the bulk solvent.** Being part of the same molecule, the charged phosphate groups are constrained by the covalent bonds to be close to each other, which is unfavorable energetically. The electrostatic repulsions between these ionized phosphate groups are so severe, and the need to screen them with counterions so great, that a number of counterions remain associated with the DNA even in the absence of these ions in the bulk solvent. These bound counterions are said to be 'condensed', in contrast to the other free ions, which form a normal diffuse ionic cloud around the DNA. The sodium ions that are condensed on B-DNA are 76% of the number of phosphate groups, independent of the bulk salt concentration, so **the net charge of the DNA is effectively decreased to −0.24 *e* per phosphate group. With divalent ions, such as Mg^{2+}, the effective net charge is −0.12 *e* per phosphate group; it is only −0.08 *e* for trivalent ions.** Although said to be condensed, these ions are fully hydrated with bound water, associated rather loosely with the DNA, and exchange readily with ions of the bulk solvent. They are located within 7 Å beyond the external surface of the DNA, some 17 Å from the helix axis. Monovalent counterions are generally not observed in oligonucleotide crystal structures and so must be very mobile; divalent cations, such as Mg^{2+}, can be observed crystallographically, but only when they are fixed in position by interacting with two or more groups.

The local concentration of monovalent cations around double-stranded DNA is approximately 1 M. The negative charges on DNA attract H^+ ions as well, so the **pH in close proximity to the DNA molecule is some 2–3 units lower than in the bulk solution.** In terms of the van der Waals dimensions of the atoms, the diameter of B-DNA is only 20 Å **but hydration and electrostatic repulsions cause the effective diameter to be 30 Å at high salt, increasing to 150 Å at 0.01 M salt.** These observations are not unique to DNA but apply generally to any polyelectrolyte of similar charge density.

Cations are usually bound to nucleic acids in a solvated state. They usually interact with DNA via water molecules in their primary solvation shell; direct ion–DNA contacts are probably rare because such interactions would need to compensate for the substantial energetic cost of removing the waters of hydration. On the other hand, Mg^{2+} ions are thought to bind directly to some phosphate groups by losing at least one water from their hydration shells. Solvated ions are usually too large to enter the minor groove, except in the case of AT tracts, where monovalent cations can partition selectively into the minor groove. Solvated ions appear to bind preferentially to guanine bases in the major groove or to the O atoms of phosphate groups. Sites where the ions bind are readily exchangeable: the same site may be occupied by any ion, including more complex molecules, such as spermine (Equation 2.4), or a water molecule. Polyamines like spermine are apparent at specific sites in crystal structures of DNA, but they appear to be disordered by most other techniques.

Divalent cations are much more effective than monovalent, and one Mg^{2+} ion has an influence on the DNA double helix comparable to that of 100–1000 Na^+ ions. Not even K^+ and Na^+ are equivalent, as they produce detectable differences in the structure of the DNA. DNA deformation may be caused by electrostatic collapse around areas of uneven cation density. The release of these counterions can be very significant when they are displaced by binding of a ligand (Section 7.2).

The hydration environment around a DNA double helix, including cation counterions, plays an important role in determining the type of conformation adopted and other properties. For example, right-handed B-DNA with a GC-rich sequence can be converted to left-handed Z-DNA simply by increasing the salt concentration (Figure 2-20).

The partial specific volumes of nucleic acids in water are relatively small, around 0.5–0.6 cm^3 g^{-1}. Nucleic acids do not occupy much extra volume in water, indicating that they interact tightly with water molecules. They are influenced remarkably by the solvent composition, especially the salt concentration, because of their high charge density, and they interact differently with the salt and water molecules of an aqueous solution. Consequently, their apparent partial specific volume differs from the true value and varies with the salt concentration (Figure 2-30). This is especially important for nucleic acids because their density (the reciprocal of their partial specific volume) is usually of interest primarily in solutions containing high concentrations of salt, especially of cesium chloride, for density-gradient centrifugation.

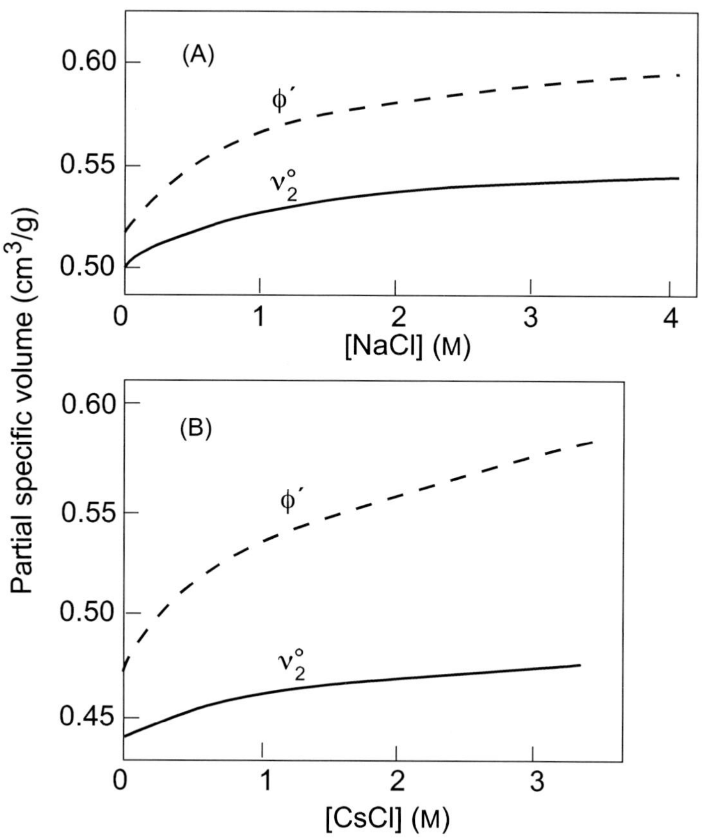

Figure 2-30. The partial specific volume of DNA as a function of the concentration of NaCl (A) or CsCl (B). The apparent partial specific volume is ϕ' (*dashed curve*), which does not account for the interactions with the salt. The true partial specific volume is given by $v°_2$. Data from H. Eisenberg (1976) *Biological Macromolecules and Polyelectrolytes in Solution*, Clarendon Press, Oxford.

The intrinsic compressibility of a nucleic acid is believed to be very small, and the compressibility measured in solution is believed to be primarily that of the hydrating solvent (Table 2-9). The measured values are negative, indicating that the hydration is dominated by interactions with the ionized phosphate groups of the nucleic acids. Differences in compressibilities of the various double helices are attributed to differences in their hydration shells. The more negative the compressibility, the stronger the hydration. The most strongly hydrated is poly(dIdC)•poly(dIdC), poly(dGdC)•poly(dGdC) exhibits intermediate hydration, and the two all-AT duplexes the weakest hydration.

Table 2-9. The apparent molar volumes, and apparent molar adiabatic compressibilities per base pair, for various synthetic DNA duplexes at 25°C

DNA duplex	Apparent molar volume (cm^3/mol of base pair)	Apparent molar adiabatic compressibility (10^{-4} cm^3/bar/mol of base pair)
Poly(A)•poly(U)	166	–52
Poly(A)•poly(dT)	168	–55
Poly(dA)•poly(U)	173	–41
Poly(dA)•poly(dT)	169	–58
Poly(dAdT)•poly(dAdT)	170	-58
Poly(dGdC)•poly(dGdC)	152	–73
Poly(dIdC)•poly(dIdC)	123	–102

Data from T. V. Chalikian *et al.* (1999) *Biopolymers* **50**, 459–471.

Role of monovalent counterions in the ultrafast dynamics of DNA. S. Sen *et al.* (2006) *J. Phys. Chem. B* **110**, 13248–13255.

The persistence length of DNA is reached from the persistence length of its null isomer through an internal electrostatic stretching force. G. S. Manning (2006) *Biophys. J.* **91**, 3607–3616.

Multiple time scales in solvation dynamics of DNA in aqueous solution: the role of water, counterions, and cross-correlations. S. Pal *et al.* (2006) *J. Phys. Chem.* **110**, 25396–25402.

The DNA structure responds differently to physiological concentrations of K^+ or Na^+. B. Heddi *et al.* (2007) *J. Mol. Biol.* **368**, 1403–1411.

Water clustering and percolation in low hydration DNA shells. I. Brovchenko *et al.* (2007) *J. Phys. Chem. B* **111**, 3258–3266.

2.4. DNA FLEXIBILITY AND DYNAMICS: CURVING, TWISTING AND STRETCHING

DNA possesses many of the properties usually associated with a linear polymer, with varying types of both mechanical rigidity and motional flexibility. DNA is unique, however, in the lengths of the chains. A natural DNA molecule will consist of millions to a billion base pairs (bp) and will be millions of times longer than it is wide. A double helix containing 100 million bp will have a covalent length of 3.3×10^8 Å, or 3.3 cm, a diameter of only 20Å (2.0 nm). This extreme elongation presents special topological problems, described in Chapter 3. A further consequence is that such long DNA molecules are readily broken physically by the shear forces generated by standard laboratory manipulations, even just pouring and pipetting; the resulting fragments usually have molecular weights of about 5×10^6, corresponding to about 7500 bp. Passing the solution through a hypodermic syringe needle will produce fragments of about 1500 bp, as will shearing in a homogenizer or blender. Fragments as small as 50–100 bp on average can be generated by sonication.

The motions of DNA in solution can be either large-scale or local: DNA is locally rather stiff, so that a short double helix can be approximated as a rigid rod, but very long DNA molecules are globally flexible. A fully extended DNA double helix might measure nearly 1 m in length, but its flexibility in solution would decrease its average end-to-end distance to only 230 μm, a contraction of more than 4×10^3-fold. DNA helices that are intermediate in length between a rigid rod and a random coil must be treated as 'worm-like chains', which take into account both the local stiffness and the long-range flexibility of the double helix.

The four types of deformation of the linear DNA double helix are curvature, bending, torsion and stretching. The global curvature and bending flexibility are normally expressed as the longitudinal **persistence length**, which is a measure of the resistance of a polymer to lateral bending. It is formally defined as the average projection of the molecular end-to-end distance vector on its initial path vector, in the limit of infinite chain length. Another useful interpretation of persistence length is the distance over which the root mean square (r.m.s.) bend angle in any particular direction is 1 radian (about 57°). A helix of the same length as the persistence length has an r.m.s. end-to-end separation that is 14% shorter than its covalent length.

The persistence length of double-stranded DNA is roughly 150–500 bp, depending upon the ionic strength and the cations present. Consequently, the helix axis of a double-stranded DNA molecule changes direction by 57° over every segment of 150–500 bp. **Normal B-DNA is considerably more flexible than either A-DNA or Z-DNA**: the persistence length of A-DNA is about 1500 bp at moderate ionic strengths. Single-stranded DNA is much more flexible, and its persistence length is only 15Å (1.5 nm) in 2 M NaCl and 30 Å (3 nm), corresponding to 4–9 nucleotides, in 25 mM NaCl.

Double-stranded DNA also displays local torsional stiffness to twisting about its long axis. This is usually expressed as a **torsional persistence length**, that length over which DNA will have an r.m.s. twist deviation of 1 radian (57°). **This torsional persistence length of DNA is about 180 bp**, approximately the same as the longitudinal persistence length. Stretching of the DNA double helix is measured by an **elastic stretch modulus**, S, which when divided by the cross-sectional area of the polymer gives the **Young modulus** of physics. The DNA double helix can be stretched more readily at low ionic strength, probably due to electrostatic repulsions between the phosphate groups.

These deformations of the DNA double helix are believed to be a result of a balance between repulsions between the ionized phosphate groups, attractions between the stacked base pairs and deformation of the covalent structure, although the details are not clear.

The relative concentrations of two sites on a DNA double helix with respect to each other, their **effective concentration**, can be measured by the tendency of the ends of the helix to form a circular molecule (Figure 3-2). The effective concentration reaches a maximum when the two groups are separated by about 500 bp, with a value of about 1×10^{-7} M. The effective concentrations of groups closer together are lower due to the stiffness of the double helix, while those further apart are less likely to encounter each other randomly. When separated by 150 bp, the effective concentration is decreased to about 2×10^{-9} M, but only to about 2×10^{-8} M when separated by 3000 bp. When the two sites are defined in terms of their position on the faces of the double helix, their effective concentrations oscillate in a sinusoidal manner with varying separations, due to the helical periodicity of the double helix and the varying tendencies of different sequences to bend. The phasing over separations of 60–200 bp may alter the effective concentration 5–10-fold.

Certain DNA sequences have abnormal mobilities in gel electrophoresis, suggesting that they have unusual shapes. For example, double-stranded DNA fragments having repeats of $(A)_n$ nucleotides (with $n \geq 4$), separated by another 4–5 nucleotides and phased with the helical repeat, migrate in the gel significantly more slowly than those having mixed sequences. The reason is that the 5′-AAAAA sequence, and its complementary 5′-TTTTT sequence, has an intrinsic bending property that can be demonstrated by the increased efficiency of cyclization of those bent DNA fragments. The molecular basis of the intrinsic bendability of the $(A)_n$ sequence may be due in part to the high propeller twist associated with the A•T base pairs in the $(A)_n•(T)_n$ sequence and the presence of only two hydrogen bonds between the bases. The bend in these sequences is relatively smooth, resulting in a curved DNA structure, and the opening (roll) of the bend is towards the minor groove. **Many proteins induce such a smooth bending when they bind to DNA** by having many small single-step bends (Section 7.2.C). **Another type of bent DNA has a sharp kink at a localized site**, usually caused and stabilized by bound proteins.

The large-scale tumbling motion of a DNA duplex of 10–100 bp occurs at a time scale of nanoseconds to microseconds (Figure 2-31). The **correlation times** for tumbling at 0°C of a cylinder the size of an oligonucleotide with 12 nucleotide base pairs are 7.5 ns for rotations about the helix axis and 14 ns for rotations perpendicular to the axis. The latter time increases roughly with the square of the length of the duplex, while the time for the parallel motion increases in proportion to the length. The correlation times at 20°C are about half the values at 0°C. This global flexibility of DNA can account for the behavior of DNA in solution and in gels.

Natural DNA molecules are often closed circles, with no termini. This introduces a further complication that is discussed in Chapter 3.

Electrostatic effects in DNA stretching. A. V. Tkachenko (2006) *Phys. Rev. E* **74**, 041801.

NMR studies of dynamics in RNA and DNA by ^{13}C relaxation. Z. Shajani & G. Varani (2006) *Biopolymers* **86**, 348–359.

Diffusion and segmental dynamics of double-stranded DNA. E. P. Petrov *et al.* (2006) *Phys. Rev. Lett.* **97**, 258101.

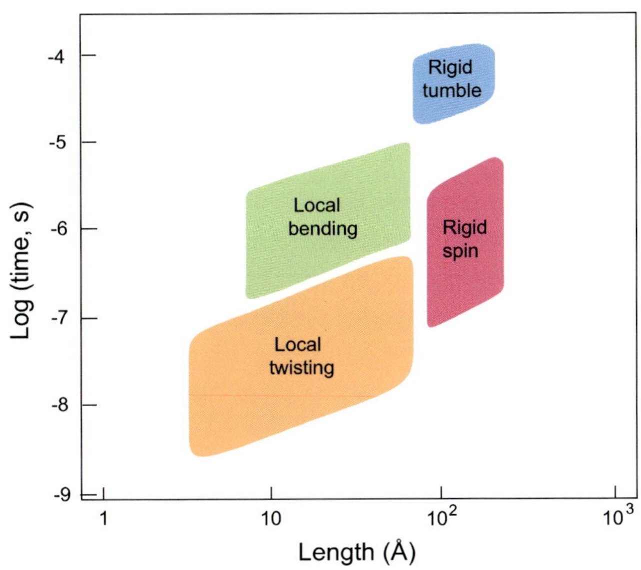

Figure 2-31. Time scales over which DNA duplexes of various lengths move rotationally in various ways. Data from M. Hogan *et al.* (1982) *Proc. Natl. Acad. Sci. USA* **79**, 3518–3522.

2.4.A. Local Flexibility

Local motions are important in many of the biological properties and functions of DNA. That various parts of the DNA structure have considerable flexibility is suggested by the highly anisotropic temperature factors observed in the X-ray crystallography structures of DNA molecules. The average bending angle between adjacent base pairs is approximately 7° under normal conditions, while the average twist angle between adjacent bases is about 3.6°. Enzymes that repair mismatched bases often bind to DNA with the offending base flipped out of the double helix and into the active site of the enzyme; whether this happens, and how frequently or rapidly, in the absence of the enzyme is not known. Many drugs and dyes intercalate into the DNA structure, which requires that adjacent DNA base pairs be separated by up to 6.8 Å (Figure 2-32). This is a large conformational change, involving many torsion angles and sugar pucker rearrangements (Figure 2-5); it is reversible, so DNA must be in constant motion.

(A) (B)

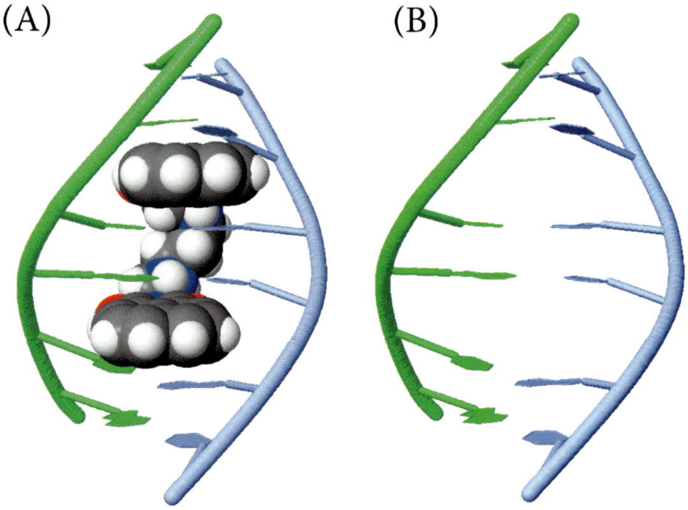

Figure 2-32. The intercalation of double-stranded DNA by the antitumor bisnaphthalimide known as LU-79553. Each strand of the DNA has the sequence 5′-d(ATGCAT). The structure was determined by NMR. (A) The intercalator is shown as a space-filling model but the DNA is depicted with just the nucleobases and a cylinder for the ribose-phosphate backbone. (B) The DNA with the intercalator deleted, illustrating how its structure is distorted. Figure generated from PDB file *1cx3* using the program Jmol.

The biological manipulation and packaging of DNA often depend crucially on local variations in both bending and torsional flexibility, and the probability of DNA untwisting is strongly correlated with a high bending flexibility. The average physicochemical properties of a long DNA molecule approximate those of a thin isotropic homogeneous rod that bends equally in all directions, but more locally DNA behaves as an anisotropic heterogeneous rod that bends primarily only in certain directions. This bending anisotropy is sequence-dependent and to a first approximation reflects both the geometry and stability of individual base steps.

The bendability of DNA is believed to depend primarily upon the stacking energies between neighboring base pairs. DNA sequences with Py–Pu steps in B-DNA are deformed most easily; therefore steps such as CpG, TpA or TpG (= CpA) appear to be more flexible than others and are the preferred binding sites for intercalators and the preferred kinked sites induced by binding of proteins (Section 7.2.C). A change from G•C to A•T base pairs renders that region of the DNA flexible and more susceptible to bending. As a result of its poor stacking, the deformability of a TpA dinucleotide step is inherently much greater than that of an ApT or ApA step. The proximities of the methyl groups of two successive thymines on the opposite strand and of the methyl groups and the phosphate backbone make the helix relatively rigid at an ApA step. In an ApT step, the stacking of the methyl group with the adjacent adenine and the intervening sugar phosphate backbone prevents bending by a roll mechanism. In a TpA step, however, the methyl group projects into the major groove without any significant stacking interactions with either the adjacent adenine or the phosphate backbone, and this step therefore displays greater deformability than the ApA- and ApT-steps. The local dynamics of the CpG site of the dodecamer CGCGAATTCGCG indicate that the deoxyribose of the ninth residue, cytosine, undergoes more motion than the other nucleotides. The greater flexibility of the CpG step has been correlated with the cleavage at this site by the restriction enzyme *Eco*R1 (Section 6.3.C). Therefore, the local dynamics of DNA clearly play important roles in DNA bending induced by proteins, in binding of intercalators, in restriction enzyme action, and undoubtedly in other biological functions.

Single phosphate groups of DNA can be replaced by the double phosphate groups of pyrophosphate or O-ethyl-substituted pyrophosphate without elongating the internucleotide distance: the additional phosphate group is simply flipped out from the sugar-phosphate backbone. This has only a minor impact on the base pairing and stacking, so there are limits to the flexibility of the double helix.

Hydrogen bonding, base stacking, and steric effects in DNA replication. E. T. Kool (2001) *Ann. Rev. Biophys. Biomol. Structure* **30**, 1–22.

Probing single-stranded DNA conformational flexibility using fluorescence spectroscopy. M. C. Murphy *et al.* (2004) *Biophys. J.* **86**, 2530–2537.

The structural basis of DNA flexibility. A. A. Travers (2004) *Philos. Trans. A* **362**, 1423–1438.

2.4.B. Hydrogen Exchange

The technique of hydrogen exchange is very useful in measuring the flexibilities of both nucleic acids and proteins. The H atoms attached to N and O atoms of the nucleobases can exchange with those of water at rates of roughly $10–10^3$ s^{-1} (i.e. on time scales of $10^{-1}–10^{-3}$ s). Even that attached to C8 of adenine and guanine bases will exchange with measurable rates at elevated temperatures. Such exchange depends upon the environment of the group, especially on whether the H atom is involved in hydrogen bonding.

The imino (–NH–) groups of uracil, thymidine and guanine are sufficiently acidic that their H atom can be removed as a proton by every encounter with a hydroxide ion (HO^-); this diffusion-controlled reaction has rate constant k_{OH} of approximately 10^{10} M^{-1} s^{-1}. Weaker bases, such as components of the buffer, can also remove these protons. Acid-catalyzed exchange proceeds by protonation of oxygen followed by deprotonation of the –NH=, to produce the iminol tautomer. Using thymine as an example:

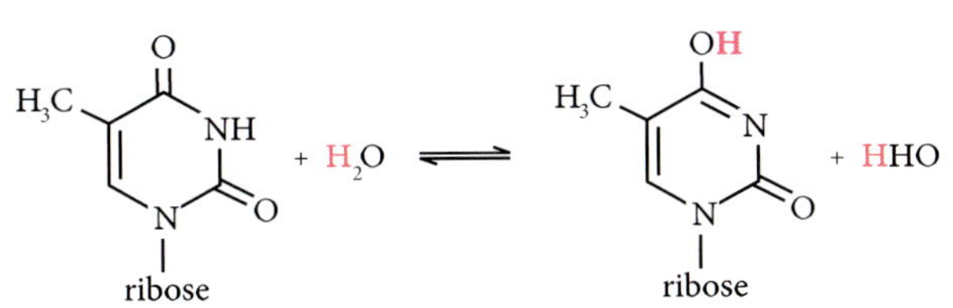

(2.7)

At low pH, acid catalysis in guanine is masked by its ionization and conversion to the conjugate acid. Exchange of an H atom of an amino group of adenine, guanine or cytosine is catalyzed by both acid and base, and there is also a pH-independent rate in the region of neutral pH. The base-catalyzed reaction involves the slow removal of an H ion from an –NH= by a hydroxide ion and its replacement by one from the solvent. The acid-catalyzed reaction involves attachment of H^+ to a ring nitrogen, followed by removal of one of the amino H atoms (as a proton) by water, to produce as a transient intermediate an abnormal tautomer:

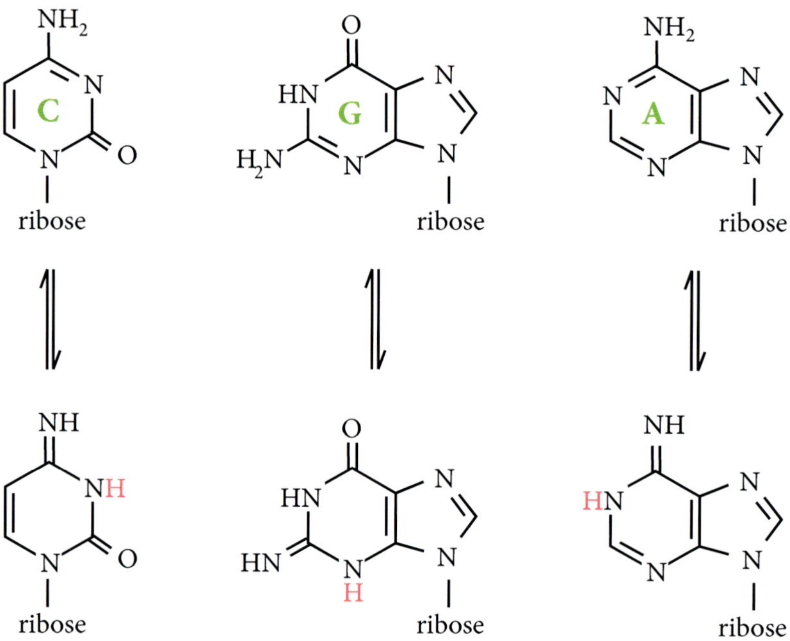

(2.8)

where the H atom from water is shown in red. In the neutral pH region, where the rate is independent of pH, the H atom of the amino group is removed by a hydroxide ion instead of water, to generate the same intermediate. At low pH in each of these cases, acid catalysis is masked by complete ionization and conversion to the conjugate acid (Figure 2-7).

The pH relevant to hydrogen exchange is that within close proximity of the DNA double helix, and this can be quite different from the bulk pH, due to the polyelectrolyte nature of DNA (Section 2.3). Such complications mean that the detailed mechanism by which exchange occurs from a DNA double helix under most circumstances is not certain, as in the case of proteins (Section 10.3). The exception is at high pH, which is possible with DNA because the double helix is stable under these conditions. The exchange rate is determined by the rate at which a group becomes accessible (known as the 'EX1 mechanism'). This has made it possible to measure the rate of base-pair opening in the DNA double helix to be between 2×10^1 and 2×10^3 s^{-1}; slower rates are indicative of unusual structural features. The rates and equilibrium constants of the opening reactions for individual base pairs at various temperatures yield the enthalpies and free energies of the barrier to opening and the open state for each base pair. Neighboring bases one or more positions removed from the opening base pair influence the temperature-dependence of the opening pathway. The rates of the reaction are related directly to the free energies of the open state, which suggests that the transition state in the opening reaction is closer to the native closed state of the base pair than to its open state.

Studies of base pair kinetics by NMR measurement of proton exchange. M. Guéron and J.-L. Leroy (1995) *Methods Enzymol.* **261**, 383–413.

A nuclear magnetic resonance investigation of the energetics of basepair opening pathways in DNA. D. Coman & I. M. Russu (2005) *Biophys. J.* **89**, 3285–3292.

2.5. BINDING OF SMALL MOLECULES

Many small molecules bind to DNA. This usually has significant physiological consequences, and many of these molecules are known as drugs and antibiotics. Most of these molecules bind much more tightly to double-stranded DNA than to single-stranded, so they stabilize the double helix, even if they perturb its structure.

Thermodynamics of nucleic acids and their interactions with ligands. A. N. Lane & T. C. Jenkins (2000) *Quart. Rev. Biophys.* **33**, 255–306.

Synthetic metallomolecules as agents for the control of DNA structure. A. D. Richards & A. Rodger (2007) *Chem. Soc. Rev.* **36**, 471–483.

Volume and hydration changes of DNA–ligand interactions. X. Shi & R. B. Macgregor (2007) *Biophys. Chem.* **125**, 471–482.

2.5.A. Binding to the Minor Groove

Most small molecules that bind to DNA do so in the narrow groove. The narrow groove at A–T sequences in B-DNA provides an excellent binding site for many drugs, due to the greater negative electrostatic potential at the bottom of the minor groove at A–T regions. The N2 amino groups of guanine nucleotides hinder drug binding, both electrostatically and sterically. **Most such drugs replace the spine of hydration water in the narrow minor groove and stabilize the DNA structure, without perturbing its overall conformation significantly** (Figure 2-19-B). The binding energy is due in part to the gain in entropy associated with the displacement of multiple water molecules by one drug molecule.

Binding of distamycin A (a pyrrole-containing antitumor antibiotic) to DNA containing A•T sequences revealed a more complex pattern. Distamycin A can bind DNA not only in a 1:1 drug:duplex complex but also in a 2:1 mode, with two distamycins bound to the minor groove in an antiparallel side-by-side manner. The 2:1 binding mode requires expansion of the minor groove to accommodate the two drug molecules. That it does so reflects the flexibility of B-DNA.

Complexes of the minor groove of DNA. B. H. Geierstanger & D. E. Wemmer (1995) *Ann. Rev. Biophys. Biomol. Structure* **24**, 463–493.

Hydration changes accompanying the binding of minor groove ligands with DNA. N. N. Degtyareva *et al.* (2007) *Biophys. J.* **92**, 959–965.

2.5.B. Intercalation

Virtually any molecule with a sufficiently large aromatic chromophore will bind to DNA double helices by insertion of the flat chromophore between the base pairs (Figure 2-32). Stacking interactions between the intercalator and the adjacent base pairs are believed to be the primary driving force; the intercalators tend to have larger and more electron-deficient chromophores than those of a DNA base pair. Intercalation disrupts the DNA double helix, forcing the base pairs apart by about 3.5 Å and unwinding the helix by 10–25°; yet the stability of the helix is increased because the intercalator binds only to the double helix. The time that different intercalators remain bound (the *residence time*) varies widely, from a few milliseconds to many hours. Binding by intercalation can also be combined with binding in the minor groove by another part of the molecule.

Most of the intercalators are mutagenic and cause frame-shift mutations in which a base pair is inserted or deleted at the site where the intercalator bound to the DNA. These mutations are believed to be caused by the presence of the intercalator during replication of the DNA molecule (Section 6.1), producing mispairing of the nucleotides. Some intercalators, such as daunomycin, are antibiotics and used in cancer chemotherapy.

Variable role of ions in two drug intercalation complexes of DNA. N. Valls *et al.* (2005) *J. Biol. Inorg. Chem.* **10**, 476–482.

Intercalation of organic dye molecules into double-stranded DNA. II. The annelated quinolizinium ion as a structural motif in DNA intercalators. H. Immels *et al.* (2005) *Photochem. Photobiol.* **81**, 1107–1115.

Quantifying force-dependent and zero-force DNA intercalation by single-molecule stretching. I. D. Vladescu *et al.* (2007) *Nature Methods* **4**, 517–522.

---

1. Ethidium Bromide

One of the best-known intercalators is **ethidium bromide** (EB):

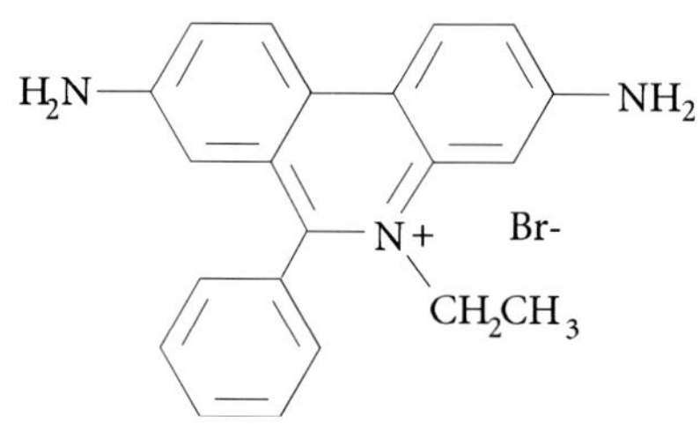

(2.9)

because it is widely used in molecular biology laboratories to detect nucleic acids. EB binds with high affinity to double-stranded DNA, with a dissociation constant of only 0.5 μM. The aromatic ring of EB intercalates within the DNA duplex, and the two amino groups hydrogen bond to the phosphate groups across the two DNA strands, which fixes the orientation. Consequently, B-DNA adopts an A-form structure locally and the deoxynucleoside sugar pucker changes from C2′-*endo* to C3′-*endo*. EB does not, however, bind to adjacent sites on the double helix, and there must be at least two base pairs between bound EB molecules. Such 'neighbor exclusion' is common with intercalation and causes binding curves to be more complex than expected with a single population of binding sites.

In aqueous solution, EB is dark yellow in appearance and possesses a broad featureless absorption spectrum, with maximum absorbance at about 480 nm. EB is only weakly fluorescent; its broad featureless fluorescence emission spectrum has a maximum at 617 nm. In aqueous solution, the fluorescence quantum yield is low and the fluorescence lifetime is short because the excited state is quenched by transfer of a proton to an adjacent water molecule. Increasing the EB concentration produces dimerization, a red shift in the absorption maximum, and a decrease in absorbance intensity; the EB dimer is nonfluorescent.

When EB binds to DNA, its absorption spectrum shifts to longer wavelengths, due to stabilization of the excited state, and the fluorescence intensity increases markedly. EB is shielded from solvent molecules and the rate of quenching decreases, increasing both the fluorescence decay time and the quantum yield. The fluorescence decay time of EB is 23 ns when bound to DNA but only about 1.8 ns in water.

The increased fluorescence when EB is bound to DNA provides the most common method for detecting and measuring DNA. The fluorescence is enhanced approximately 30-fold when excitation is at 520 nm and emission at 600 nm. This enhancement is practically independent of the base composition of the DNA. Single-stranded DNA and RNA also stimulate the fluorescence of EB, but to a lesser extent than does duplex DNA.

Appropriately designed dimers of EB show binding affinities and fluorescence enhancements even greater than those of the monomer. One intercalates at a ratio of 1 dimer per 4–5 bp and can detect and quantify picogram amounts of DNA.

EB is a powerful mutagen, extremely toxic by inhalation, ingestion and skin contact, and a suspected carcinogen and reproductive toxin. EB originally found fame as a drug useful in combating trypanosomal, microbial, bacterial and viral infections. All these biological effects are a direct consequence of EB binding to DNA and inhibiting its replication.

Ethidium DNA agarose gel electrophoresis: how it started. P. Borst (2005) *IUBMB Life* **57**, 745–747.

On the electronic structure of ethidium. N. W. Luedtke *et al.* (2005) *Chemistry* **11**, 495–508.

Local conformational changes induced in B-DNA by ethidium intercalation. J. M. Benevides & G. J. Thomas (2005) *Biochemistry* **44**, 2993–2999.

The complex of ethidium bromide with genomic DNA: structure analysis by polarized Raman spectroscopy. M. Tsuboi *et al.* (2007) *Biophys. J.* **92**, 928–934.

2. Psoralen Photo Cross-linking

The natural reagent **psoralen**:

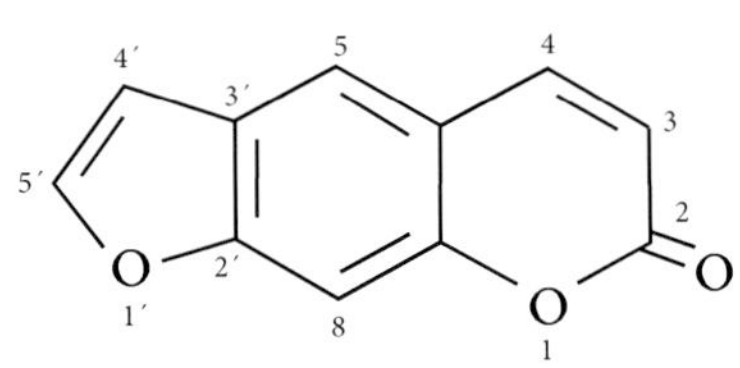

(2.10)

intercalates within the DNA double helix and can then undergo chemical reaction with adjacent pyrimidine bases upon being illuminated with UV light. The three-ring structure absorbs UV light with maxima at 220, 250 and 300 nm. Upon irradiation, the 4′–5′ double bond of the psoralen can react with the 5–6 double bond of an adjacent pyrimidine ring, to generate a cyclobutane ring like that depicted in Equation 2.3. This adduct can still absorb UV light in the wavelength range 320–400 nm, and its 3–4 double bond can similarly react with the 5–6 double bond of another adjacent pyrimidine ring. This pyrimidine will be on the opposite strand of the DNA double helix, so the two strands will be cross-linked covalently:

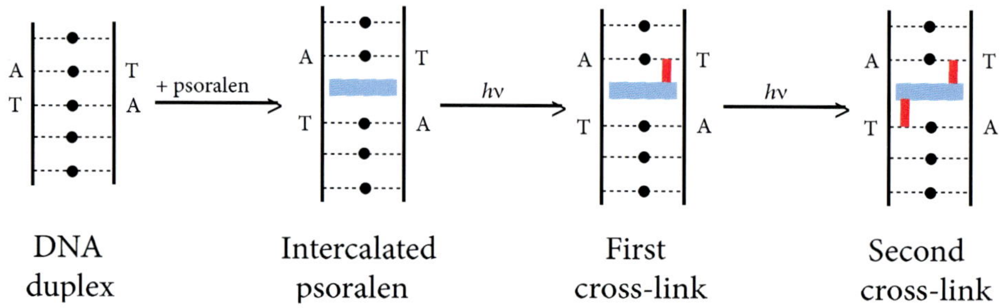

(2.11)

No reaction occurs in the absence of UV light. This reaction can be used in cells, making it useful for studying the *in vivo* structure of DNA.

The most useful forms of psoralen have methyl groups at the 4, 5′ and 8 positions, plus a hydroxymethyl or amino methyl group at the 4′ position. **Anglecins** are angular psoralens that have the five-membered ring fused to the 5–6 bond of psoralen, rather than the 6–7 (or 2′–3′). This angular structure cannot cross-link two bases, but it is a useful reagent for studying the individual cross-linking reactions.

Quantitative analysis of DNA interstrand cross-links and monoadducts formed in human cells induced by psoralens and UVA irradiation. C. Lai *et al.* (2008) *Anal. Chem.* **80**, 8790–8798.

Photoreactivity of furocoumarins and DNA in PUVA therapy: formation of psoralen–thymine adducts. J. J. Serrano-Pérez *et al.* (2008) *J. Phys. Chem.* **112**, 14002–14010.

Modulation of psoralen DNA crosslinking kinetics associated with a triplex-forming oligonucleotide. D. H. Oh *et al.* (2008) *Photochem. Photobiol.* **84**, 727–733.

2.6. CHEMICAL MODIFICATION AS A PROBE OF STRUCTURE

Chemical modification is often used to probe the DNA structure, to determine whether a particular group is accessible to the reagent. Most groups are much less reactive when they are part of a double-helical structure, especially when they are involved in base pairing or bound to proteins (Section 7.1.C).

Diethyl pyrocarbonate (DEPC) carboxyethylates purines at the N7 position when they are single-stranded or in the Z conformation:

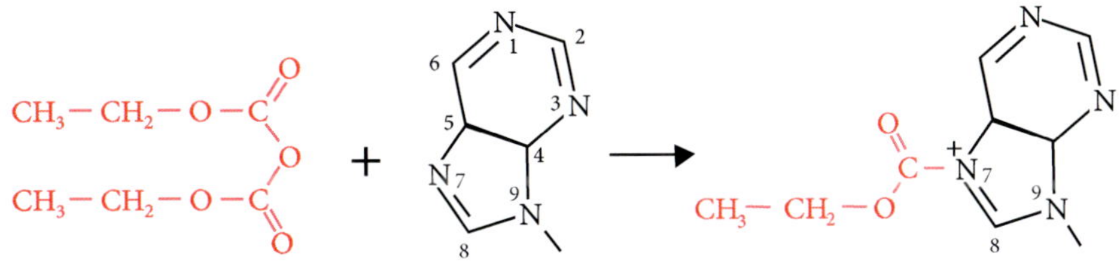

Diethylpyrocarbonate Purine (2.12)

The reaction can proceed further at N9 and open the purine ring by expelling C8. **Dimethyl sulfate** methylates the N7 position of guanines (Figure 2-33-C) when in double- or single-stranded structures, but not if it is involved in Hoogsteen hydrogen bonding (Figure 2-21). **Formaldehyde** reacts reversibly with amino groups, just as it does in proteins, so long as they are not involved in base pairing; but it also produces cross-links (Figure 2-33). **Hydroxylamine** (H_2N–OH) or **ethoxyamine** (H_2N–O–CH_3) converts an amino group to a hydroxyamino (–NHOH) or methoxyamino (–NH–O–CH_3) group (Figure 2-33-B). **Nitrous acid** converts amino groups to hydroxyls (Figure 2-33-A) but also produces cross-links (Figure 2-34).

Bisulfite (HSO_3^-) deaminates cytosine nucleotides to uracil if they are single-stranded (Figure 2-33-D). **Permanganate** or **osmium tetroxide** (OSO_4) will oxidize the C5–C6 double bond of thymine or cytidine residues:

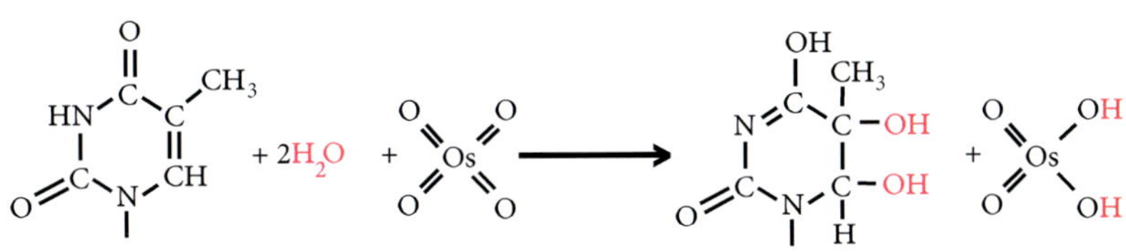

Thymine (2.13)

when they are present in single strands. Pyridine increases the rate of reaction with osmium tetroxide and produces an adduct of osmium and pyridine across the double bond. **Bromo-** or **chloro-acetaldehyde** forms fluorescent cyclic etheno derivatives with the base pairing positions of adenines, cytosines and, to a lesser extent, guanines, so long as they are not involved in base pairing:

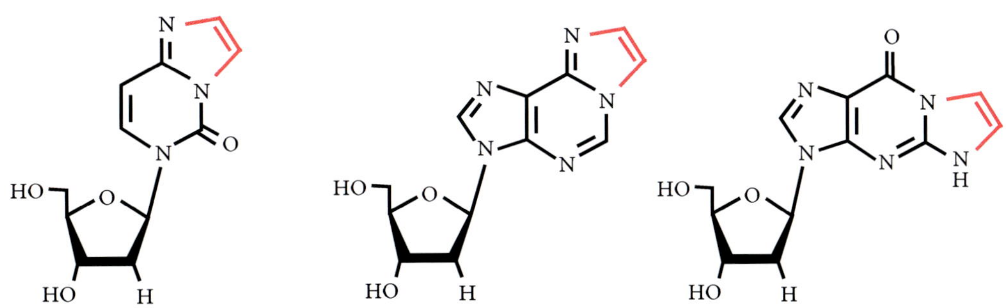

Cytosine Adenosine Guanosine (2.14)

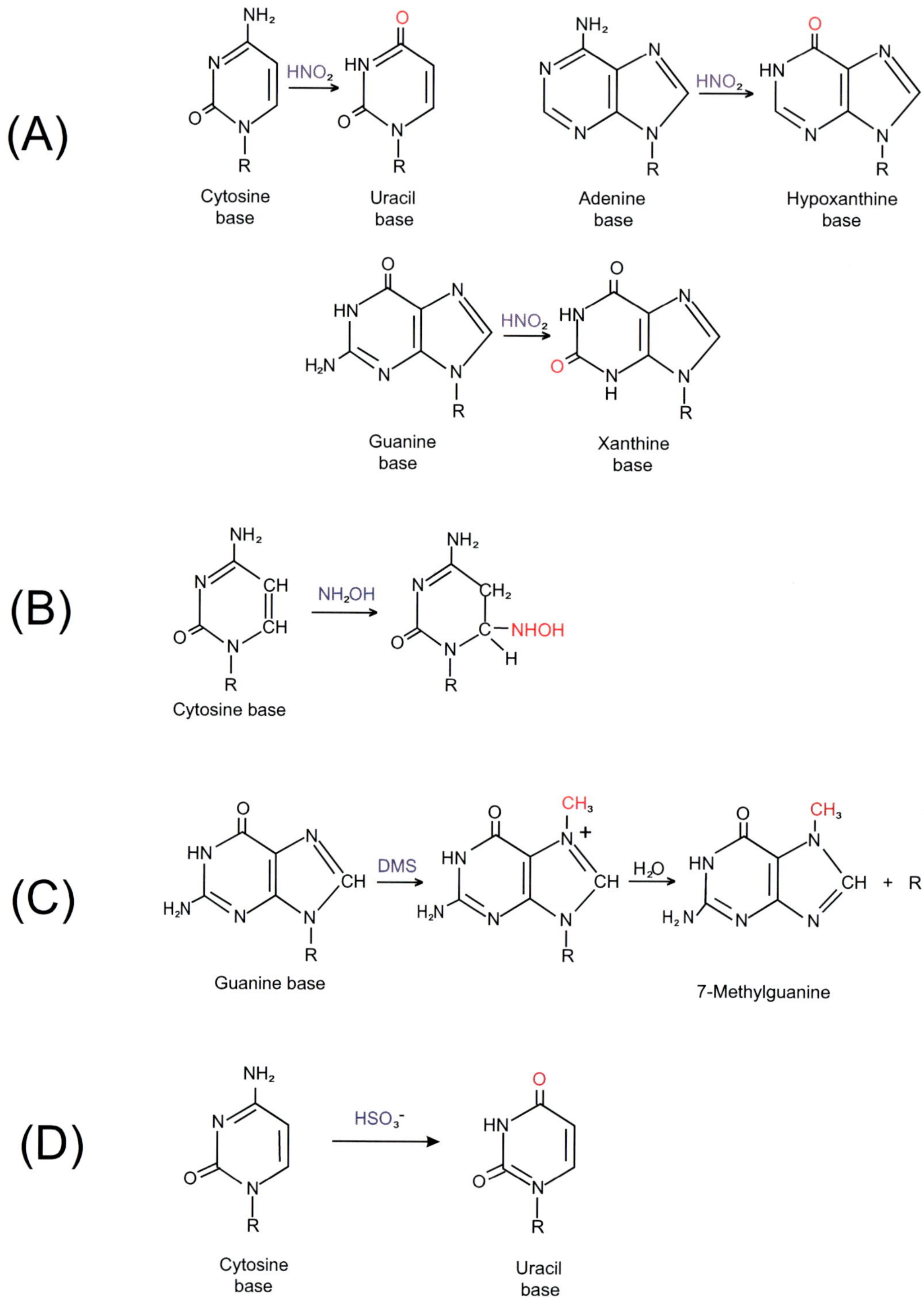

Figure 2-33. Common chemical modifications of DNA bases. The ribose group is depicted by R. (A) Deamination by nitrous acid coverts cytosine to uracil, adenine to hypoxanthine, and guanine to xanthine. (B) Reaction with hydroxylamine, as illustrated here with cytosine. (C) Alkylation of guanine by dimethyl sulfate (DMS); generating the quaternary N atom destabilizes the glycosidic bond and releases the modified base from the deoxyribose sugar. (D) Deamination of cytosine to uracil by bisulfite.

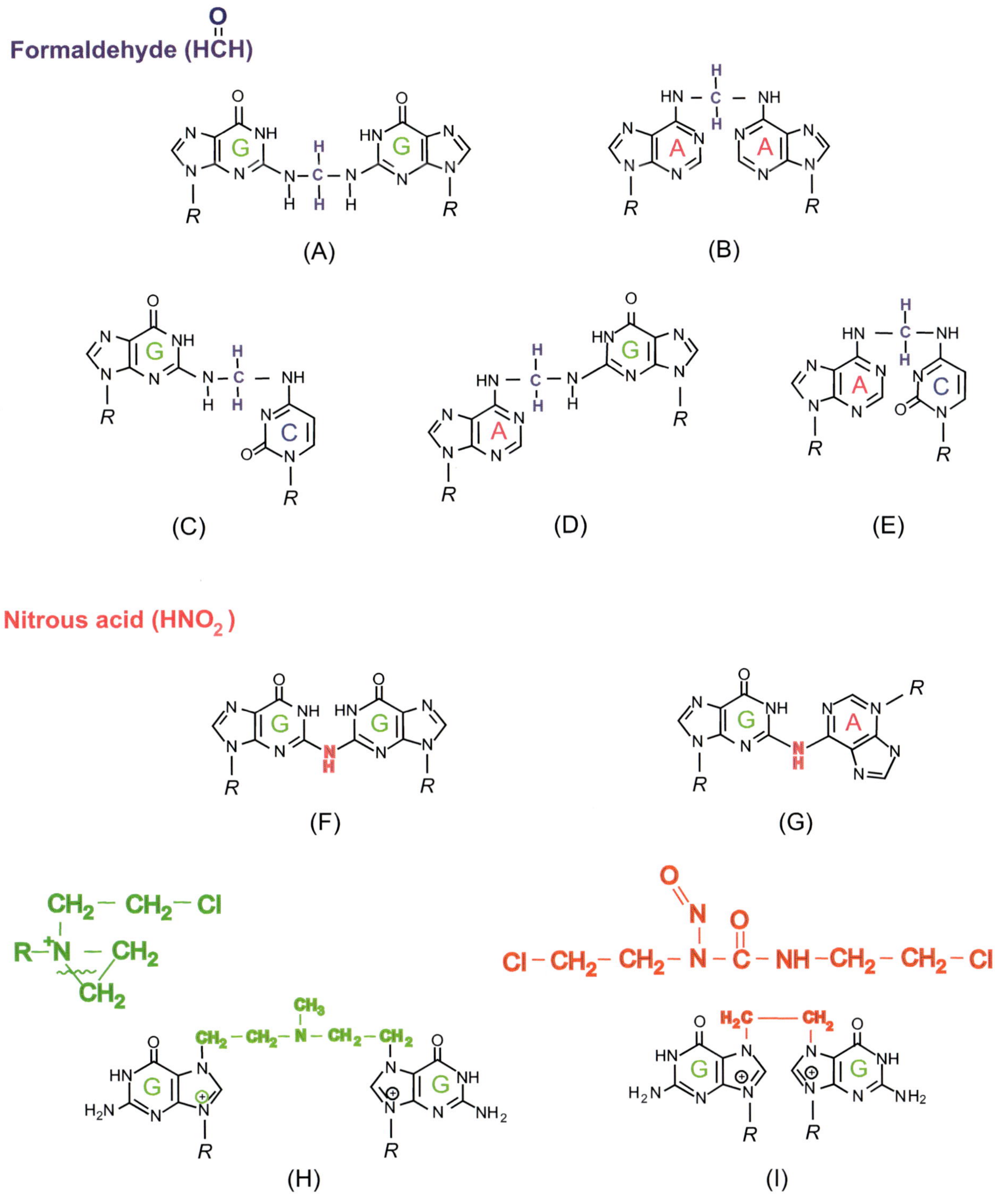

Figure 2-34. Cross-links produced in double-stranded DNA by formaldehyde ($H_2C{=}O$), nitrous acid (HNO_2) and two N mustards. The ribose group is depicted by R. Adapted from B. Singer & J. T. Kusmierek (1982) *Ann. Rev. Biochem.* **51**, 655–693.

Nitrosoureas:

$$\begin{array}{ccccccc} & & R & & O & & \\ & & | & & \| & & \\ O{=}N & - & N & - & C & - & NH_2 \end{array} \qquad (2.15)$$

will transfer the R group to any accessible O and N atoms of nucleic acids, on the bases, ribose ring and phosphate group.

Chemical modification by **acylation** of amino groups introduced at the 2′-position reports on the local nucleotide flexibility, because unconstrained 2′-amino nucleotides more readily reach a reactive conformation in which the amide-forming transition state is stabilized by interactions between the amine nucleophile and the adjacent 3′-phosphodiester group. The acylation reaction requires that the base lie outside of the helix and in the major groove. Bulged 2′-amine-substituted cytidine nucleotides react approximately 20-fold more rapidly than do nucleotides constrained by base pairing. In contrast, those base-paired but flanked by a 5′- or 3′-bulge react two- or six-fold more rapidly, respectively, than the perfectly paired duplex. The relative lack of 2′-amine reactivity for nucleotides adjacent to a DNA bulge suggests that the structural perturbations of the bulge do not extend significantly into the flanking duplex structure.

Some of these chemical modifications make their adjacent phosphodiester backbones labile to cleavage by the base **piperidine**, so the positions of the modifications in a DNA molecule of defined size can be determined by the lengths of the DNA fragments produced, similar to the approach used in chain-cleavage DNA sequencing (Section 6.4.C). Some chemical probes, as well as some naturally occurring drugs, can cleave DNA by an oxidative mechanism in which an H atom is abstracted from the deoxyribose ring, producing a radical that reacts further with oxidants such as O_2 or H_2O_2, causing chain cleavage. The site of the cleavage can be determined readily from the lengths of the fragments produced. **Hydroxyl radicals** cleave the sugar-phosphate backbone nonspecifically; they are generated spontaneously by ferrous ions in the presence of oxygen and a reducing agent. A complex of ferrous iron with the chelating agent EDTA (Figure 12-7) behaves similarly. Other reagents use bound transition metals to carry out similar reactions. Nonspecific reagents of this type are useful when the DNA molecule is not universally accessible to the reagent, due to being bound to a surface or a protein; the cleavage pattern reflects which nucleotides are accessible, which can indicate the helical periodicity of the double helix. Such nonspecific reagents can be made specific by attaching them covalently to polynucleotides that base-pair with complementary sequences or to agents that intercalate at specific sites; their local concentration can then be much greater than that of the bulk solvent.

Some enzymes are specific for the structure of a polynucleotide. For example, **S1 nuclease** hydrolyzes the backbones of both DNA and RNA when they are single-stranded. It will cleave the single-stranded regions in the loops of cruciforms, in the strand released after triple-strand formation, and in internal loops. It does not cleave where there is a mismatch between a single pair of bases.

Chemical modifications are especially useful in identifying the sites on DNA molecules where specific proteins bind, and thereby protect the DNA from modification (Section 7.1.C). Many of these reactions also produce changes in base-pairing during replication of the DNA and consequently cause genetic mutations, so the reagents responsible are often mutagens and must be handled with care.

Mechanics of DNA flexibility visualized by selective 2′-amine acylation at nucleotide bulges. D. M. John *et al.* (2004) *J. Mol. Biol.* **337**, 611–619.

Photochemical determination of different DNA structures. Y. Xu *et al.* (2007) *Nature Protoc.* **2**, 78–87.

2.6.A. Cross-linking

Chemical groups in DNA that are close in space can be cross-linked using reagents with two reactive groups (Figure 2-34). If two different DNA groups react with the same reagent molecule, they must be in the proximity required by that reagent. This can be a relatively simple probe of the overall structure of the DNA molecule. The reagents used most commonly are *N*-acetyl-*N*′-(*p*-glyoxylbenzoyl)cystamine, bis(2-chloroethyl)-methylamine (also known as 'nitrogen mustard'), *p*-azidophenyl acetimidate and iminothiolane.

~ CHAPTER 3 ~

DNA TOPOLOGY

The linear double-helical structure of DNA described in Chapter 2 is only the first level of DNA structure. DNA molecules can be extremely long, approximately 30 cm for a billion nucleotides, yet they must be accommodated within the nucleus of a cell, the dimensions of which (typically less than 0.1 mm in diameter) are only a tiny fraction of the length of the DNA molecules it must contain; the linear DNA molecules must be compacted extensively. Compaction introduces a second problem: how such a compacted DNA double helix is opened up to serve as a template for its transcription into messenger RNA (Section 6.2). A third problem is even more severe, concerning DNA replication (Section 6.1): how can such a compacted linear molecule be unraveled and separated into its two strands, so that they can be replicated to produce two identical molecules that can then be separated and packaged into two separate nuclei during and after cell division? If the 3 billion base pairs of DNA in the haploid human genome were in one piece (it is actually divided into 23 chromosomes), it would be a linear DNA molecule 1 m long and 20 Å in diameter. To illustrate the physical complexity of such long molecules, consider that this is comparable to a normal piece of string that is 1 mm in diameter being 500 km long! The well-known everyday problem of unraveling a compacted string (even only a few meters in length) illustrates adequately the topological problems that could be encountered within a cell in order to express and duplicate its DNA. Further complications occur with the many DNA molecules that are circular *in vivo* because the termini are joined in normal phosphodiester bonds. Cells have developed very sophisticated machinery to manage these problems of raveling and unraveling DNA. Here we will consider the basic problem of DNA packing: its topology.

Duplex DNA acquires topological properties if organized into a **topological domain**, which is a closed region in which the two strands are linked. The two types of elementary topological domains are (1) **closed circular duplex DNA** (ccDNA), in which the two strands are each covalently continuous, and (2) DNA loops that are closed by the binding of protein molecules (Figure 3-1). Protein-sealed DNA loops can occur locally in parts of DNA molecules that are not closed circular. The principles of circular DNA topology are directly applicable to DNA loops defined by bound protein molecules, and of even greater importance (Chapter 7), but protein molecules can dissociate and then reassociate (Section 7.2). Consequently, most experimental and theoretical work on DNA topological domains has used ccDNA, because it is defined by its covalent bonds. Circular DNA molecules exist in nature as the genomes of bacteria and as plasmids, or they can be prepared easily by using a **DNA ligase** enzyme to link covalently the two ends (Figure 3-2). This occurs much more readily if the two ends of the linear DNA molecule have a few unpaired nucleotides at each end that are complementary to

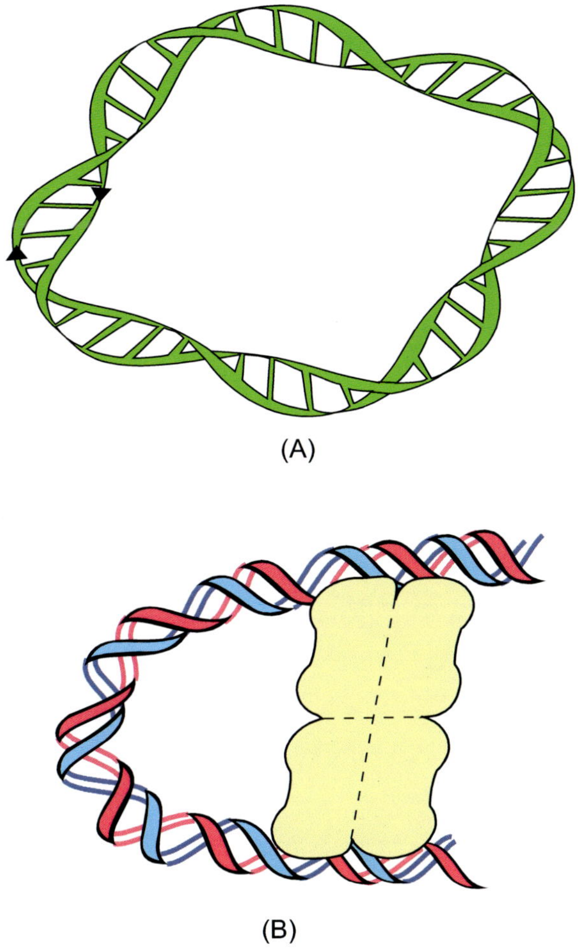

Figure 3-1. Two kinds of elementary topological domains of double-stranded DNA. (A) A closed duplex DNA, in this case a relaxed duplex with no supercoiling. The polynucleotide backbones of the two strands are depicted as the two continuous bands, linked by the ladder-like base pairs. Both strands in this DNA are covalently continuous and the linking number is an integer. There are eight nodes in the projection shown and $Lk = +4$ (Section 3.2). (B) A DNA loop formed by binding of a tetrameric protein. Only the two polynucleotide backbones of the DNA duplex are indicated. In this case, $Lk \approx 6$. The exact value of Lk depends upon the precise location at which the two strands are held together by the protein, and Lk consequently need not be an integer.

each other, known as **sticky ends** (Section 6.3.A). The topology of the resulting circular double helix is altered if one of the ends is rotated relative to the other by one or more full 360° turns before linking the ends.

The reference molecule for DNA topology is the **nicked circular duplex DNA**, in which at least one phosphodiester bond of ccDNA has been hydrolyzed. This destroys the topological domain, because free rotation can take place about the site of the chain scission; such species are designated as **open duplex DNAs**.

Circular DNA molecules like those illustrated in Figures 3-1 and 3-2 are elementary topological domains in which the circular DNA is topologically monomeric and the axis of the DNA is not itself linked to that of any other DNA. DNA can also be organized into higher order, nonelementary topological domains, in which the duplex DNA axes of one or more circular DNAs are themselves linked (Figure 3-3). **Knotted DNA** represents a class in which the duplex axis of a single DNA forms a knot, and it can be produced by site-specific genetic recombination. **Catenated DNA** has the axes of two or more ccDNA submolecules linked and contains one or more higher order topological domains. The submolecules might consist of individual elementary topological domains.

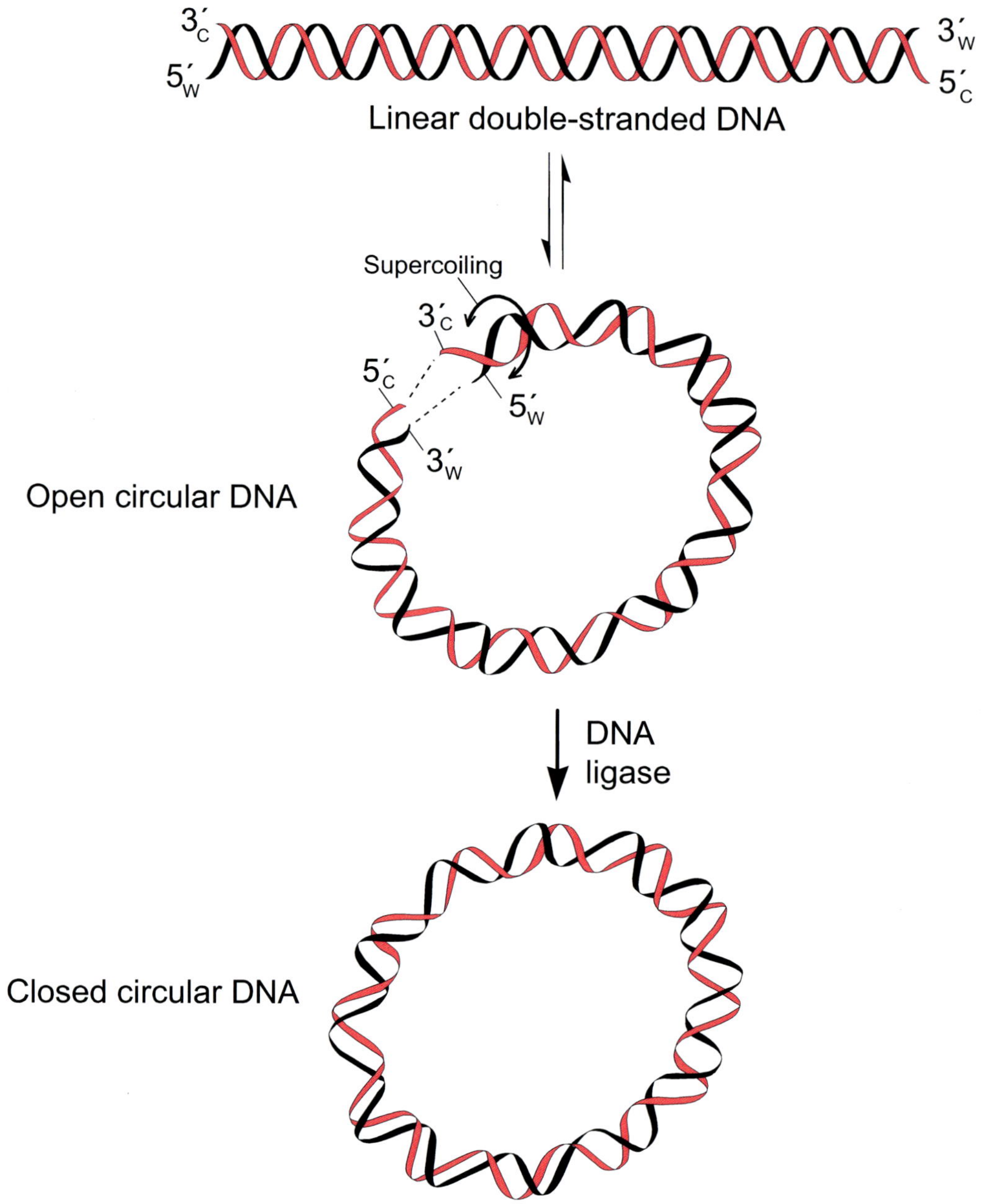

Figure 3-2. Converting a linear double-stranded DNA molecule to a circle. The individual strands in the linear molecule are labeled as W (Watson, *black*) or C (Crick, *red*). Circularization links the opposite ends of each of the W and C strands. It is assisted if the two ends have complementary single strands, known as 'sticky ends'. Supercoiling results if one end is rotated by one or more full turns before the ends are linked. Each twist of 360° changes the linking number by 1. The phosphodiester backbones are completed by the action of a DNA ligase enzyme.

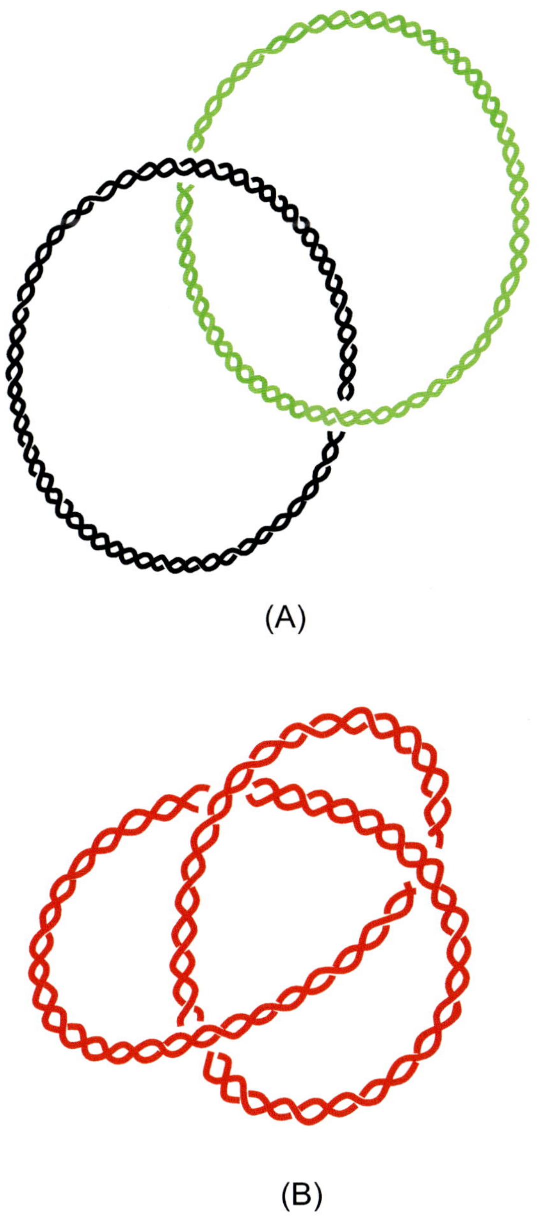

Figure 3-3. Two types of higher order topological domains. (A) Two closed circular DNAs are linked once to form a catenane. The higher order linking number (catenation number) has a value of 1 in this case, because the axes of the two DNA molecules are linked once. (B) The axis of a single DNA molecule is knotted into a trefoil.

Practical methods for supercoiled plasmid DNA production. J. Ballantyne (2006) *Methods Mol. Biol.* **127**, 311–337.

DNA supercoiling and bacterial gene expression. C. J. Dorman (2006) *Sci. Prog.* **89**, 151–166.

DNA topology: dynamic DNA looping. A. Travers (2006) *Curr. Biol.* **16**, R838–840.

Dynamic state of DNA topology is essential for genome condensation in bacteria. R. L. Ohniwa *et al.* (2006) *EMBO J.* **25**, 5591–5602.

Strong effects of molecular topology on diffusion of entangled DNA molecules. R. M. Robertson & D. E. Smith (2007) *Proc. Natl. Acad. Sci. USA* **104**, 4824–4827.

3.1. SUPERCOILING AND SUPERHELICES: TOPOISOMERS

Imagine in Figure 3-2 that one end of a linear DNA duplex is twisted by one or more full turns before the ends are joined. Each twist will introduce and build up elastic strain in the DNA circle that is generated, much as occurs in a rubber band. Consequently, this supercoiling increases the free energy of the DNA. Supercoiling can be either left-handed or right-handed. The number of times the two strands of a ccDNA are coiled about each other cannot be altered without first cleaving at least one of its polynucleotide strands; this can be demonstrated easily with a buckled belt, in which each edge of the belt represents a strand of DNA. The two DNA backbone strands whose interwinding defines a topological domain contain no chemical bonds in common; nevertheless, the two strands cannot be separated except by breaking a covalent bond. This type of connection is sometimes called a **topological bond**.

Circular DNAs often have the duplex coiled about itself, which generates a peculiar twisted appearance. This is a result of **supertwisting**, or **superhelicity**, of the DNA duplex (Figure 3-4); it is often referred to as DNA's 'tertiary structure'. Such superhelical structures arise to lessen the strain introduced in a ccDNA from the way the two individual DNA strands are twisted about each other. Two types of superhelical structure occur and can be reversibly interconverted without strand scission (Figure 3-5). First is the **plectonemic** or **interwound** superhelix, in which the DNA axis winds solenoidally up and down a virtual cylinder, forming a double helix, with smoothly curved connection at the ends of the cylinder. It occurs with purified, monomeric closed circular DNA and is further illustrated in Figure 3-4. Second is the **toroidal** superhelix, in which the DNA axis winds solenoidally around the surface of a torus or a segment of a torus, forming a single helix; it is primarily encountered in DNA associated with other DNA molecules (catenated DNA) or wrapped on a protein surface. These two superhelical forms are different conformations and are interconvertible, but they have opposite handedness. The most simple parameter to describe supercoiling is n, the number of superhelical turns that are apparent in its structure when the molecule lies flat in a plane.

Supercoiled DNA may be converted to a **relaxed circle** simply by cleaving one of its strands at a single site; the DNA backbone of the opposite strand is then free to swivel about its backbone bonds so as to relieve any strain by changing the supercoiling. **The relaxed state of a DNA circle has the lowest free energy and is not considered to be supercoiled**, even though the two strands are wound round each other.

Naturally occurring DNA circles generally have negative supercoiling, i.e. they are **underwound** by about 6%. This is believed to be a result of the way the DNA is packaged (Section 3.7) and it prepares the DNA for physiological processes that require separation of the DNA strands, such as their replication or transcription.

Analysis of chemical and enzymatic cleavage frequencies in supercoiled DNA. H. Tsen & S. D. Levene (2004) *J. Mol. Biol.* **336**, 1087–1102.

DNA axial rotation and the merge of oppositely supercoiled DNA domains in *Escherichia coli*: effects of DNA bends. V. A. Stupina & J. C. Wang (2004) *Proc. Natl. Acad. Sci. USA* **101**, 8608–8613.

Influence of supercoiling on the disruption of dsDNA. S. M. Chitanvis & P. M. Welch (2005) *J. Chem. Phys.* **123**, 124901.

Effect of supercoiling on formation of protein-mediated DNA loops. P. K. Purohit & P. C. Nelson (2006) *Phys. Rev. E* **74**, 061907.

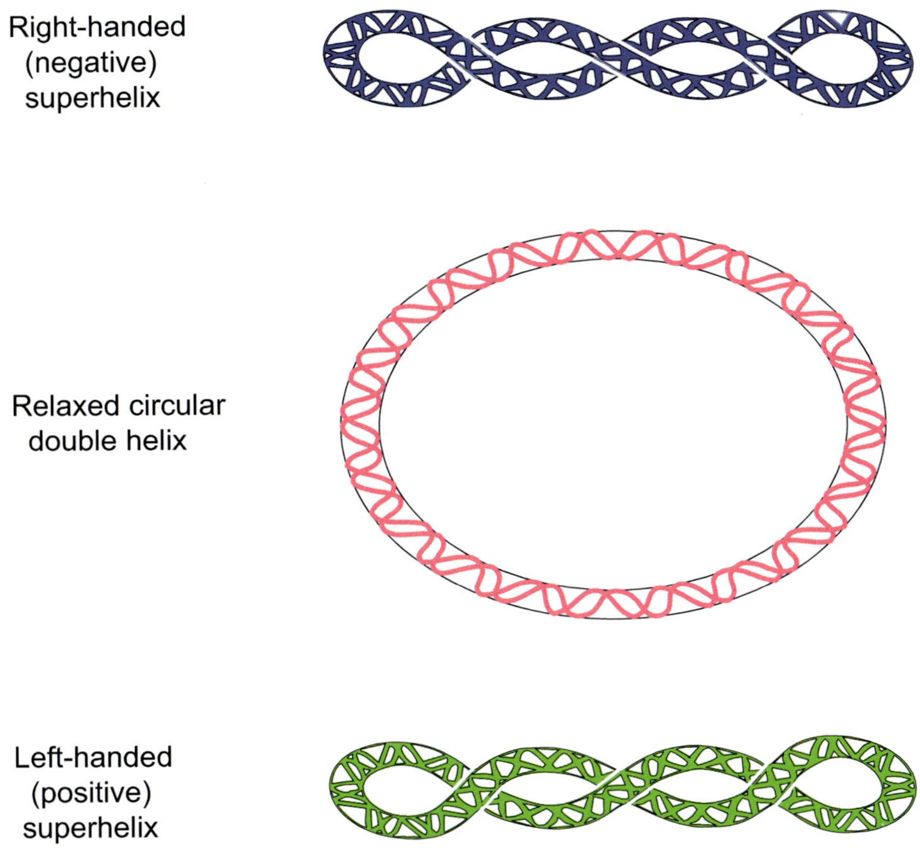

Figure 3-4. Relaxed and supercoiled DNA. The three structures differ in the number of times the two DNA strands are wound about each other (they have different linking numbers) and they cannot be interconverted without breaking one of the strands. The superhelix conformation with the double helices wound around each other is adopted to minimize the strain in the circular double helix. Negatively supercoiled DNA is the form normally present in cells, although the positively supercoiled form is generated transiently during replication and recombination.

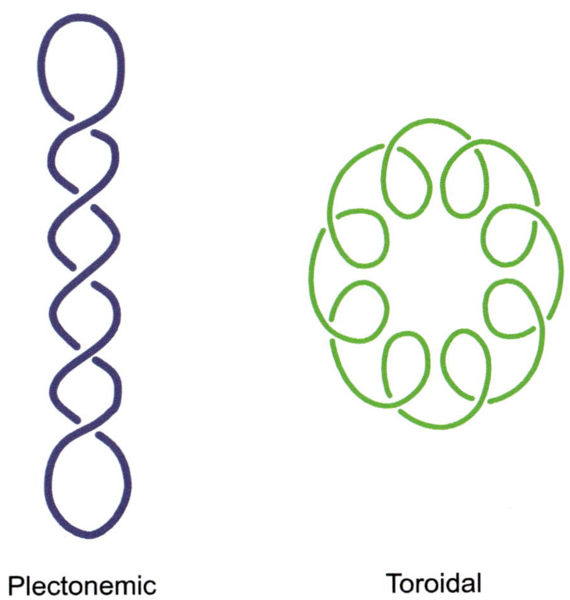

Figure 3-5. The two fundamental types of superhelix winding in superhelical DNA: a plectonemic (or interwound) superhelix (*left*) and a toroidal (or solenoidal) superhelix (*right*). The plectonemic form is the more stable in isolation but the toroidal form is generated by binding to some proteins.

3.2. LINKING NUMBER, *Lk*

Supercoiling is described quantitatively by the **linking number**, *Lk*, which is the number of times that one DNA strand winds about the other and is the most important parameter of DNA elementary topological domains. ***Lk* is a measure of the total number of complete revolutions that either strand makes about the other**; so long as the strands remain intact, *Lk* is a fixed quantity. Two closed duplex DNAs that differ only in *Lk*, but are otherwise identical, are **topoisomers**. They contain exactly the same nucleotide sequence and covalent connections; nevertheless topoisomers with different *Lk*s are distinct and have different properties that depend strongly upon *Lk*.

The linking number has several simple and very useful properties. (1) ***Lk* is an integer for superhelical DNA** because both strands must be closed curves. It need not be an integer for protein-sealed domains, where the geometry can be more flexible. (2) ***Lk* is constant so long as the topological domain remains intact**. *Lk* can be altered only if a topological domain is broken, by cleaving one or both strands of double-stranded DNA or, if appropriate, by the disruption of links to the protein sealing the domain. (3) **The linking number is independent of the DNA geometry** because it is a topological quantity. In particular, *Lk* does not vary with deformation of the trajectory of either strand or with changes in the characteristic duplex geometric quantities (pitch, roll, twist, tilt, propeller twist, etc.). (4) **The linking number is independent of the ordering of the two curves**; consequently, for two DNA strands W and C, $Lk(W,C) = Lk(C,W)$. (5) By convention, **the *Lk* of a closed circular DNA formed by a right-handed double helix is positive**. (6) Tightening the double helix increases *Lk* and is considered to be positive supercoiling.

A topological approach to nucleosome structure and dynamics: the linking number paradox and other issues. A. Prunell (1998) *Biophys. J.* **74**, 2531–2544.

Purification and use of DNA minicircles with different linking numbers. G. Camilloni *et al.* (1999) *Methods Mol. Biol.* **94**, 51–60.

Finite-element analysis of the displacement of closed DNA loops under torsional stress. J. H. White & W. R. Bauer (2004) *Philos. Transact. A* **362**, 1335–1353.

3.2.A. Linking Difference, ΔLk

The **linking difference** is a measure of how the winding of an elementary topological domain of closed duplex DNA deviates from that of its nicked, relaxed circular counterpart. The chemically and biologically interesting properties of a topological domain are actually determined by ΔLk, rather than the linking number *Lk* itself. The linear double helix in Figure 3-2 has the two strands wound around each other in a normal double helix, but supercoiling is introduced by rotating one end relative to the other by one or more turns. The number of turns, which can be negative or positive, determines ΔLk.

The corresponding nicked circular or linear DNAs strictly speaking do not have a linking number but they can be assigned a **pseudolinking number**, Lk_0, which is the number of times the two DNA strands are wound around each other. In contrast to *Lk*, Lk_0 need not be an integer, and generally is not, even for homogeneous duplex molecules, as the two strands are not continuous and can have incomplete turns at their ends. Lk_0 for relaxed DNA is determined simply by the number of base pairs,

N, and the DNA helical repeat, h_0 base pairs per turn (bp/turn):

$$Lk_o = N/h_0 \tag{3.1}$$

For linear B-DNA in dilute NaCl at 37°C, $h_0 = 10.5$ bp/turn on average, although the local value depends upon the base composition. It is also different for noncanonical DNA structures, such as Z-DNA (Section 2.2.B.3) and H-DNA (Section 2.2.C.3), and locally denatured DNAs.

The linking difference is defined as the difference between Lk and Lk_o:

$$\Delta Lk = Lk - Lk_o \tag{3.2}$$

If $\Delta Lk \neq 0$, the DNA is under a stress that deforms the double helix axis into a superhelix. A relaxed closed duplex tends to be a flat circular molecule, but increasing the supercoiling to $|\Delta Lk| > 1.5$ causes it to adopt a figure 8 type of conformation (Figure 3-4).

1. Relaxed Duplex DNA

Relaxed duplex DNA is that topoisomer closest to nicked circular DNA, whose linking number is the nearest integer to Lk_o. Because Lk_o itself is generally not an integer, a relaxed DNA species usually has a value of ΔLk that is small and fractional, ε, which is simply the difference between the exact value of Lk_o from Equation 3.1 and the nearest integer, so $-0.5 \leq \varepsilon \leq 0.5$. For example, DNA of the plasmid pBR322 has $N = 4363$ and $Lk_o = 415.52$ if $h_0 = 10.5$ base pairs/turn, so the relaxed duplex DNA topoisomer is the one with $Lk = 416$ and $\varepsilon = -0.48$. In physical terms, ε represents the minimum rotation that is required to bring the 5′ and 3′ ends of a nicked circular DNA together to permit covalent joining by a DNA ligase (Figure 3-2). The value of ε can be determined experimentally from the Boltzmann distribution of topoisomers that forms upon thermal equilibration in the presence of a topoisomerase (Section 3.6.A). This distribution fits well to a Gaussian curve of topoisomer frequency versus ΔLk, and ε is the separation between the center of this distribution and the location of the nearest, most prominent, topoisomer (Figure 3-6).

All naturally occurring closed circular DNAs have $\Delta Lk < \varepsilon$, when the ccDNA is said to be **underwound**. The linking difference is consequently sometimes called the **linking deficiency**, although the term 'linking excess' would be more appropriate for cases in which $\Delta Lk > \varepsilon$ and the DNA is **overwound.** If $\Delta Lk = \varepsilon$, the DNA is said to be **relaxed,** even though it might still contain a small linking difference.

Changing either Lk or Lk_o can change both the sign and the magnitude of ΔLk. The value of Lk can be changed most readily by treatment with a DNA topoisomerase (Section 3.3). In contrast, Lk_o can be changed by altering the temperature, type or concentration of the salt, or by the addition of an intercalating drug (Section 2.5.B). A relaxed duplex DNA can therefore be made to supercoil simply by making a change in its environment, so it is necessary to specify the conditions under which Lk_o is measured. The standard conditions are usually 37°C, 0.2 M NaCl, in the absence of any reagents that affect the DNA twist, when the value of Lk_o is described as Lk_o^0.

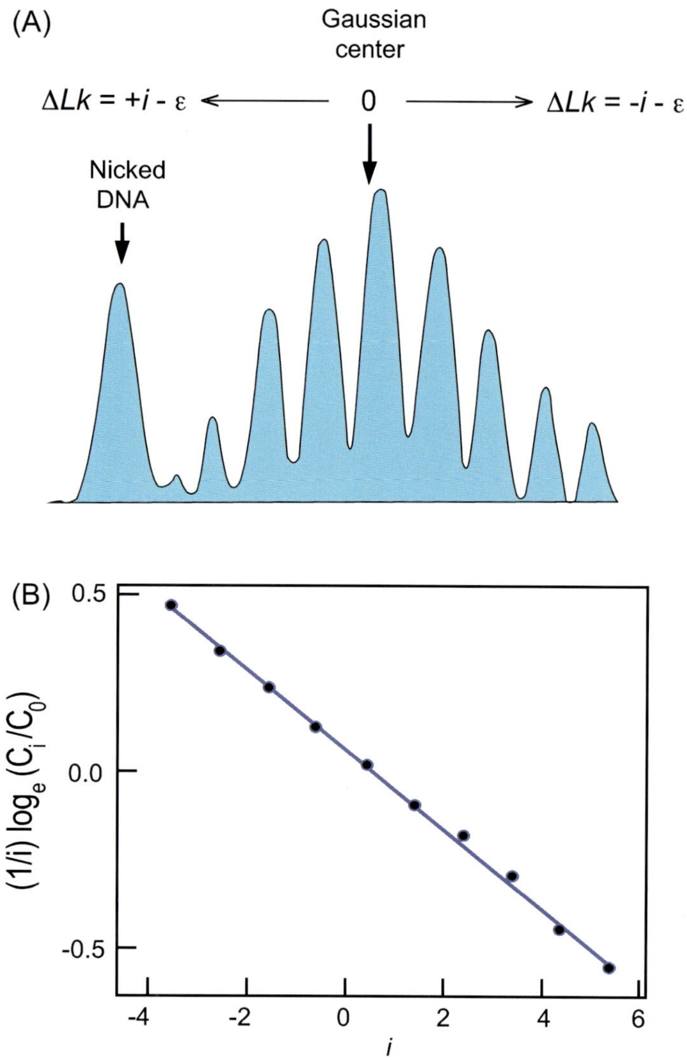

Figure 3-6. Hypothetical analysis of the distribution of topoisomers by gel electrophoresis. (A) Densitometer tracing of the gel, showing the intensity profile as a function of distance. The location of the nicked DNA is indicated, and the other peaks are the topoisomers with different linking numbers. The approximate location of the Gaussian center of the distribution is indicated. (B) Analytical plot of the topoisomer distribution according to Equation 3.15. The ΔLk-values of the topoisomers are $i + \varepsilon$, where $i = 0, \pm 1, \pm 2$, etc.

3.2.B. Superhelix Density, σ

The values of Lk, Lk_o and ΔLk are generally proportional to the DNA length. The supercoiling of two DNA molecules of different lengths can be compared by defining a normalized quantity, the **superhelix density** or **specific linking difference**, σ:

$$\sigma = \frac{Lk - Lk_o}{Lk_o} \tag{3.3}$$

Most naturally occurring circular duplex DNAs are underwound and have values of σ between 0 and −0.1 under standard conditions, but DNAs having σ as low as −0.17 have been prepared *in vitro*. Circular duplex DNAs with relatively large positive σ-values can be prepared using DNA reverse gyrase to increase the linking number (Section 3.3). Both σ and Lk_0 are readily measured, so the easiest way to measure Lk is to combine Lk_0 with the measured value of either ΔLk or σ. Using the example of pBR322 DNA again, the natural DNA has an average value of σ = −0.06. In this case, the average linking number for this native DNA is 390.58. The linking number of the nearest (most prevalent) topoisomer is 391.

As in the case of ΔLk, the value of σ can be altered by changes in environmental conditions alone, without any change in *Lk*, such as by changing the temperature. **The temperature coefficient of σ is $\Delta\sigma/\Delta T = 3.1 \times 10^{-4}$ deg^{-1}**. If the temperature is changed from 5° to 37°C, the change in σ is +0.01, which would change ΔLk by +4.16 turns for pBR322 DNA with 4363 base pairs. An increase in temperature thus causes a relaxed DNA to supercoil in the positive sense (a left-handed, interwound superhelix) and reduces the supercoiling of a naturally occurring (underwound) superhelical DNA. Changing the cation species and concentration has similar effects. The parameter pX^+ is defined as the negative logarithm of the cation concentration, analogous to the pH. Over the range of ionic strength from 0.05 to 0.3 M, **$\Delta\sigma/\Delta pX^+ = 4.47 \times 10^{-3}$ for the ions Na^+, K^+, Li^+ and NH_4^+ and 6.70×10^{-3} for Rb^+, Cs^+ and Mg^{2+}**. For example, if the potassium ion concentration were to be increased from 0.01 M to 1.0 M, the change in σ should be –0.013, which is comparable to a decrease in temperature of 42°C.

Separation of supercoiled DNA using capillary electrophoresis. D. T. Mao & R. M. Lautamo (2001) *Methods Mol. Biol.* **162**, 333–344.

Separation of large circular DNA by electrophoresis in agarose gels. K. D. Cole & C. M. Tellez (2002) *Biotechnol. Prog.* **18**, 82–87.

Early melting of supercoiled DNA topoisomers observed by TGGE. V. Viglasky *et al.* (2000) *Nucleic Acids Res.***28**, E51.

3.3. TOPOISOMERASES

DNA topoisomerases are enzymes that catalyze the interconversion of DNA topoisomers. In the process, the DNA strands in a duplex DNA move through each other, to change the linking number. The changes can occur to duplex DNA segments in the same or different molecules, so topoisomerases can also interconvert catenanes and knotted DNA molecules (Figure 3-3). Type I DNA topoisomerases can change the linking number by ±1, by introducing a transient break in one strand, passing the other strand through the break, and resealing the cleaved strand. The topologies of catenanes and knots can be altered by these enzymes only if at least one of the strands contains a nick. Type II topoisomerases break and rejoin both strands at the same time, causing double-stranded DNA to pass through a double-strand break and thereby changing *Lk* by ± 2. The topology of a catenane or knot can be changed by these enzymes even if the chains are intact.

In the absence of an energy input, topoisomerases catalyze the net interconversion of topoisomers only if new topoisomers have lower free energies than the original; in the presence of such a topoisomerase, the topoisomers become relaxed and the spectrum generated reflects their intrinsic free energies (Figure 3-6). Other topoisomerases can use the energy from hydrolysis of ATP to introduce negative or positive supercoiling, even if this increases the free energy of the ccDNA. **DNA gyrase** is such a type II DNA polymerase and introduces negative supercoiling. In contrast, **reverse gyrase** introduces positive supercoiling, increasing the value of *Lk*.

The topoisomerases perform these topological tricks by binding the DNA molecule so that the hydroxyl group of a tyrosine residue attacks a DNA phosphate group; this cleaves the backbone of the strand and attaches the cleaved strand to the enzyme. One or both strands are passed through the break, then the cleavage of the polynucleotide chain is reversed, and the intact DNA molecule dissociates from the topoisomerase.

Topoisomerases are crucial components of the cellular machines that catalyze DNA replication and transcription. Without them, the DNA molecules would become hopelessly entangled.

Crystal structure of an intact type II DNA topoisomerase: insights into DNA transfer mechanisms. M. Graille *et al.* (2008) *Structure* 16, 360–370.

DNA topoisomerases: harnessing and constraining energy to govern chromosome topology. A. J. Schoeffler & J. M. Berger (2008) *Quart. Rev. Biophys.* **41**, 41–101.

Structural studies of type I topoisomerases. N. M. Baker *et al.* (2009) *Nucleic Acids. Res.* **37**, 693–701.

Topoisomerase II: a fitted mechanism for the chromatin landscape. N. Roca (2009) *Nucleic Acids. Res.* **37**, 721–730.

3.4. TWIST AND WRITHE

The individual conformations of a single topoisomer can be described by two further parameters, the molecule's **twist**, *Tw* (Section 3.4.A) and its **writhe**, *Wr* (Section 3.4.B). The twist is the number of complete revolutions that one polynucleotide strand makes about the duplex axis, while the writhe is the number of turns that the duplex axis makes about the superhelix axis. The sum of the two numbers must be the same as the linking number:

$$Lk = Tw + Wr \tag{3.4}$$

This is the most fundamental equation describing DNA topology. It indicates that an increase in twist must be accompanied by a decrease in its writhe (Figure 3-7). A belt wound around a cylinder has a large writhe number but a very small twist; extending the belt decreases the writhe but increases the twist. The linking number of a closed duplex DNA molecule must be an integer, but not the twist and writhe numbers.

The linking number is a topological quantity that remains constant, irrespective of the exact structure and orientation of the molecule, so long as the DNA backbone remains intact. In contrast, the parameters *Tw* and *Wr* are geometrical conformational quantities that may change as the DNA bends, twists or kinks.

3.4.A. Twist, *Tw*

The twist is the number of complete revolutions that one polynucleotide strand makes about the duplex axis in any particular conformation. Twist is a general property of two curves that coil about each other. The twist of one curve (W) about another (C), *Tw*(W,C), is usually measured by how often W spins about C as one advances along C. In contrast to the linking number, the twist depends upon the ordering of the two curves; except for certain special cases, $Tw(W,C) \neq Tw(C,W)$. By convention, *Tw* is positive for right-handed duplex turns.

If the DNA axis is a straight line, or if the axis is closed and lies entirely in a plane, the twist is known as Tw_0. For B-DNA in solution, Tw_0 is normally the number of base pairs divided by 10.5, the number of base pairs per turn. In this case, B-form DNA containing 30 base pairs would have a twist of Tw_0 =

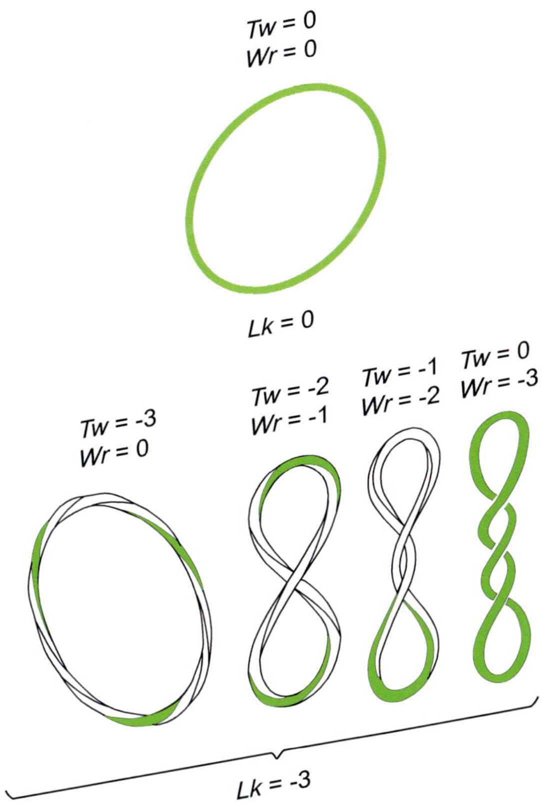

Figure 3-7. Simple illustration of the differences between the linking number (*Lk*), the twist (*Tw*) and the writhe (*Wr*) of a closed duplex DNA molecule. It is represented as a rubber ring of square cross-section, with one surface *colored* and the other three *white*. The twist is apparent by the number of times the visible surface changes color. Adapted from C. R. Calladine & H. R. Drew (1992) *Understanding DNA: the molecule and how it works*, Academic Press, London.

$N/h_0 = 30/10.5 = 2.857$. DNA rarely has a truly linear axis, however, and any out-of-plane motion of the helical axis changes the value of *Tw*. The extent of the deviation depends upon the amount of axial torsion. The twist of any structure can be calculated but it is not straightforward.

Chirality and surface twist of DNA wrapped on protein surfaces. J. H. White & W. R. Bauer (1995) *J. Biomol. Struct. Dynamics* **12**, 815–826.

Is a small number of charge neutralizations sufficient to bend nucleosome core DNA onto its superhelical ramp? G. S. Manning (2003) *J. Am. Chem. Soc.* **125**, 15087–15092.

DNA twisting flexibility and the formation of sharply looped protein–DNA complexes. T. E. Cloutier & J. Widom (2005) *Proc. Natl. Acad. Sci. USA* **102**, 3645–3650.

B-DNA under stress: over- and untwisting of DNA during molecular dynamics simulations. S. Kannan *et al.* (2006) *Biophys. J.* **91**, 2956–2965.

Role of tension and twist in single-molecule DNA condensation. K. Besteman *et al.* (2007) *Phys. Rev. Lett.* **98**, 058103.

Terminal twist-induced writhe of DNA with intrinsic curvature. K. Hu (2007) *Bull. Math. Biol.* **69**, 1019–1030.

3.4.B. Writhe, *Wr*

The writhe (*Wr*), or writhing number, is the number of turns that the duplex axis makes about the superhelix. Writhe is a property of a single closed space curve; in the case of DNA, the appropriate space curve is the duplex axis of the double helix. Writhe **can be calculated only for a closed curve, so *Wr* is not applicable to nicked circular or linear DNAs**. The magnitude of the writhe is directly proportional to n, the number of superhelical turns. **The writhe is zero when the DNA's duplex axis is constrained to lie in a plane**. In all other cases some of the winding is always converted into changes in Tw and the absolute value of Wr is always less than n.

The writhe is not a topological quantity so its calculation is *not* independent of the projection chosen; this clearly distinguishes Wr from Lk. Wr is the average of the projected writhing number over all possible projections. Deforming the axis changes the values of the directed writhing numbers, and hence of Wr. The easiest way to calculate the writhe is from the difference between the linking number and the twist, using Equation 3.4. This method is indirect but considerably easier than direct methods because both Lk and Tw are measured much more readily than Wr.

A topological invariant to predict the three-dimensional writhe of ideal configurations of knots and links. C. Cerg & A. Stasiak (2000) *Proc. Natl. Acad. Sci. USA* **97**, 3795–3798.

Computation of writhe in modeling of supercoiled DNA. K. Klenin & J. Langowski (2000) *Biopolymers* **54**, 307–317.

Writhe distribution of stretched polymers. S. Sinha (2004) *Phys. Rev. E* **70**, 011801.

Evaluating changes of writhe in computer simulations of supercoiled DNA. R. de Vries (2005) *J. Chem. Phys.* **122**, 064905.

3.5. DNA TOPOLOGY AND GEOMETRY

For any process in which Lk is unchanged, any changes in the twist must be matched exactly by changes in the writhe of opposite sign (Equation 3.4 and Figure 3-7):

$$\delta Tw = -\delta Wr \text{ (for constant } Lk) \tag{3.5}$$

A nicked circular DNA has no writhe, so $Tw_0 = Lk_o$. Introducing this into Equation 3.4 gives:

$$\Delta Lk = \Delta Tw + Wr \tag{3.6}$$

where $\Delta Tw = Tw - Tw_0$ and represents the deviation of the twist from the open circular value for B-DNA. This equation indicates that there are two differences between a relaxed duplex DNA and a nicked circular duplex DNA.

(1) A nicked circular duplex DNA has $Wr = \Delta Tw = 0$, but for the relaxed duplex DNA $Wr + \Delta Tw = \varepsilon$. Consequently, a relaxed duplex DNA can be distorted slightly by some combination of changes in Tw and Wr. For a perfect elastic rod, all the distortion would go into twist, but it all goes into writhe in the case of DNA.

(2) The twist and writhe remain coupled in a relaxed duplex DNA, as in Equation 3.5. In contrast, for nicked circular DNA ΔLk is not defined and Equation 3.6 does not apply, so in this case ΔTw

and Wr are uncoupled and may fluctuate independently. These differences explain why nicked circular DNA often migrates more slowly than relaxed circular DNA in both gel electrophoresis and sedimentation experiments under identical solution conditions.

The important structural parameters for superhelical DNAs have been estimated for closed circular DNAs of 3.5×10^3 and 7.0×10^3 base pairs. The superhelix radius r, measured in Å, depends upon the superhelix density σ (Section 3.2.B) according to the empirical equation:

$$\frac{1}{r} = 0.00153 - 0.268\sigma \qquad (3.7)$$

The superhelix pitch is observed to be 54–56°, nearly constant and independent of σ. The length of the superhelix axis is independent of σ but proportional to the contour length of the DNA. The ratio of the superhelix axis length to DNA length is 0.41. The number of supercoils, n, is proportional to the absolute value of ΔLk: $n = 0.89\,|\Delta Lk|$. The ratio between the twist increment and the writhe is constant, and $\Delta Tw/Wr \approx 3$. Individually, $\Delta Tw = 0.28\,\Delta Lk$ and $Wr = 0.72\,\Delta Lk$, indicating how changes in ΔLk are proportioned between the twist and the writhe.

Some experiments indicate that the superhelix radius is highly salt-dependent, with the opposing duplex segments approaching self-contact at moderate counterion concentrations, but others indicate that the structure remains much more relaxed at high salt, with no evidence of contacts between adjacent duplexes.

The plectonemic form is generally more stable than the toroidal form (Figure 3-5) for ccDNA in isolation. A relaxed circular form generally adopts a figure of 8 structure when ΔLk reaches an absolute value of 1.5–1.9. Thereafter, each increase of ΔLk by 1 usually introduces one more superhelical turn. Supercoiling can also induce large-scale conformational changes, such as local denaturation of the double helix, change from B-DNA to Z- or H-DNA (Section 2.2), or formation of a cruciform structure (Section 2.2.C.7). Increased negative superhelicity can induce adoption of the Z-DNA structure (see Figure 3-11), which made it possible to measure the free energy difference between B- and Z-DNA.

Mathematical modelling of interwound DNA supercoils. D. M. Stump *et al.* (2000) *J. Biomech.* **33**, 407–413.

3.5.A. Experimental Characterization of DNA Topology

Supercoiling affects the overall structure of a DNA molecule and is apparent in this way.

1. Electron Microscopy

The large sizes of DNA molecules make it possible to observe their topologies directly with an electron microscope. However, this requires that the molecules be spread out so that they can be absorbed on the support film in an extended nonaggregated form (Figure 3-8). This is usually accomplished by mixing the negatively charged DNA molecule with a basic protein, such as cytochrome *c*, or low-molecular weight substances, such as benzyl-dimethyl-alkyl-ammonium chloride and ethidium bromide (Section 2.5.B.1). This mixture is spread to form a molecular monolayer at the liquid–air interface, which is then adsorbed onto the support film, stained with uranyl acetate, rinsed in ethanol and air-dried. To improve contrast, the preparation can be rotary-shadowed, usually with platinum and palladium, at an angle of 6–9° from the plane of the metal vapor. Metal shadowing greatly exaggerates the thickness of the DNA molecules but the topology is evident (Figure 3-8).

The linking number of a closed duplex DNA can simply be counted from its structure when the molecule lies flat in a plane. In that case, the writhe is zero, so $Lk = Tw$, and Lk may be measured simply by counting the DNA's duplex turns. Nonplanar structures, however, require other procedures.

Polylysine-coated mica can be used to observe systematic changes in the supercoiled DNA conformation by scanning force microscopy in solution. M. Bussiek *et al.* (2003) *Nucleic Acids Res.* **31**, e137.

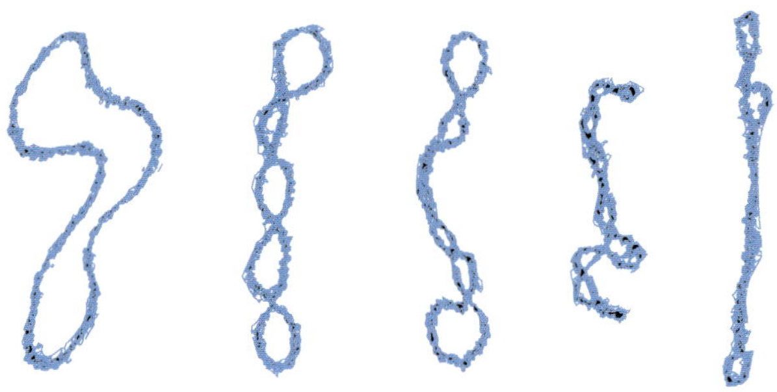

Figure 3-8. Typical results of electron microscopy after rotary shadowing of circular duplex DNA molecules differing in their conformations, from no supercoiling on the *left* to tightly supercoiled on the *right*. The background has been removed.

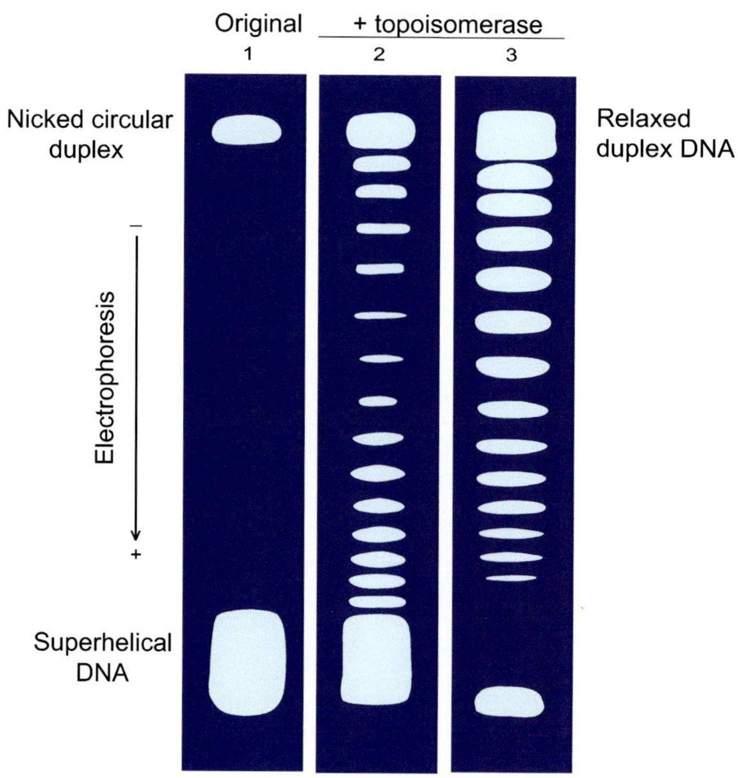

Figure 3-9. Schematic diagram of a typical separation of DNA topoisomers by electrophoresis in an agarose gel. The three lanes show the stepwise relaxation of closed circular DNA by incubation with a topoisomerase. Lane 1 shows the original mixture of superhelical (fastest band) and nicked circular duplex (slowest band) species. Lanes 2 and 3 show the formation of various topoisomers following different times of incubation with the topoisomerase. The final product is relaxed duplex DNA. Adjacent bands contain topoisomers that differ by unity in linking number.

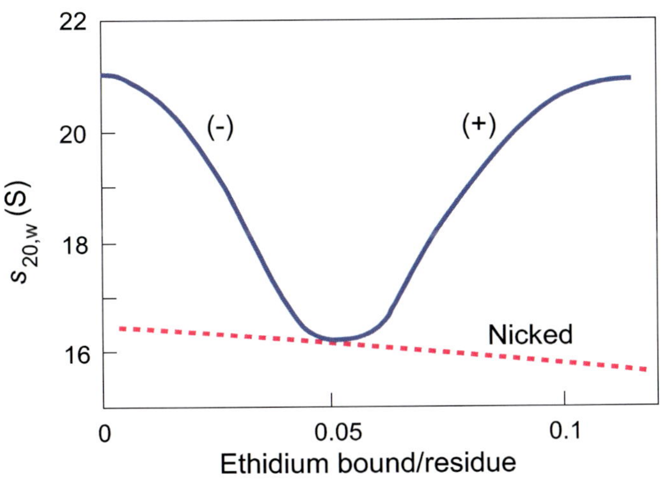

Figure 3-10. Effects on the sedimentation rate of the binding of ethidium bromide to an underwound closed circular DNA from SV40 virus and to the relaxed, nicked circular DNA. The ethidium intercalates between the bases, unwinds the helix, expands the hydrodynamic volume of the molecule and decreases its rate of sedimentation. The *s*-value of the initially underwound ccDNA approaches that of the relaxed molecule but then increases again as positive supercoils are formed. Data from W. Bauer & J. Vinograd (1968) *J. Mol. Biol.* **33**, 141–171.

2. Gel Electrophoresis to Separate Topoisomers

A supercoiled DNA molecule is more compact than a relaxed DNA molecule of the same length, so supercoiled DNA generally migrates faster when subjected to gel electrophoresis (Figure 3-9) or sedimentation (Figure 3-10). In a population of molecules with a distribution of linking numbers, the various topoisomers will migrate during gel electrophoresis as individual bands (Figure 3-9). Generating all possible values of *Lk* by the action of a topoisomerase permits the value of *Lk* to be assigned to each band by simple counting, so long as all the topoisomers can be resolved.

3. Intercalation by Ethidium Bromide

Intercalating agents such as ethidium bromide alter a circular DNA's degree of superhelicity because they cause the DNA double helix to unwind by about 26° at the site of binding (Section 2.5.B). The linking number is unchanged but the decrease in the negative twist of the supercoil is compensated by an increased writhe. Intercalation in a typical underwound ccDNA therefore causes an expansion of its hydrodynamic volume and a decrease in its sedimentation coefficient (Figure 3-10). As more ethidium binds, the sedimentation coefficient decreases to that of relaxed circular DNA, then it increases again as positive supercoils are introduced. If the number of ethidium molecules bound per base at the minimum sedimentation rate is ν, the unwinding angle per ligand bound is φ (= 26°) and the average helix repeat in the relaxed form is h_0 (= 10.5 bp/turn), the value of the superhelix density, σ, will be given by:

$$\sigma = -h_0\left(\frac{\phi}{360^\circ}\right)(2\nu) = -1.52\nu \qquad (3.8)$$

The molecule described in Figure 3-10 with $\nu = 0.05$ therefore has $\sigma = -0.076$.

4. Two-dimensional Gel Electrophoresis

Introducing a second dimension and intercalation by ethidium bromide expands the range of topoisomers that can be resolved by electrophoresis and provides information about conformational changes in the DNA molecule. The DNA is first subjected to gel electrophoresis as in Figure 3-9 in a

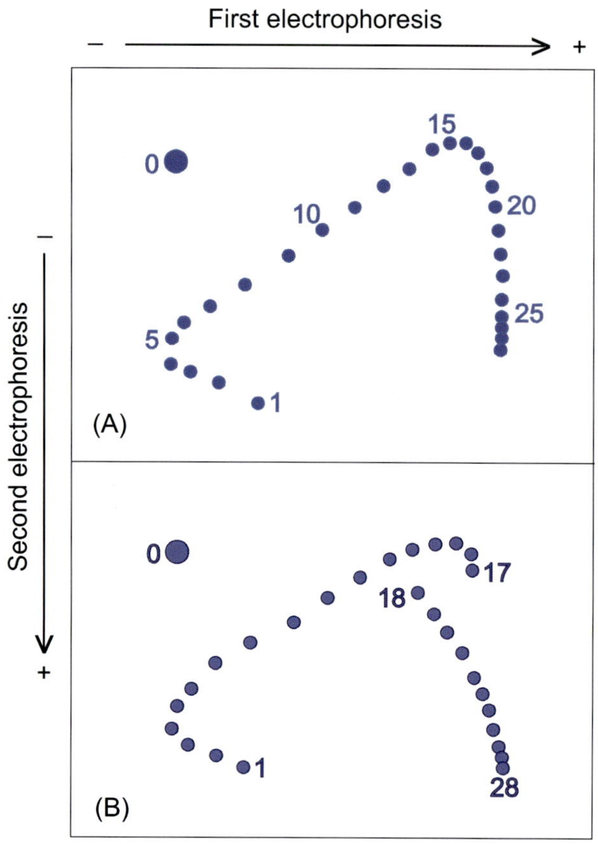

Figure 3-11. 2-D gel electrophoresis of a supercoiled plasmid. The spectrum of topoisomers was generated by incubating a supercoiled plasmid ccDNA with a topoisomerase. The plasmid used in (A) had the polynucleotide $d(GC)_{16}$ inserted in (B). The first electrophoresis was from *left* to *right*. After incubation with ethidium bromide, the second electrophoresis was from *top* to *bottom*. The nicked circular DNA migrates in spot 0. The other spots are numbered from 1 to 28 in order of decreasing *Lk*. Data from J. C. Wang.

slab gel. The separated DNA within the gel is then incubated with ethidium to permit intercalation and reduce the negative writhe (Figure 3-10). The intercalated DNA molecules in the gel are then subjected to electrophoresis at a right angle to that of the first electrophoresis (Figure 3-11). All the topoisomers should be resolved, and their relative mobilities in the first and second dimensions provide information about their topologies.

The two-dimensional (2-D) gel shown in Figure 3-11-A resolved 28 topoisomers that were generated by relaxing a plasmid DNA with a topoisomerase; they are numbered from 1 to 28 in order of decreasing *Lk*. Spot 0 is the nicked circle. The electrophoretic mobilities of isomers 1–4 decreased in both dimensions, indicating that they were positively supercoiled initially. In the original mixture, the relaxed state occurred between topoisomers 4 and 5. The remainder of the topoisomers had increasingly negative supercoiling. The second dimension shows the reduction of negative writhe that results from intercalation. After binding ethidium, topoisomers 1–14 had positive writhe, while 16–28 were still negative. The smooth curve on which these 28 spots lie indicates that no abrupt transitions occurred as a result of the different linking numbers or the change in writhe resulting from intercalation.

The same plasmid was used in Figure 3-11-B, except that it contained a $d(GC)_{16}$ insert, with 32 base pairs alternating between G•C and C•G. The most striking difference is the abrupt change in mobilities of topoisomers 18–28 in the first dimension. A substantial change in overall conformation occurred when *Lk* changed by –1 from that of topoisomer 17. Topoisomers 13–16 migrated in the first dimension at the same approximate rate as those indexed six higher, i.e. 19–22. Their similar migrations suggest that they have the same *Wr*. Therefore, they differ in *Tw* by six, because (Tw_{14}

$-\ Tw_{20}) = (Lk_{14} - Lk_{20}) - (Wr_{14} - Wr_{20}) = 6 - 0$. This is close to the difference expected if the 32-bp GC insert flipped from right-handed B-DNA, with 10.5 bp/turn, in topoisomers 1–17, to a left-handed Z-helix, with 12 bp/turn, in topoisomers 18–28: the difference in *Tw* would be expected to be (32/10.5) + (32/12) = 5.7.

3.6. ENERGETICS OF SUPERCOILING

A topological domain has associated with it a specific free energy associated with the topological strain, called either the **topological domain free energy** or the **free energy of superhelix formation.** This quantity generally increases with the extent of supercoiling of either chirality; it is generally proportional to the square of ΔLk. The equilibria of all binding reactions that involve changes in either the DNA twist or in the writhe are shifted by this free energy term. Examples are the binding of intercalative drugs, such as ethidium bromide (Section 2.5.B.1), and of proteins (Chapter 7).

The free energy of the topology must be included in any chemical or physical reaction in which the topology changes. The linking number for a DNA topological domain is constant and places an additional constraint upon all chemical reactions and other changes of state; it causes significantly altered thermodynamic quantities, thermal denaturation patterns, transitions to other duplex structures and binding of various reagents. Two opposing tendencies must generally be balanced: (1) minimizing the deviation of the twist, *Tw*, from that of the corresponding nicked duplex DNA, to reduce the superhelix radius, *r*, and (2) minimization of the bending distortion that accompanies superhelix formation; electrostatic repulsions between the phosphate groups tend to repel the opposing duplex strands and increase *r*. The net free energy change due to a change in linking difference, ΔLk, results from some combination of these two factors and is known as the **free energy of superhelix formation, ΔG(σ,T)**. Any change in the composition or concentrations of salts affects both factors: the stiffness decreases with increasing ionic strength, but so do the electrostatic repulsions within the duplex. The diameter of the van der Waals surface of B-DNA is only 20 Å, but hydration and electrostatic repulsions cause the effective diameter to be 30 Å at high salt concentrations, increasing to 150 Å at 0.01 M salt. The responses of superhelical structure to changing ionic conditions or temperature are complex.

The free energy due to the topology has been measured, with very similar results, by three independent methods: (1) using the perturbation of the binding constant of intercalating drugs with changes in σ, which can measure ΔG over the entire accessible range of ±σ; (2) using the width of the distribution of topoisomers when a closed circular DNA is equilibrated in the presence of a topoisomerase (Figure 3-6), which measures ΔG over a narrow range of ±σ in the vicinity of $\sigma = 0$; and (3) relating the extent of the linking difference to local denaturation of the DNA duplex, which can measure ΔG over a wide range of values of $\sigma < 0$.

The free energy of superhelix formation is described in terms of the superhelix density σ by:

$$\Delta G(\sigma,T) = \frac{NRT}{h_o^2} q(T)\sigma^2 \tag{3.9}$$

where *N* is the number of base pairs in the DNA, h_0 is the helical repeat of nicked circular or linear duplex DNA (10.5 bp/turn) and $q(T)$ is the free energy coefficient. The last quantity for chains longer than 2400 bp is 1160 bp at 37°C, independent of *N* and σ. For shorter chains, its value depends upon *N*:

$$q(37°, N) = 3939 \text{ bp} - 1.1\, N \quad (3.10)$$

It decreases at higher temperatures:

$$q(\text{T}) = [(0.968 \times 10^6/\text{T}) - 1920] \text{ bp} \quad (3.11)$$

The entropy and enthalpy of superhelix formation are independent of temperature but quadratic functions of the superhelix density. For DNA molecules with about 5×10^3 bp, the enthalpy change per mole of base pairs ($\Delta H/N$) has been measured to be 17.3 σ^2 kcal/mol, and the entropy change per mole of base pairs ($\Delta S/N$) is 35 σ^2 cal/mole/deg. The free energy change per mole base pair ($\Delta G/N$) is 6.4 σ^2 kcal/mol at 37°.

The free energy of a supercoiled DNA molecule is usually at a minimum when about 70% of any change in *Lk* is expressed in *Wr* and 30% in *Tw*. Consequently, a decrease in *Lk* causes both right-handed negative supercoiling of the DNA axis and unwinding of the duplex. The energies of the twist deformability are virtually the same for all 10 possible base pair steps but their response is highly nonuniform. In particular, pyrimidine/purine steps are much more flexible than purine/purine steps, followed by purine/pyrimidine steps.

On the origin of the temperature dependence of the supercoiling free energy. J. J. Delrow *et al.* (1997) *Biophys. J.* **73**, 2688–2701.

Effect of polyethylene glycol on the supercoiling free energy of DNA. A. N. Naimushin *et al.* (2001) *Biopolymers* **58**, 204–217.

Link, twist, energy, and the stability of DNA minicircles. K. A. Hoffman *et al.* (2003) *Biopolymers* **70**, 145–157.

3.6.A. Energy Distribution of Topoisomers

The free energy of superhelix formation is best measured with the thermal distribution of topoisomers that results at equilibrium in the presence of a DNA topoisomerase. A circular duplex DNA can be incubated with a topoisomerase, or the appropriate nicked circular DNA can be incubated in the presence of DNA ligase (Figure 3-2), until equilibrium is established. After inactivation of the enzyme, the equilibrium distribution of topoisomers that was produced is separated by gel electrophoresis (Figure 3-6). The energy distribution is determined by the free energy difference between each topoisomer *i* and the hypothetical completely relaxed species at the center of the distribution, with $\Delta Lk = \varepsilon$. The distribution should be described by:

$$c_i = \left(\frac{1}{\text{Q}}\right) \exp\left[\frac{\Delta \text{G}(1+\varepsilon)}{\text{RT}}\right] \quad (3.12)$$

where c_i is the concentration of topoisomer *i*. The factor Q is the partition function, which in this case is simply a constant. Incorporating Equation 3.9 and using the definition of σ, the Boltzmann energy distribution becomes:

$$c_i = \left(\frac{1}{Q}\right)\exp\left[\frac{q}{N}(1+\varepsilon)^2\right] \tag{3.13}$$

It is convenient for analytical purposes to express this equation as a ratio of the concentration of topoisomer i relative to that of the most prominent topoisomer, with $i = 0$:

$$\frac{c_i}{c_0} = \exp\left[\left(\frac{q}{N}\right) i\,(i+2\varepsilon)\right] \tag{3.14}$$

The separation between the maximum of this distribution and the location of the nearest topoisomer should be equal to ε, the difference in winding between a nicked circular DNA and the nearest relaxed, closed DNA. Taking the logarithms of both sides of Equation 3.14 produces:

$$\left(\frac{1}{i}\right)\log_e\left(\frac{c_i}{c_0}\right) = \left(\frac{q}{N}\right)(i+2\varepsilon) \tag{3.15}$$

Plotting the relative concentrations of the various topoisomers on a linear scale should give the expected Gaussian distribution (Figure 3-6-B); the slope gives q/N and the intercept at $i = 0$ yields ε.

Performing such an analysis at different temperatures demonstrates that the distribution of topoisomers is shifted at different temperatures, towards less negative linking numbers at higher temperatures. For example, increasing the temperature from 14°C to 21°C increases the ΔLk of the predominant topoisomers by about 2. The data demonstrate that each base pair of the double helix unwinds by about 0.012° per °C increase in temperature.

Equilibrium distributions of topological states in circular DNA: interplay of supercoiling and knotting. A. A. Podtelezhnikov *et al.* (1999) *Proc. Natl. Acad. Sci. USA* **96**, 12974–12979.

3.6.B. Topology-Dependent Binding of Ligands

The free energy for the binding of ν ligands per base pair to a closed circular DNA, ΔG(ν,σ), is the sum of two contributions: the intrinsic binding free energy, ΔG°(ν), and the associated change in the free energy of superhelix formation, ΔΔG°(σ). The change in free energy upon binding is related to the binding constant K by the usual relationship ΔG° = –RT $\log_e$ (K). K′ is the intrinsic binding constant, which can be measured by binding to a nicked circular or linear duplex DNA, so ΔG°′ = –RT $\log_e$ (K′). The change in the superhelix free energy term ΔG°(σ) is:

$$\Delta\Delta G°(\sigma) = [d\Delta G°(\sigma)/d\sigma]\Delta\sigma \tag{3.16}$$

A ligand that alters the duplex rotation by f turns upon binding will produce $\Delta\sigma = h_0 f/N$. Combining these relationships demonstrates that the binding constant depends upon the superhelix density and the temperature:

$$K(\sigma,T) = K'(T)\exp\left[-q(T)\,\sigma^2\,\frac{2f}{h_0}\right] \tag{3.17}$$

This analysis illustrates how the topology can have very significant effects on K. For example, a protein that binds to a DNA of $\sigma = -0.07$ and removes one duplex turn in the process will bind with $K = 3K'$. In other words, the affinity of such a closed circular DNA for the protein will be approximately three times that of a comparable nicked circular or linear DNA under the same environmental conditions. This effect can be even more pronounced with a supercoiled DNA that has a very high (negative) superhelix density. For example, if $\sigma = -0.17$ and one duplex turn is removed in the binding process, $K = 600\ K'$. Consequently, if two otherwise identical DNA molecules of low and high (negative) superhelix density compete for binding to the same protein, the closed DNA of very negative σ will have a 600-fold advantage. Of course, if $\sigma > 0$ the effect is reversed and the closed DNA of low σ will bind the more tightly.

The classic example of a reaction that involves a change in the average duplex rotation is the intercalation of ethidium bromide, which unwinds the DNA duplex by 26°, or 0.072 turns of the duplex (Section 2.5.B.1 and Figure 3-10).

Closed circular DNA as a probe for protein-induced structural changes. J. H. White *et al.* (1992) *Trends Biochem. Sci.* **17**, 7–12.

3.7. DNA WRAPPED AROUND THE NUCLEOSOME

The DNA in the chromosomes of eukaryotic organisms is greatly compacted by being wrapped around units of a complex of eight histone proteins: a $H3_2$–$H4_2$ tetramer, flanked by two H2A–H2B dimers. About 166 bp of DNA are wound around the wedge-shaped histone octamer in two left-handed superhelical turns, to form one nucleosome (Figure 3-12). It is a wedge-shaped disk, about 110 Å in diameter and about 55 Å high, with a pseudodyad axis of symmetry close to where the

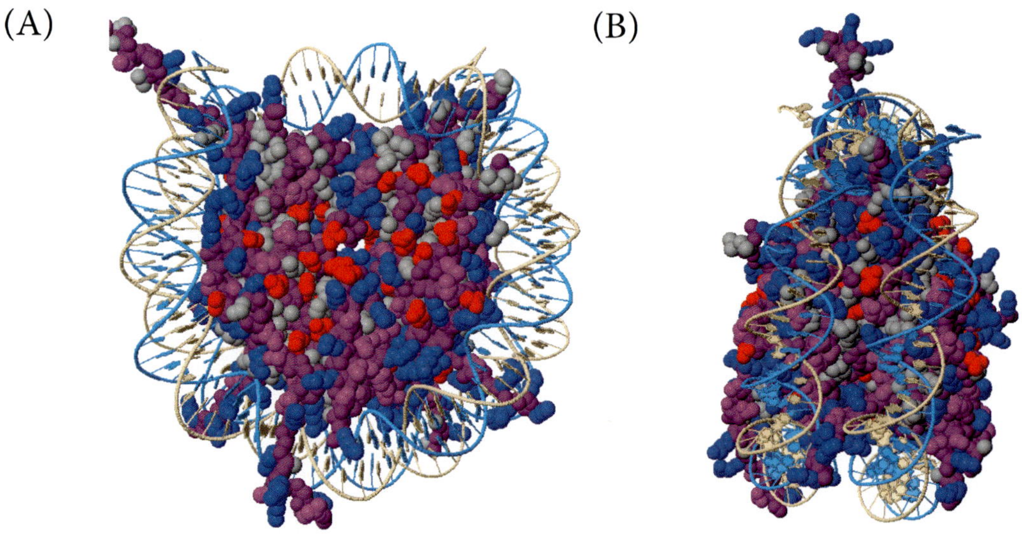

Figure 3-12. The structure of the nucleosome core particle determined by X-ray crystallography. To obtain a homogeneous structure, the histones were synthesized in bacteria to avoid the many post-translational modifications they normally experience, and a synthetic DNA molecule was used so that all the particles had the same DNA sequence. (A) A view down the superhelix axis; (B) a perpendicular view. The DNA is depicted as a skeletal model, whereas the histone proteins are depicted as van der Waals spheres for their atoms. Acidic groups are *red*, basic groups are *blue*, whereas polar but uncharged groups are *mauve*. Figure generated from PDB file *1aoi* using the program Jmol.

bound DNA segment enters and exits. Adjacent nucleosomes are linked by up to 74 bp of DNA, known as the **linker DNA**, plus one molecule of linker histone H1.

The overall helical periodicity of the bound DNA is 10.2 bp/turn but it varies in detail along the DNA. The DNA superhelix is not uniformly bent, due to interactions with the histones; the sharpest distortions are at positions 15 and 40–50 bp on either side of the dyad. Each of the histone dimers has three independent DNA-binding sites (Section 7.3.F). These binding interactions must overcome the free energy cost of bending DNA into the solenoid shape.

The ends of the DNA emerging from the nucleosome can be fixed, which defines a topological domain (Figure 3-13). It is then clear that the DNA wrapped twice around one nucleosome has $\Delta Tw = 0$ and $Wr = -2$. If the histone core is removed and the DNA pulled straight, the writhe is converted to twist, and $\Delta Tw = -2$ and $Wr = 0$.

The helical repeat of 10.2 bp/turn implies that the 146 bp of DNA wrapped on the nucleosome have $Tw = 14.14$. If the DNA duplex had been linearly extended in solution, $Tw_0 = 146/10.54 = 13.85$. Therefore, the surface wrapping and resultant curvature of the DNA axis about the nucleosome increases the twist by only about 2%.

Wrapping DNA in a nucleosome compacts it considerably, but even further compaction occurs when nucleosomes that are adjacent on a DNA molecule interact to form higher ordered structures.

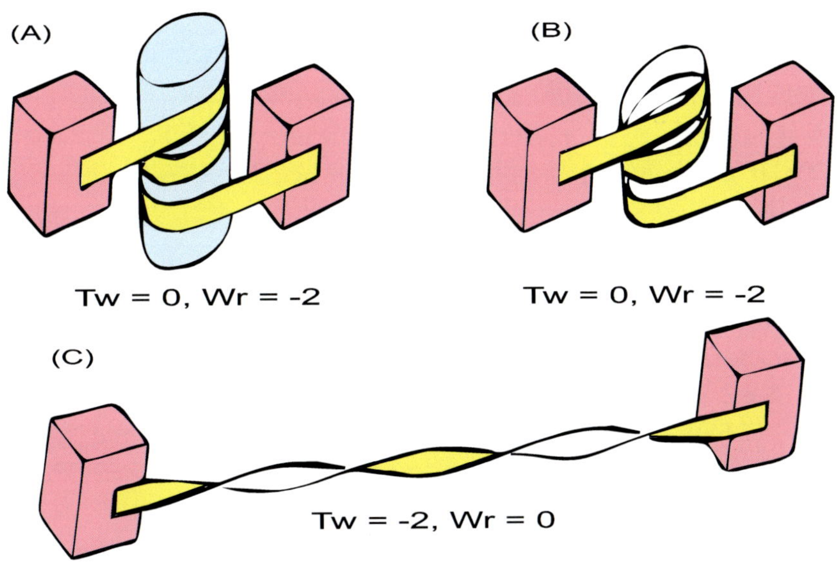

Figure 3-13. The topology of wrapping a molecule like DNA around a histone core to form a nucleosome. In this schematic illustration, the DNA double helix is represented as a simple ribbon, which ignores the double-helical nature of the DNA molecule. (A) The histone core is depicted as the central cylinder, wound with two turns of ribbon, while the two blocks on the *left* and *right* serve to fix the ends of the ribbon. The ribbon is not twisted but has $Wr = -2$. (B) The central cylinder is removed. (C) The ribbon is extended, demonstrating the twist of −2 that results as the writhe goes to zero. Adapted from C. R. Calladine & H. R. Drew (1992) *Understanding DNA: the molecule and how it works,* Academic Press, London.

~ CHAPTER 4 ~

RNA STRUCTURE

The functions of RNA molecules in living organisms range from the storage of genetic information in RNA viruses, defining the structures of macromolecular machines like ribosomes (Section 7.4.F), transferring and regulating the genetic information to the ribosomes, to the enzyme-like activities of ribozymes (Chapter 8). RNA molecules adopt diverse structures in response to these many functional requirements.

RNA has a covalent structure (Figure 4-1) very similar to that of DNA, the only differences being the change from a 2′-deoxyribose sugar in DNA to a ribose sugar in RNA, which involves only a change of a 2′-H atom to a 2′-OH group, and from a methyl group in thymine to an H atom in uracil. The backbone of RNA still consists of 3′ ⟶ 5′ phosphodiester bonds, even though an alternative 2′⟶ 5′ linkage is possible. The alternative linkage is used transiently in some of the functions of RNA, such as the removal and splicing of internal segments during processing of messenger RNA. **The presence of the 2′-OH group makes the phosphodiester bonds of RNA much more susceptible to hydrolysis than those of DNA**, especially at high pH (Figure 4-2). The 2′-hydroxyl group also increases dramatically the variety of three-dimensional (3-D) structures that RNA can adopt. **The bases in RNA molecules are often modified in nature**, much more so than in DNA, and the most common outcomes are illustrated in Figure 4-3. Although the polynucleotide backbone has considerable conformational flexibility, only a relatively small number of conformations are observed in biological RNAs. The 2′-hydroxyl group means that there are vicinal 2′- and 3′-hydroxyl groups at the 3′-ends of RNA molecules that can be oxidized by periodate (IO_4^-) to the 2′,3′-dialdehyde. The dialdehyde is not stable but reacts readily with other groups, especially amines. This reaction can be used to attach the 3′-ends of RNA molecules to a solid support.

The role of DNA in storage and transmittal of genetic information imposes the double-helical structure on most DNA molecules, **whereas RNA molecules are naturally double-stranded only in the RNA viruses**, where they have the same genetic role as DNA. Otherwise, RNA molecules are synthesized as single strands, by transcription of the DNA genome, and they remain single-stranded. Such RNA molecules are not replicated. In fact, **double-stranded DNA is a signal to the cell of a virus infection**, so cells have a special mechanism, known as **RNA interference**, of searching for double-stranded RNA molecules and then responding appropriately by destroying them. **Many RNAs function as unstructured, single-stranded molecules.** For example, the single polynucleotide strand of messenger RNA must be unfolded for the genetic message to be translated; stable structures in the coding region generally inhibit protein biosynthesis, although some structures are important in the regulation of translation (Figure 4-4-C).

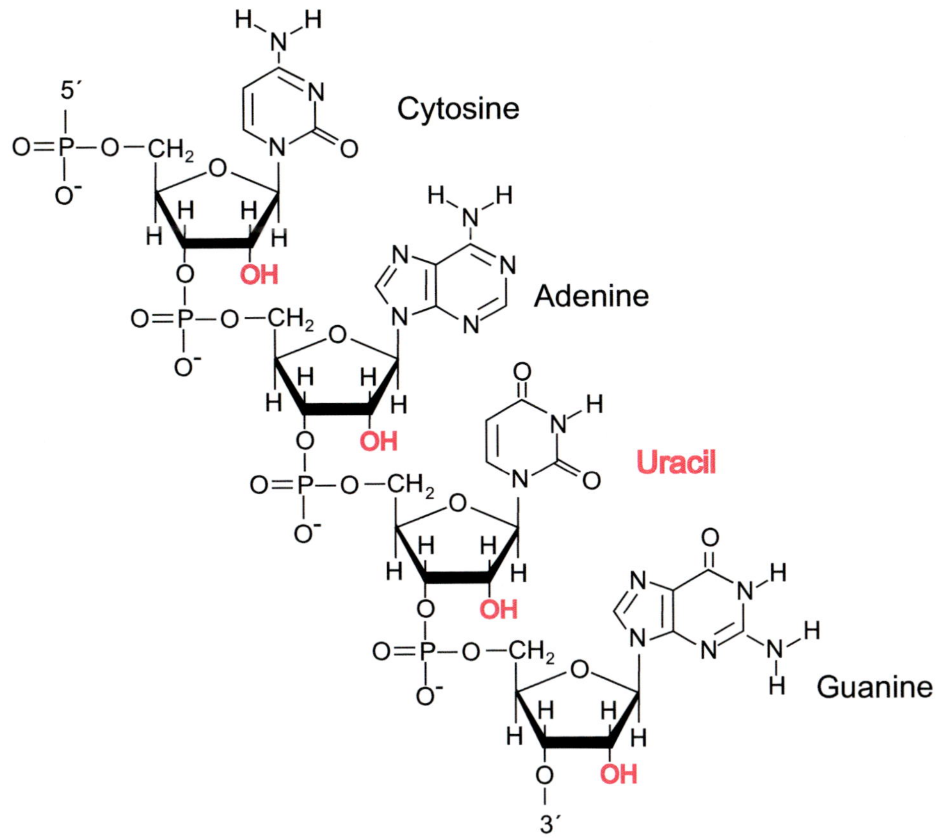

Figure 4-1. The chemical structure of the four primary nucleotides of RNA, showing each of the four bases. The differences from DNA are indicated in *red*.

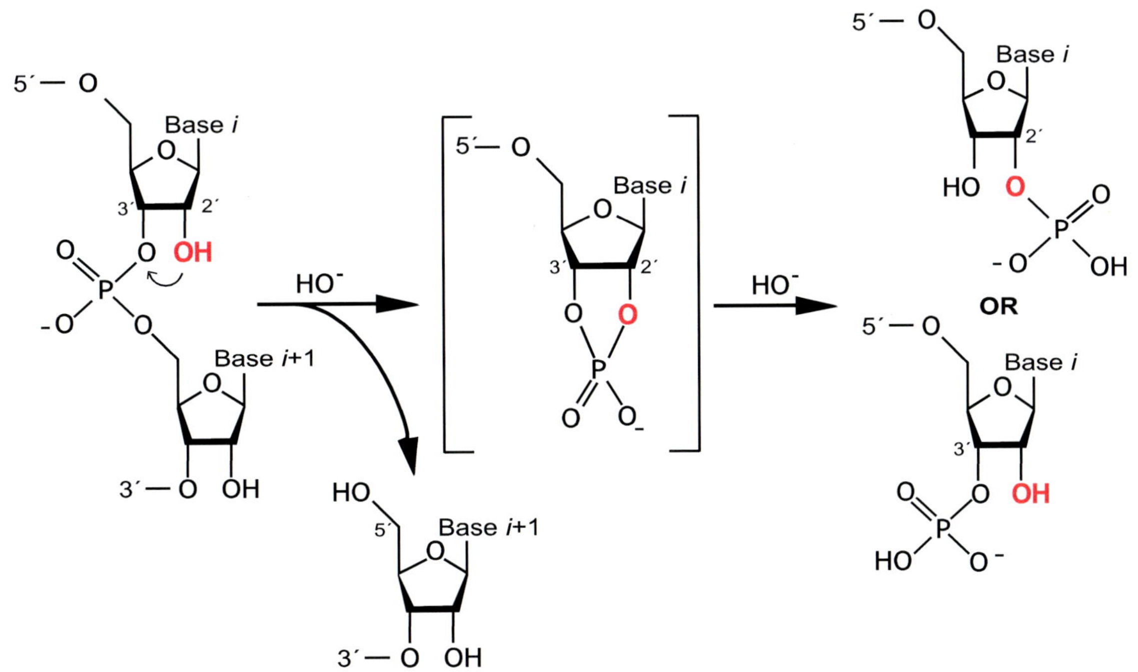

Figure 4-2. Cleavage of the RNA internucleotide linkage under basic conditions. The 2′-hydroxyl group attacks the phosphate group attached to the adjacent 3′-O atom, to produce the 2′,3′-phospho intermediate (in brackets). After alkaline hydrolysis of the intermediate, the phosphate group can end up on either the 2′- or 3′- O atom.

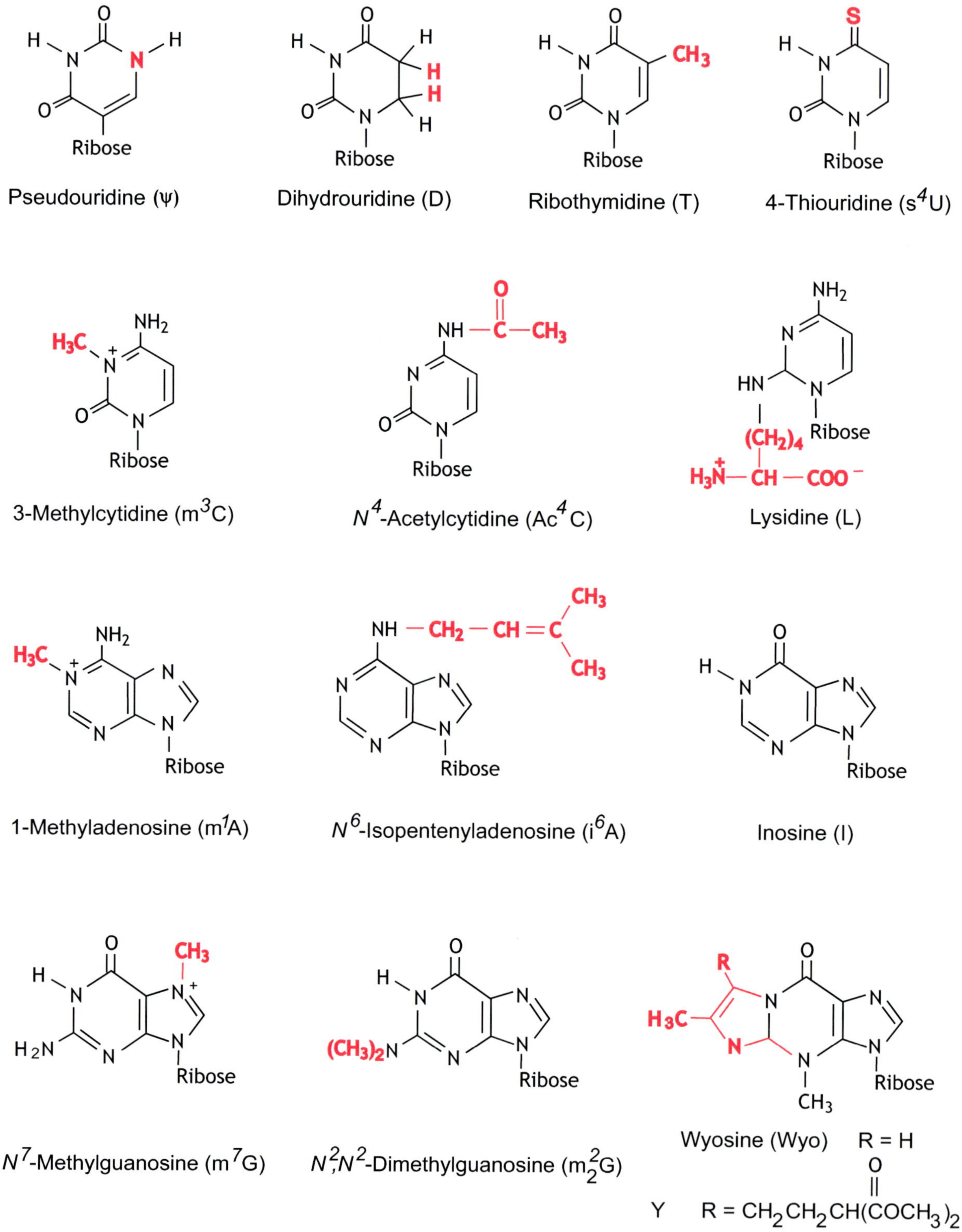

Figure 4-3. The most common modified forms of the nucleotides in tRNA. They may also be methylated at their ribose 2′-positions, and are then indicated by an 'm' after the nucleotide abbreviation, such as Cm, Gm, etc.

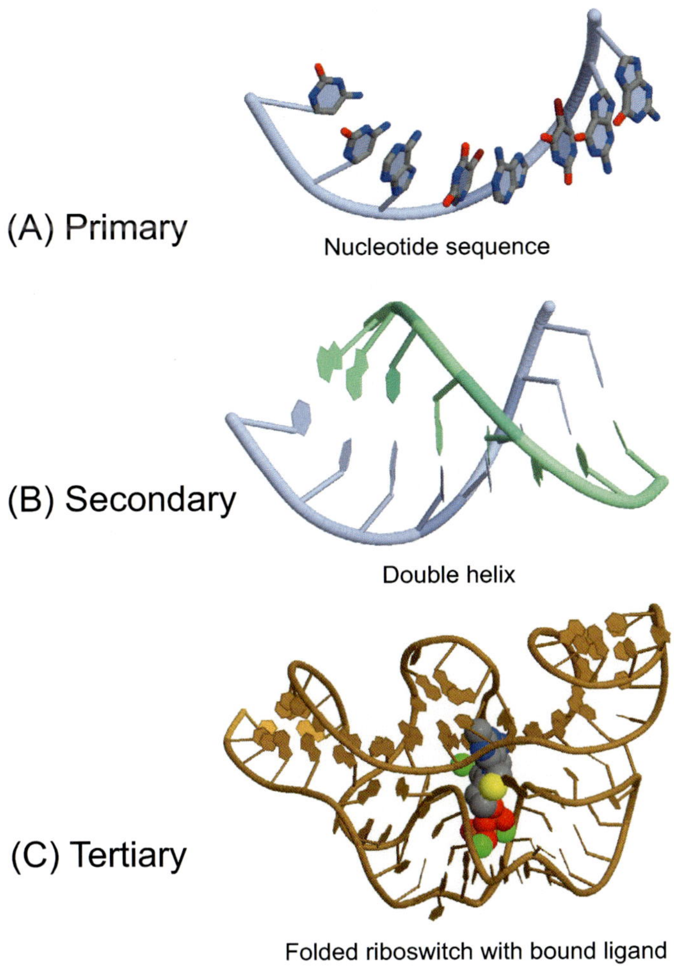

Figure 4-4. Hierarchy of RNA folding. (A) A single strand of RNA and its nucleotide sequence: its primary structure. (B) The A-form double helix of RNA: its secondary structure. (C) The 3-D structure of the thiamine pyrophosphate-specific riboswitch, with bound thiamine pyrophosphate: the tertiary structure. Figure generated from PDB files *115d* and *2gdi* using the program Jmol.

Other single-stranded RNA molecules adopt an array of structures to rival proteins in their complexity and functions; some can bind small molecules specifically (Figure 4-4-C) and some even catalyze biochemical reactions (Chapter 8). Many of the coenzymes that are involved in catalysis by proteins are related to the substituents of RNA. **It is thought that RNA was the basis of the first type of living organisms, performing the dual functions of holding the genetic information and catalyzing reactions.** RNA may have performed the present-day functions of both DNA and proteins, which arose later and largely replaced RNA because they are more specialized and better suited to their more specific functions.

It is convenient to describe folded RNA structures in hierarchical terms (Figure 4-4), comparable to those used in describing protein structure: primary, secondary, tertiary and quaternary structures. The **primary structure** refers to the sequence of an RNA molecule; **secondary structure** is the regular local structure formed, usually an antiparallel double helix; **tertiary structure** involves interactions between elements of the secondary structure and other groups distant in the sequence; **quaternary structure** is the association of multiple RNA molecules. In many cases, the folding of the RNA molecule involves interactions with one or more protein molecules, which can produce very complex structures, as in ribosomes (Section 7.4.F). In these cases, the common motifs to be described here are incorporated into those structures.

Details of virtually all the known structures of RNA can be obtained from the **Protein Data Bank** (**www.rcsb.org**) and from **http://ndbserver.rutgers.edu**.

The RNA World. R. F. Gesteland & J. F. Atkins, eds (1993) Cold Spring Harbor Press, NY.

NMR spectroscopy of RNA. B. Furtig *et al.* (2003) *Chembiochem.* **4**, 936–962.

RNA structural motifs: building blocks of a modular biomolecule. D. K. Hendrix *et al.* (2005) *Quart. Revs. Biophys.* **38**, 221–243.

NMR methods for studying the structure and dynamics of RNA. M. P. Latham *et al.* (2005) *Chembiochem.* **6**, 1492–1505.

Dynamics-based amplification of RNA function and its characterization by using NMR spectroscopy. H. M. Al-Hashimi (2005) *Chembiochem.* **6**, 1506–1519.

4.1. SECONDARY STRUCTURE OF RNA

The secondary structure of RNA is dominated by the formation of antiparallel double helices stabilized by Watson–Crick base pairs between complementary segments with complementary sequences, comparable to those in DNA (Figure 2-13). The main difference is that uracil (U) bases replace thymine (T), so the canonical base pairs in RNA are C•G, G•C, U•A and A•U. As in DNA, the more stable G•C and C•G base pairs predominate (Table 4-1). Noncanonical base pairs occur much more frequently than in DNA; the most frequent (Table 4-1) is the U•G base pair:

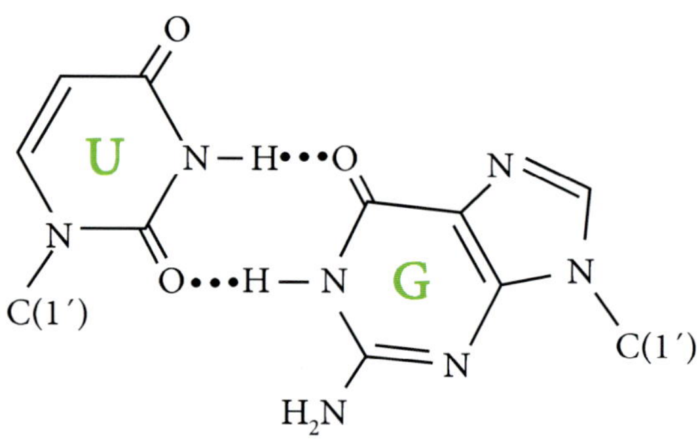

(4.1)

It introduces slight distortions into a double-helical structure that are recognized by proteins and other RNAs. RNA structures tend to be more complex than just antiparallel double helices.

RNA double helices, like those of DNA, have an antiparallel right-handed helical conformation. **RNA double helices usually adopt the A-form structure**, however, which differs significantly from canonical B-form DNA double helices in the conformation of the sugar and displacement of the bases from the helical axis (Figure 2-18). In an A-form RNA helix, base-stacking results in nearly perfect parallel orientations of all the bases in the helix. The lengths of hydrogen bonds in RNA are shorter than in DNA and sequence-dependent.

The 2′-hydroxyl group is the major determinant of the conformational and thermodynamic differences between RNA and DNA. Its presence on the ribose sugar in RNA hinders formation of a B-type helix. A-type helices have a deep, narrow major groove that is relatively inaccessible and does

Table 4-1. Frequencies of base pairs in RNA sequences with conserved secondary structures

Base pair	Frequency (%)
G • C	34.3
C • G	28.2
A • U	11.8
U • A	16.9
G • U	3.9
U • G	3.5
Others	1.4

Frequencies in 1081 base pairs in ribosomal and tRNAs and RNase-P, compiled in Table 1 of P. G. Higgs (2000) *Quart. Rev. Biophys.* **33**, 199–253.

not facilitate specific interactions with other molecules. The minor groove exposes the ribose 2′-OH groups, which are good hydrogen bond acceptors and donors. It is also shallow and broad, making it accessible to ligands and other groups, although the bases are not readily distinguished. A network of water molecules spans the minor groove, forming hydrogen bonds to the 2′-OH group. The RNA duplex is less tightly hydrated than the corresponding DNA helix, consistent with the A-type double helix of DNA being favored only at low water concentrations.

RNA double helices are more stiff than those of DNA, with a persistence length (Section 2.4) some 1.5- to two-fold longer than that of DNA. The flexibility patterns of B-DNA and A-RNA are quite different: a few soft essential movements explain most of the natural flexibility of A-RNA, whereas many are necessary for B-DNA. DNA is generally more flexible, but A-RNA is easier to deform in certain ways. The structures of RNA duplexes are not uniform, although they vary less than those of DNA. These variations depend on the sequence and structural context of the helix relative to the global 3-D structure. These local differences lead to very diverse shapes, with profound consequences for recognition by proteins and other ligands.

Unlike DNA, and except for the double-stranded RNA genomes of viruses, **the double helices in natural RNAs are primarily intrastrand**, generated by the single strand doubling back upon itself. The helices are also relatively short, seldom more than 10 base pairs in length, and are interrupted by single-stranded nucleotides forming loop elements: hairpins, bulges and internal loops (Figure 4-5). These, together with the helical junction that is formed when more than two double helices come together, are the secondary structure motifs and the building blocks upon which most complex RNA structures are built.

In many natural folded RNAs, more than half of all nucleotides are incorporated into double-stranded helices (Figure 4-6). **Duplex regions can form through very long-range interactions, between segments distant in the primary structure**, and these interactions are crucial to determine and

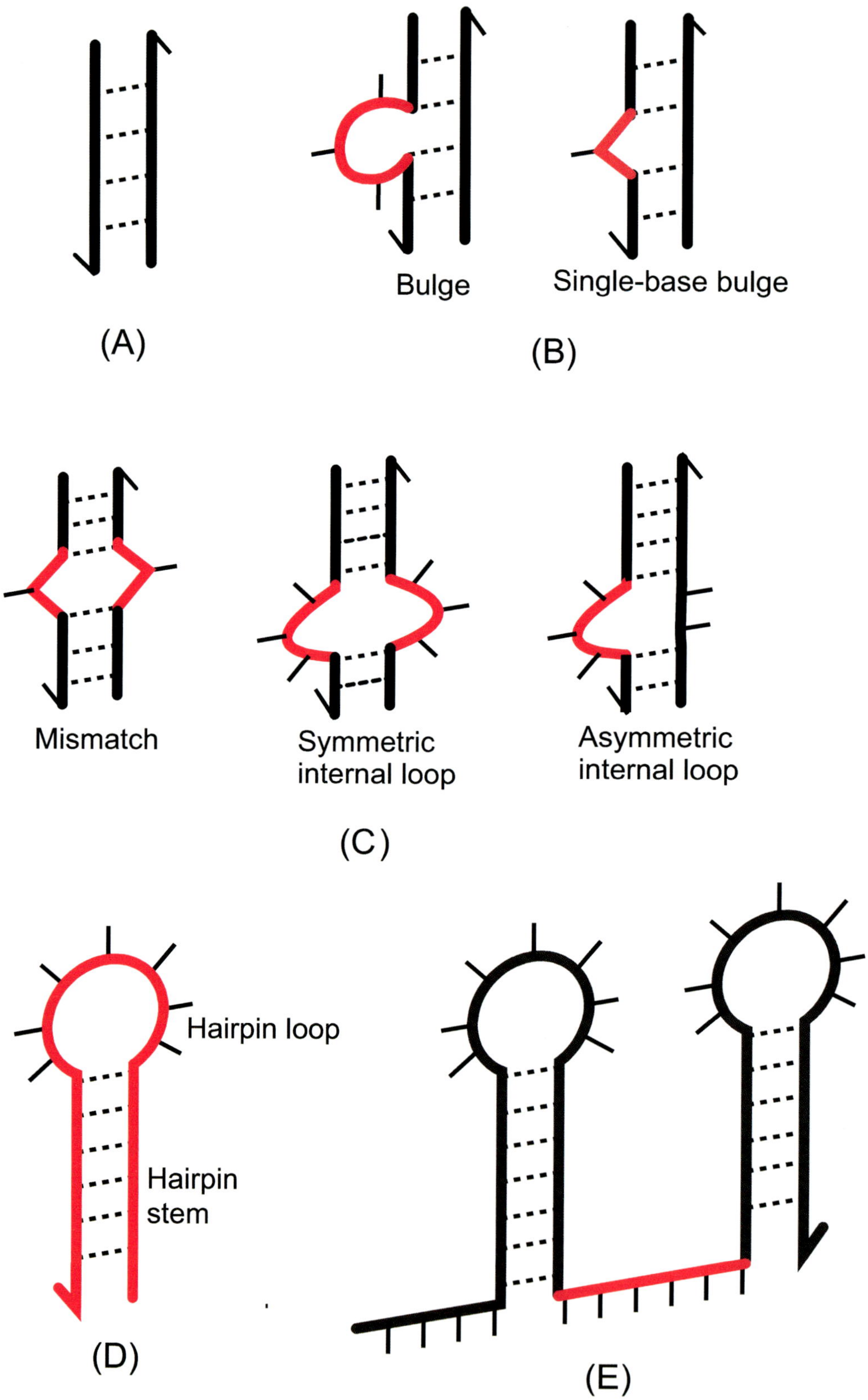

Figure 4-5. Secondary structure motifs of RNA. The backbone of the RNA molecule is depicted as a *thick line*; *dashed lines* indicate complementary antiparallel base pairs, while *thin lines* are unpaired bases. The *half-arrow* indicates the 3′-end of the oligonucleotide. (A) Intramolecular antiparallel double helix; (B) bulges; (C) internal loops within double-stranded segments; (D) hairpin; (E) single-stranded regions between double-stranded hairpins.

stabilize the overall fold of an RNA molecule. For example, opposite strands in double helices within ribosomal RNA are separated by as many as 2000 nucleotides.

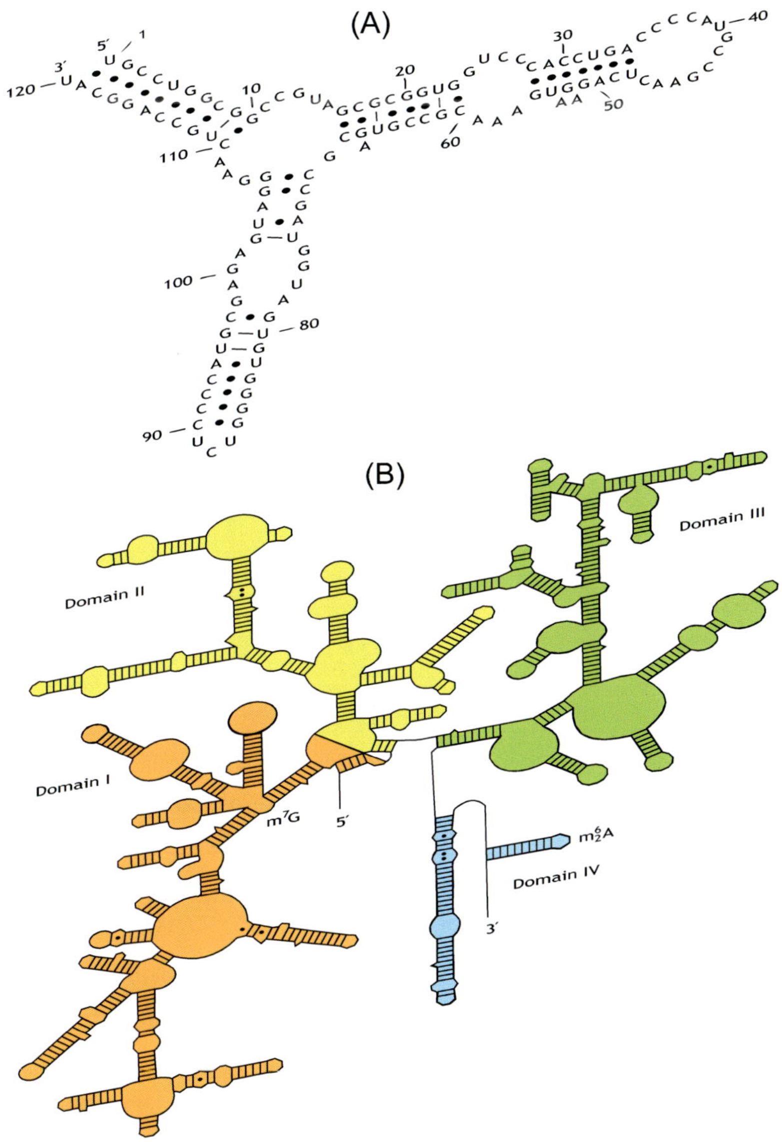

Figure 4-6. Secondary structures of two RNA molecules. (A) The 120-nucleotide 5S RNA of *Escherichia coli*. Watson–Crick base pairs are indicated by *dots*, the common G•U pairs by *dashes*. Adapted from R. R. Gutell *et al.* (1985) *J. Mol. Biol.* **32**, 183. (B) The 1542-nucleotide 16S RNA from *E. coli* ribosomes. The secondary structure was divided up into four domains, indicated by different colors. Two bases that have been modified are indicated. The 3-D structure is shown in Figure 7-37. Adapted from H. F. Noller (1984) *Ann. Rev. Biochem.* **53**, 134.

The relative flexibility of B-DNA and A-RNA duplexes: database analysis. A. Perez *et al.* (2004) *Nucleic Acids Res.* **32**, 6144–6151.

Influence of the 2′-hydroxyl group conformation on the stability of A-form helices in RNA. J. Fohrer *et al.* (2006) *J. Mol. Biol.* **356**, 280–287.

RNA secondary structure design. B. Burghardt & A. K. Hartmann (2007) *Phys. Rev. E* **75**, 021920.

4.1.A. Hairpin Loops

The most common element of RNA secondary structure is the **hairpin** or **stem–loop** (Figure 4-5). For example, bacterial 16S RNA of ribosomes contains approximately 30 hairpins that are phylogenetically conserved (Figure 4-6-B). A hairpin forms when the phosphodiester backbone folds back on itself to form a double-helical tract, called the **stem**, leaving unpaired nucleotides at the turn to form a single-stranded region, called the **loop**. The primary driving force is the base pairing in the stem, which must overcome the opposing requirements to bend the polynucleotide backbone and fix its conformation. Such hairpins can be stable in isolation (Table 4-2).

Table 4-2. The stabilities of isolated RNA hairpins. The nucleotides in the turn are in bold

Sequence	T_m (°C)	ΔG° (37°C) (kcal/mol)	ΔH° (kcal/mol)	ΔS° (cal/mol/°C)
rAC**UUCG**GU[a]	36.8	0.0	−28.9	−93.1
rGGC**ACAA**GCC[b]	62.7	−2.2	−29.2	−86.9
rGGC**GCAA**GCC[b]	71.0	−3.0	−30.3	−88.0
rGGAG**UUCG**CUCC[c]	60.1	−3.4	−48.6	−145.7
rGGAC**GAAA**GUCC[c]	65.9	−4.2	−49.1	−145.0
rGGAC**UUCG**GUCC[c]	71.7	−5.7	−56.5	−163.9
rGGAC**UUCG**GUCC[d]	76.2	−6.3	−55.9	−159.9

[a] In 10 mM sodium phosphate buffer, 0.01 mM EDTA. M. Molinaro & I. Tinoco (1995) *Nucleic Acids Res.* **23**, 3056–3063.

[b] In 100 mM NaCl, 10 mM sodium phosphate buffer, 0.5 mM EDTA. J. SantaLucia *et al.* (1992) *Science* **256**, 217–219.

[c] In 10 mM sodium phosphate buffer, 0.1 mM EDTA. V. P. Antao & I. Tinoco (1992) *Nucleic Acids Res.* **19**, 5901–5905.

[d] In 1 M NaCl, 10 mM sodium phosphate buffer, 0.1 mM EDTA. V. P. Antao & I. Tinoco (1992) *Nucleic Acids Res.* **20**, 819–824.

Most single-stranded loops in natural RNAs vary in length from two to 14 nucleotides. The ribose unit is almost always in the extended 2′-*endo* conformation, rather than the usual 3′-*endo* geometry of A-form RNA helices, presumably to increase the distance that can be spanned by each individual nucleotide. A precise geometry is defined for many loops by hydrogen bonding between nucleotides not involving Watson–Crick base pairing and between bases and backbone functional groups, as well as by stacking interactions between the loop nucleotides. Small hairpin loops contain a high degree of structure, whereas longer loops containing more than seven to eight unpaired nucleotides, such as those present in transfer RNA (Section 4.2.B), preserve extensive base stacking interactions but are generally more poorly structured and thermodynamically less stable. Most loops in ribosomal RNA are four to nine nucleotides in length, perhaps reflecting the need for thermodynamic stability.

Hairpin loops are often recognized by other molecules, especially proteins (Section 7.4). Their binding is often unusual in having a strong entropic contribution to the total binding energy, possibly due to the expulsion of bound water molecules.

Freely diffusing single hairpin ribozymes provide insights into the role of secondary structure and partially folded states in RNA folding. G. Pljevaljcic *et al.* (2004) *Biophys. J.* **87**, 457–467.

Hairpin RNA: a secondary structure of primary importance. P. Svoboda & A. Di Cara (2006) *Cell. Mol. Life Sci.* **63**, 901–908.

Biochemical and thermodynamic characterization of compounds that bind to RNA hairpin loops: toward an understanding of selectivity. J. R. Thomas *et al.* (2006) *Biochemistry* **45**, 10928–10938.

4.1.B. Tetraloops

Tetraloops are hairpin loops consisting of four nucleotides; they are exceptionally common in cellular RNAs. There are two very common families: UNCG and GNRA, where N is any of the four nucleotides and R is a purine. For example, 70% of all tetraloops in 16S ribosomal RNAs from all organisms belong to either of these two families. Other classes include the CUUG, ANYA and (U/A) GGN tetraloops. Tetraloop structures have high thermodynamic stabilities; UNCG is the most stable; capping a double-helical tract with a UNCG tetraloop is equivalent thermodynamically to extending the double-helical stem by two base pairs.

In spite of the differences in their sequences, the tetraloop families have extensive structural similarities (Figure 4-7). In each case, the first and fourth unpaired nucleotides close the loop by forming nonWatson–Crick base pairs, which reduces the number of unpaired nucleotides to only two. Unusual conformations of the backbone are stabilized by hydrogen bonds between its phosphate groups and a base.

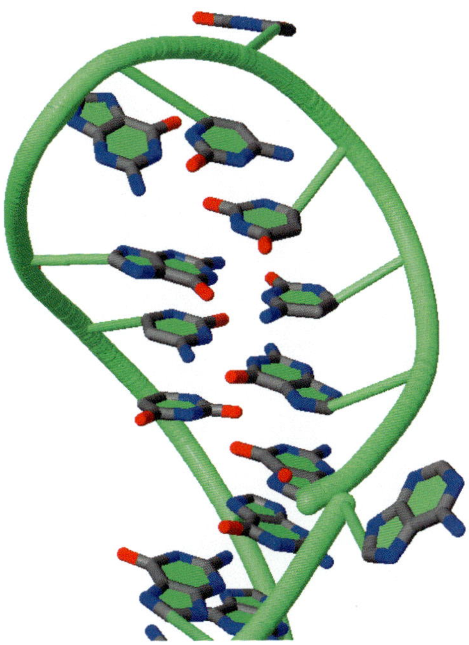

Figure 4-7. The solution structure of a UUCG tetraloop. The nucleobases are depicted in detail, but the ribose–phosphate backbone is depicted only as a cylinder. Figure generated from PDB file *1f7y* using the program Jmol.

Residue specific ribose and nucleobase dynamics of the cUUCGg RNA tetraloop motif by NMR ^{13}C relaxation. E. Duchardt & H. Schwalbe (2005) *J. Biomol. NMR* **32**, 295–308.

Solution structure of a GAAG tetraloop in helix 6 of SRP RNA from *Pyrococcus furiosus*. K. Okada *et al.* (2006) *Nucleosides Nucleotides Nucleic Acids* **25**, 383–395.

4.1.C. Bulges and Internal Loops

Extra unmatched nucleotides (**single-base bulges**) are common structural motifs in the helices of folded RNA molecules, and they often participate in interactions with other molecules and in the formation of tertiary structure. Bulges and internal loops occur when two double-helical tracts are separated by one or more unpaired nucleotides on either one or both strands (Figure 4-5); the former generates a **bulge**, whereas the latter generates an **internal loop**. Equal numbers of bases on each strand produce a **symmetric** internal loop, whereas different numbers of bases generate an **asymmetric loop**. A single base mismatch generates a symmetric internal loop of two nucleotides. Symmetrical internal loops of three nucleotides have the potential of a canonical base pair (middle pair) flanked by noncanonical base pairs on both sides (loop-terminal pair). An **extruded helical single strand** is two or three unpaired bases that are extruded from the main double-helical stack, forming an independent stack of bases. In a **cross-strand stack**, a base on one strand stacks with a base on the opposing strand, rather than stacking with the adjacent bases on its own strand.

The presence of an internal loop or bulge usually reduces the thermodynamic stability of the double helix, but **many internal loops contain G•A base pairs that are noncanonical but contribute to the stability**. Other molecules often recognize such sites, presumably because the unpaired nucleotides are more readily accessible. Such interactions are often associated with conformational transitions in the bulge region, such as flipping out of the bulge base from an intrahelical stack. NonWatson–Crick base pairs readily form within internal loops, while unpaired nucleotides within a bulge may stack within the helix or be bulged outside. **The presence of an internal loop or bulge can induce bending in an RNA helix**; the degree of bending depends on the RNA sequence within the loop and can change upon ligand binding. These motifs consequently are ideal sites for conformational switches, where ligand binding can result in long-range conformational changes.

The NMR structure of an internal loop from 23S ribosomal RNA differs from its structure in crystals of 50S ribosomal subunits. N. Shankar *et al.* (2006) *Biochemistry* **45**, 11776–11789.

The UAA/GAN internal loop motif: a new RNA structural element that forms a cross-strand AAA stack and long-range tertiary interactions. J. C. Lee *et al.* (2006) *J. Mol. Biol.* **360**, 978–988.

Conformational transitions in RNA single uridine and adenosine bulge structures: a molecular dynamics free energy simulation study. A. Bathel & M. Zacharias (2006) *Biophys. J.* **90**, 2450–2462.

4.2. TERTIARY STRUCTURE OF RNA

RNA is much more adept than DNA at forming complex tertiary interactions because of the presence of the 2′-hydroxyl group, which provides each nucleotide with additional hydrogen-bonding potential. In combination with the many polar groups on the bases, a multitude

of hydrogen bond interactions are possible. In addition to the unique 2′-hydroxyl group of the RNA backbone, tertiary interactions often involve noncanonical base pairs, unpaired bases and the negatively charged phosphate groups. Nevertheless, secondary structure is the predominant feature of RNA tertiary structures, and RNA secondary structure elements tend to maintain their 3-D structures when excised from very complex tertiary structures. **Two or more secondary structure elements tend to interact, to generate an RNA tertiary structure and define the overall folding of the RNA molecule.** The secondary structure of RNA can orient key residues into positions appropriate for tertiary interactions. For example, the L-shape of all transfer RNAs (Section 4.2.B) is generated by the geometry of the four-way junction that, combined with a conserved length of the double-helical regions, helps to position nucleotides from the T- and D-loops in close proximity to facilitate the tertiary loop–loop base pairs. During formation of the tertiary structure, unpaired bases can twist or flip out of a helical segment and generate unique structural features for recognition by other RNA segments.

Energetically, the secondary structure is the main component of RNA architecture, while the tertiary structure contributes much less to the Gibbs free energy stability of the native folded state. A folded RNA molecule with definite secondary and tertiary structures will normally display two melting peaks, the first, low-temperature, transition corresponding to melting of the tertiary structure, and the second, high-temperature, transition corresponding to melting of the secondary structure (Section 5.2). **RNA structures are less cooperative than are those of proteins.** Therefore, determination of the secondary structure is an essential step in the study of structure–function relationships of an RNA molecule (Section 4.4.A). In other cases, such as the ribosome, the RNA architecture involves a large number of proteins (Section 7.4.F).

Base stacking and hydrogen bonding are often involved in tertiary structure interactions. Base triplets form when an existing base pair becomes involved in another set of hydrogen bonds with a third base (Figure 4-11); this can occur on either the minor groove or major groove side of an RNA double helix. Hydrogen bonds between bases and backbones are observed when double helices are packed together in compact structures (Section 4.2.A.6).

Tertiary structures are often formed by unpaired nucleotides within a secondary structure motif interacting with unpaired nucleotides from another secondary structure module. Any secondary structure motif, hairpin loops, internal loops and bulge loops, can be involved in these interactions. The tertiary contacts can involve intercalation of the bases, base triplets and Watson–Crick base pairing between complementary loop sequences. This type of loop–loop interaction is best illustrated by the tertiary structure of transfer RNA (Section 4.2.B). A transfer RNA initiates folding by forming four hairpin loops using Watson–Crick base pairing, to produce a four-way helical junction. Helical stacking tertiary interactions generate two extended helical domains, which brings into proximity all unpaired residues in the loop regions and facilitates formation of the tertiary base pairing and base triplets that lock the transfer RNA into the L-shaped 3-D structure.

Ions influence the folding of RNA into specific tertiary structures, in that different salts affect the free energy change upon folding. Ions also determine the structure by the way that they are arranged on or near an RNA molecule and occupy various environments. Packing double helices together is opposed by the strong electrostatic repulsions between the negatively charged phosphate groups, unless the negative charges are screened by cations. Divalent Mg^{2+} ions are very effective at stabilizing tertiary structures in RNAs. In most cases, folding of an RNA molecule is so strongly coupled to its interactions with Mg^{2+} that it is difficult to separate the free energies of Mg^{2+}–RNA interactions from the intrinsic free energy of RNA folding. Group I self-splicing introns (Section 4.2.C) have a core of five Mg^{2+} ions that facilitates the close packing of loops into the minor groove of a double helix. The metal-binding sites are generated by base-stacking and hydrogen-bonding interactions between nucleotides.

Non-canonical base pairs and higher order structures in nucleic acids: crystal structure database analysis. J. Das *et al.* (2006) *J. Biomol. Struct. Dyn.* **24**, 149–161.

RNA folding during transcription. T. Pan & T. Sosnick (2006) *Ann. Rev. Biophys. Biomolec. Structure* **35**, 161–175.

Preparation and crystallization of RNA. B. L. Golden (2007) *Methods Mol. Biol.* **363**, 239–257.

NMR studies of RNA dynamics and structural plasticity using NMR residual dipolar couplings. M. Getz *et al.* (2007) *Biopolymers* **86**, 384–402.

4.2.A. Common Structural Motifs

The structures of RNAs can be very variable but certain structural motifs are observed in many different RNA structures.

Three-dimensional motifs from the SCOR, structural classification of RNA database: extruded strands, base triples, tetraloops and U-turns. P. S. Klosterman *et al.* (2004) *Nucleic Acids Res.* **32**, 2342–2352.

Annotation of tertiary interactions in RNA structures reveals variations and correlations. Y. Xin *et al.* (2008) *RNA* **14**, 2465–2477.

1. Pseudoknots

Pseudoknots form when a single-stranded region interacts with complementary primary sequences of a hairpin or internal loop by Watson–Crick base pairing (Figure 4-8). A pseudoknot creates an extended helical region through helical stacking of the hairpin double-helical stem with the newly formed loop–loop interaction helix. The pseudoknot is only marginally more stable than the two separate hairpins but its stability can be increased by tertiary interactions such as base triplets between unpaired nucleotides in the bridging loops and base pairs within the extended helix.

The presence of a stable pseudoknot structure within the coding region of a messenger RNA, when adjacent to a special heptanucleotide sequence described as 'slippery', causes a shift in the frame with which the sequence of the nucleotide triplets is being read and translated into amino acid residues. This type of frame shifting is indispensable in the replication of numerous viral pathogens and is also exploited in the expression of several cellular genes.

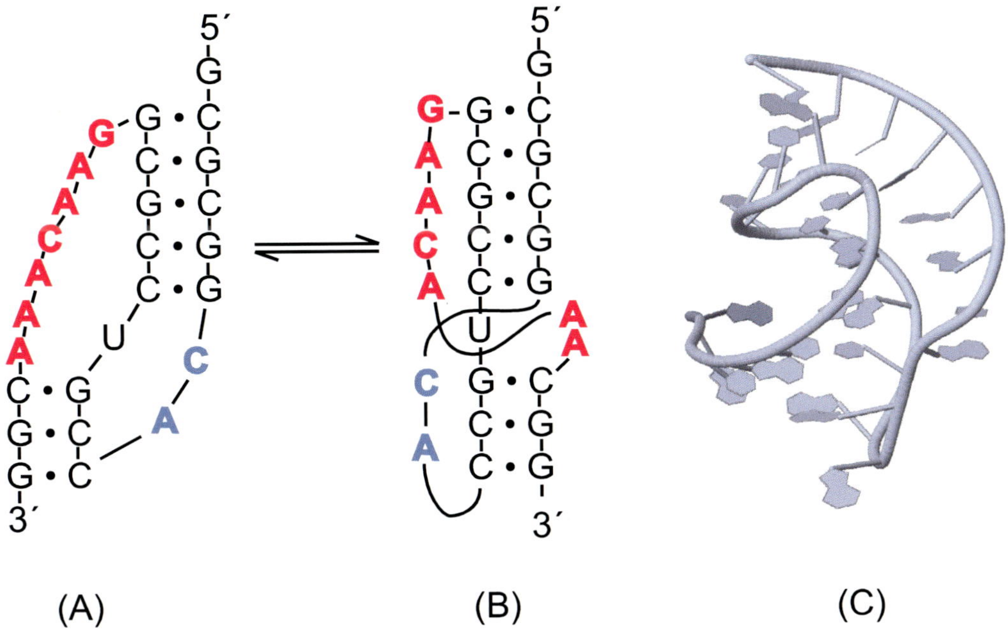

Figure 4-8. An example of a pseudoknot (A), with a schematic illustration of the type of 3-D structure that can be generated (B). An actual example of such a structure is displayed in (C), generated from PDB file *1l2x* using the program Jmol.

Predicting RNA pseudoknot folding thermodynamics. S. Cao & S. J. Chen (2006) *Nucleic Acids Res.* **34**, 2634–2652.

Pseudoknots in RNA secondary structures: representation, enumeration, and prevalence. E. A. Rodland (2006) *J. Comput. Biol.* **13**, 1197–1213.

Tertiary structure of an RNA pseudoknot is stabilized by 'diffuse' Mg^{2+} ions. A. M. Soto *et al.* (2007) *Biochemistry* **46**, 2973–2983.

2. Coaxial Helices: Interhelical Stacking

Nucleotide bases from two separate helices often stack and align their axes to form a pseudocontinuous coaxial helix (Figure 4-9). More than half of RNA structures exhibit statistically significant coaxial packing of helices. Coaxial helices dominate several large RNA structures and are believed to be highly stabilizing. The stacking between helices can occur via a single base or a base-pair bridge between the two helices. Multiple helices can stack in this manner to produce a nearly continuous single helical structure.

Coplanar and coaxial orientations of RNA bases and helices. A. Laederach *et al.* (2007) *RNA* **13**, 643–650.

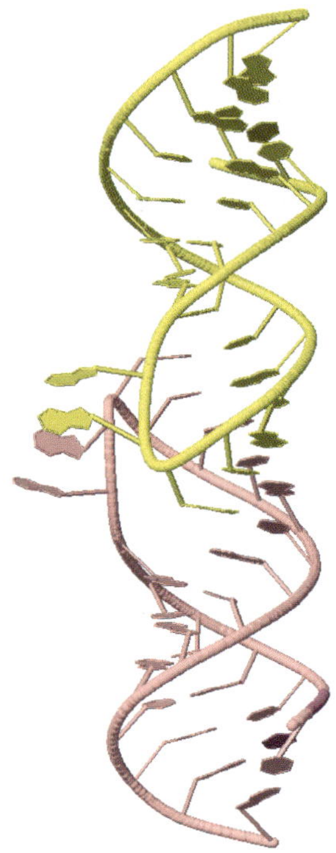

Figure 4-9. An example of two coaxial helices, formed in this case between two hairpin loops that are depicted in different colors. Figure generated from PDB file *1zci* using the program Jmol.

3. A-minor Motif

One of the most abundant and ubiquitous structural motifs stabilizing RNA tertiary structures is known as the **A-minor motif**. It involves the insertion of the smooth minor-groove edge of an adenine base into the minor groove of a neighboring double helix (Figure 4-10). This occurs preferentially at C•G base pairs, where the fit is especially snug. The base can form hydrogen bonds with one or both of the 2′-hydroxyl groups of those pairs. A-minor motifs stabilize contacts between RNA helices, interactions between loops and helices, and the conformations of junctions and tight turns.

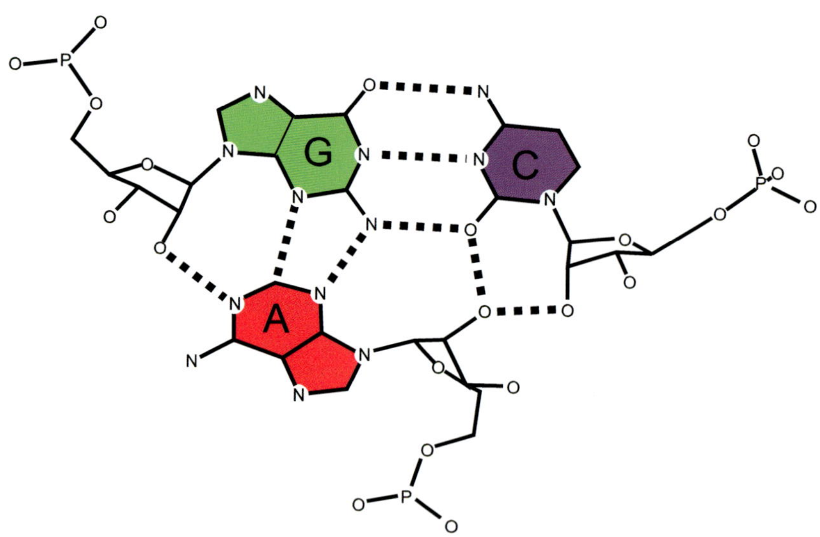

Figure 4-10. Example of an A-minor motif. The adenine nucleotide at the bottom is interacting with a G•C base pair of a double helix, in the minor groove. Besides the hydrogen bonds, there is a very close stereochemical fit between the interacting groups. Only O, N and P atoms are indicated.

RNA tertiary interaction in the large ribosomal subunit: the A-minor motif. P. Nilssen *et al.* (2001) *Proc. Natl. Acad. Sci. USA* **98**, 4899–4903.

Biochemical identification of A-minor motifs within RNA tertiary structure by interference analysis. S. A. Strobel (2002) *Biochem. Soc. Trans.* **30**, 1126–1131.

4. Dinucleotide Platform

A common example of a structural module built upon base-stacking interactions is provided by the **adenosine platform motif** (Figure 4-11). It can also occur with other bases and is known more generally as the **dinucleotide platform**. Two consecutive bases in the same strand at the end of a double helix can form a nonWatson–Crick base pair when within an internal loop, with the bases roughly coplanar. The backbone connecting the two bases is almost perpendicular to the axis of the helix they cap. Consequently, the dinucleotide platform extends the double helix by approximately one unit. The minor groove is enlarged, which facilitates interactions with other parts of the molecule. This generates a platform capable of mediating long-range tertiary interactions, especially the packing of helices in close proximity.

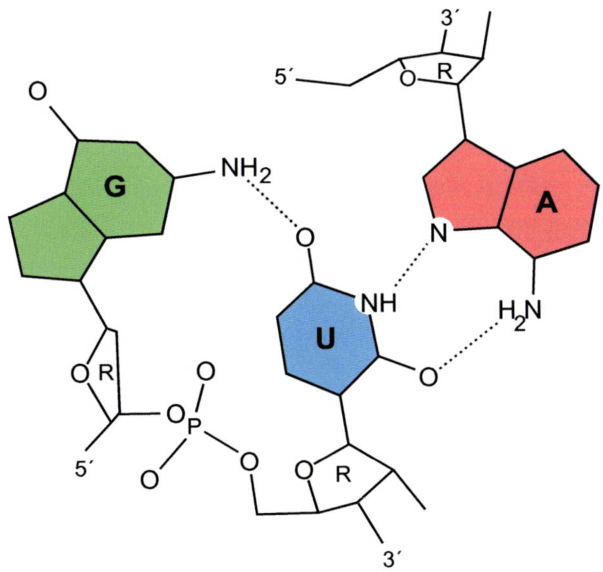

Figure 4-11. The structure of a dinucleotide platform that is also involved in a base triple as observed crystallographically. Hydrogen bonds are indicated by the *dashed lines*; only some of the hydrogen-bonding groups of the bases are shown. R is ribose. The G and U nucleotides are adjacent in the sequence of the polynucleotide chain in the lower *left* and form the dinucleotide platform. The U base is also hydrogen-bonded to an A base from another part of the sequence, to generate a base triple.

5. 'Kissing' Hairpin Loops

Nucleotides in two single-stranded loops can interact by base pairing if their sequences are complementary (Figure 4-12). This structure in 3-D can generate coaxial helices.

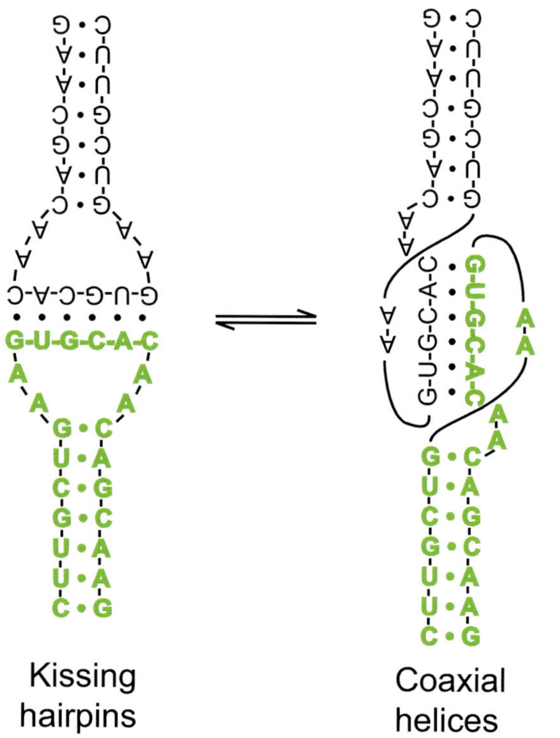

Figure 4-12. 'Kissing' of two hairpin loops when their single-stranded loops have complementary sequences. Both hairpins have the same sequence here, but that is not necessary. One strand is indicated in *color*. A coaxial stack of helices can be formed by base pairing between the hairpins and their Watson–Crick stems.

6. Ribose Zipper

If two antiparallel strands come into close proximity without being base paired, the ribose rings can interact through hydrogen bonds between consecutive backbone ribose 2′-hydroxyl groups (Figure 4-13). The hydrogen of the 2′-hydroxyl group of the 5′-nucleotide in both strands forms a hydrogen bond to the oxygen of the 2′-hydroxyl and to the base of the 3′-nucleotide on the opposite strand. The hydrogen bond is to the N3 atom when the base is a purine and to the O2 atom when a pyrimidine, so there is no sequence specificity. The zipper is usually limited to two or three nucleotides. Eleven types of ribose zippers can be distinguished on the basis of differences in their ribose–base interactions.

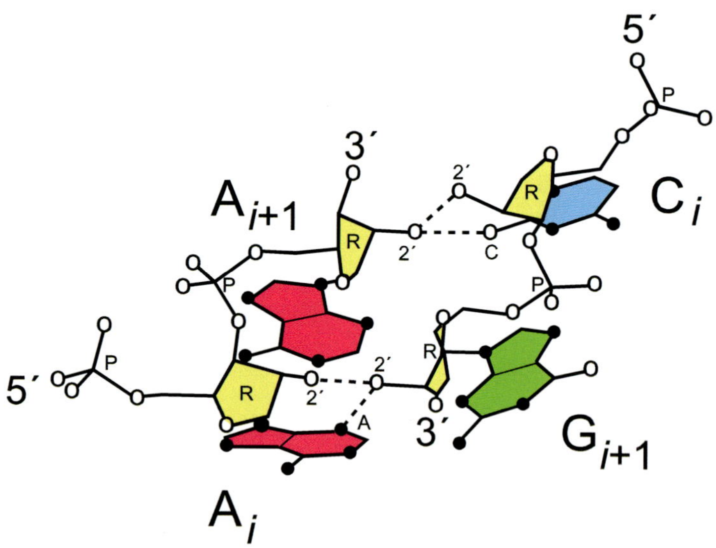

Figure 4-13. Canonical structure of a 'ribose zipper' between nucleotides A–A on one strand and C–G on the other. O atoms are indicated by *open circles*, N atoms by *black circles*. R indicates the ribose ring, while P is the phosphorous atom. The hydrogen bonds are indicated between O2′ of the 3′-nucleotide of one strand and O2′ and N3 of the base of the 5′-nucleotide of the opposite strand.

Almost all ribose zippers link Watson–Crick double-stranded stems, or stem-like base-paired structures, with loop segments that can be external, internal or junction. Ribose zippers tend to have similar sequences because stacked base triples are formed between consecutive base pairs on the stem or stem-like segment with bases (often adenines) from the loop-side segment.

About two-thirds of the ribose zippers in ribosomes interact with proteins of the complex (Section 7.4.F). Most of these ribosomal proteins bind to the ribose zipper chain segments using basic amino acid residues that form hydrogen bonds to the RNA backbone.

Sequence and structural conservation in RNA ribose zippers. M. Tamura & S. R. Holbrook (2002) *J. Mol. Biol.* **320**, 455–474.

7. Uridine Turn

A sharp bend occurs in the phosphate–sugar backbone between the first and second nucleotides of a **uridine turn**, which is also known as a **U turn** or **pi turn** (Figure 4-14). The second base then stacks on a third base. In some cases there is a hydrogen bond between the first and third nucleotides. These motifs serve a variety of functions in different RNA molecules.

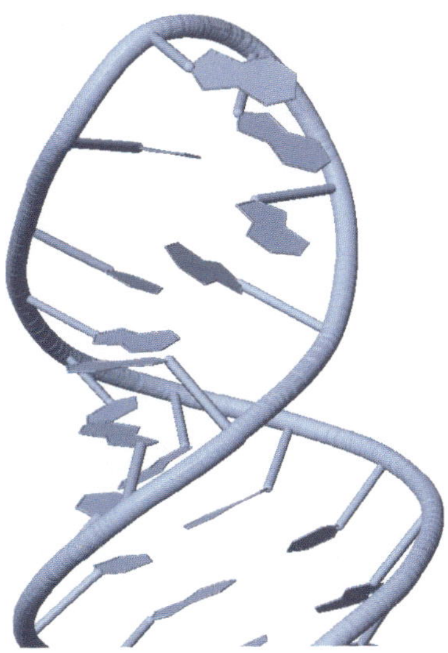

Figure 4-14. Structure of a uridine turn at the top of a hairpin. Figure generated from PDB file *1mt4* using the program Jmol.

A U-turn motif-containing stem–loop in the coronavirus 5′-untranslated region plays a functional role in replication. P. Liu *et al.* (2007) *RNA* **13**, 763–780.

The genomic HDV ribozyme utilizes a previously unnoticed U-turn motif to accomplish fast site-specific catalysis. J. Sefcikova *et al.* (2007) *Nucleic Acids Res.* **35**, 1933–1946.

8. *Tetraloop/Receptor Interactions*

Tetraloops (Section 4.1.B) often interact, via Mg^{2+} ions, with a 'receptor' structure that consists of roughly 10 ribonucleotides in a distorted double helix (Figure 4-15). Interaction of the GAAA tetraloop with its 11-nucleotide receptor is one of the most frequently occurring long-range tertiary interactions in RNAs, and it is important for the stability of the tertiary structure.

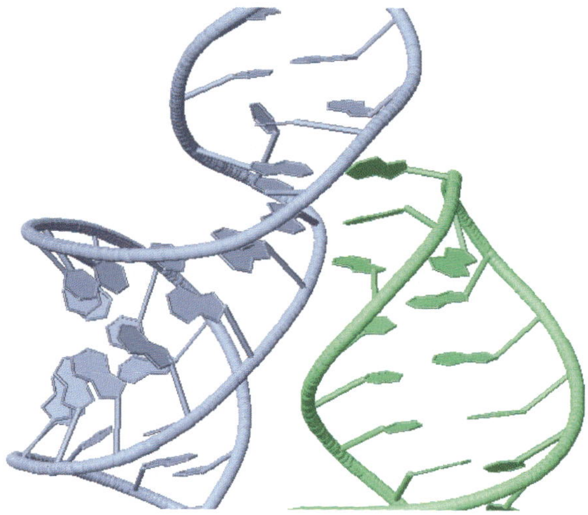

Figure 4-15. A tetraloop interacting with its 'receptor'. The tetraloop is *green*, whereas its receptor is *blue*. Figure generated from PDB file *2adt* using the program Jmol.

Site-directed spin labeling studies reveal solution conformational changes in a GAAA tetraloop receptor upon Mg^{2+} -dependent docking of a GAAA tetraloop. P. Z. Qin *et al.* (2005) *J. Mol. Biol.* **351**, 1–8.

Role of metal ions in the tetraloop–receptor complex as analyzed by NMR. J. H. Davis *et al.* (2007) *RNA* **13**, 76–86.

9. *Roles of Ions*

RNA is a polyelectrolyte just like DNA (Section 2.3). In a complex and compact 3-D RNA structure, the sugar–phosphate chain folds several times on itself, and negatively charged phosphate groups inevitably come into close contact. Positively charged ions are necessary to relieve the resulting electrostatic repulsions. Biologically, the most prevalent and efficient cations are magnesium ions and polyamines, such as putrescine, spermidine and spermine (Equation 2.4). Magnesium cations are 'hard' ions, with well-defined electronic structures, that interact favorably with the 'hard' negatively charged O atoms of the phosphate groups. Polyamines like spermine always carry positively charged amino groups, and their flexibilities and small sizes allow them to snuggle into helical grooves and in between helical sugar–phosphate backbones.

Two types of metal-ion interactions occur with RNA: (1) diffuse ions that accumulate near RNA due to the electrostatic field while retaining their hydration sphere, and (2) chelated ions that are in direct contact with RNA at a specific location and may have some waters of hydration displaced by coordination with polar RNA atoms. Such interactions are usually with the ionized phosphate groups but they may also occur with polar groups on the bases. The A-form double helix of RNA has a deep and narrow major groove, and cations are usually observed to be present in both the major and minor grooves. In group I self-splicing introns (Section 4.2.C), an organized core containing five magnesium ions constructs an exterior surface that facilitates the close packing of noncanonical loops into the minor groove of a double helix. Within the interior of this ion core, base-stacking and hydrogen-bonding interactions between nucleotides form specific metal-ion binding sites. Monovalent cations bind to adenosine platforms and to the RNA equivalents of the guanine quartet (Section 2.2.C.4). In many cases the ion-binding site is formed even in the absence of the ion, although it is much less stable, whereas in a few cases large structural rearrangements occur on metal binding.

Varying the magnesium ion concentration during thermal unfolding (Section 5.2) distinguishes between melting of the secondary and tertiary structures. Decreasing concentrations of magnesium ions cause the tertiary structure to melt at decreasing temperatures, whereas the melting temperature of the secondary structure does not change.

Ions and RNA folding. D. E. Draper *et al.* (2005) *Ann. Rev. Biophys. Biomolec. Structure* **34**, 221–243.

Mg^{2+}–RNA interaction free energies and their relationship to the folding of RNA tertiary structures. D. Grilley *et al.* (2006) *Proc. Natl. Acad. Sci. USA* **103**, 14003–14008.

Simulation of Ca^{2+} and Mg^{2+} solvation using polarizable atomic multipole potential. D. Jiao *et al.* (2006) *J. Phys. Chem. B* **110**, 18553–18559.

4.2.B. Transfer RNA Structures

Transfer RNAs (tRNAs) are single RNA chains of 75–93 nucleotides that are present in the cytosol and organelles of all living cells, serving as adapters between the nucleic acid and protein primary structures. Each tRNA is covalently attached to one particular amino acid, and it recognizes the codons of messenger RNA for that amino acid; consequently, the attached amino acid is inserted into the growing polypeptide chain when the charged tRNA is bound to the appropriate codon of the messenger RNA that is being translated. A superscript is used to indicate the amino acid accepted and used in protein biosynthesis by each tRNA. Transfer RNAs need to recognize and interact with a number of components of the protein biosynthesis machinery, so they consist of complex folded structures in which some parts have common functions and others are specific for their particular amino acid. Probably because of their complex structures and functions, tRNAs contain many modified versions of the classic nucleotides; more than 80 have been characterized. All such modifications occur after synthesis of the RNA molecule and may be modifications of the base (methylation, thiolation, etc.) or ribose (methylation). Further modifications of the tRNA molecules occur after their biosynthesis (Figure 4-16). Segments can be removed from both ends of the primary transcript, and internal sequences (*introns*) may also be removed. The CCA sequence at the 3′-end, which is linked to the appropriate amino acid, is usually added by a specific enzyme.

Guided by their primary structures, **tRNA molecules fold into cloverleaf-like secondary structures** with well-defined stems and loops that make up the acceptor arm, D-arm and -loop, anticodon arm and loop, and T-arm and -loop (Figure 4-16). The acceptor stem always has seven base pairs and four single-stranded nucleotides, including the absolutely conserved CCA sequence at the 3′-end. The D-stem and loop are of variable length in different tRNAs, whereas the anticodon stem always has five base pairs and the anticodon loop seven nucleotides. The variable region usually has four to five nucleotides but can contain up to 24. Finally, the T-stem always has five base pairs and the T-loop seven nucleotides.

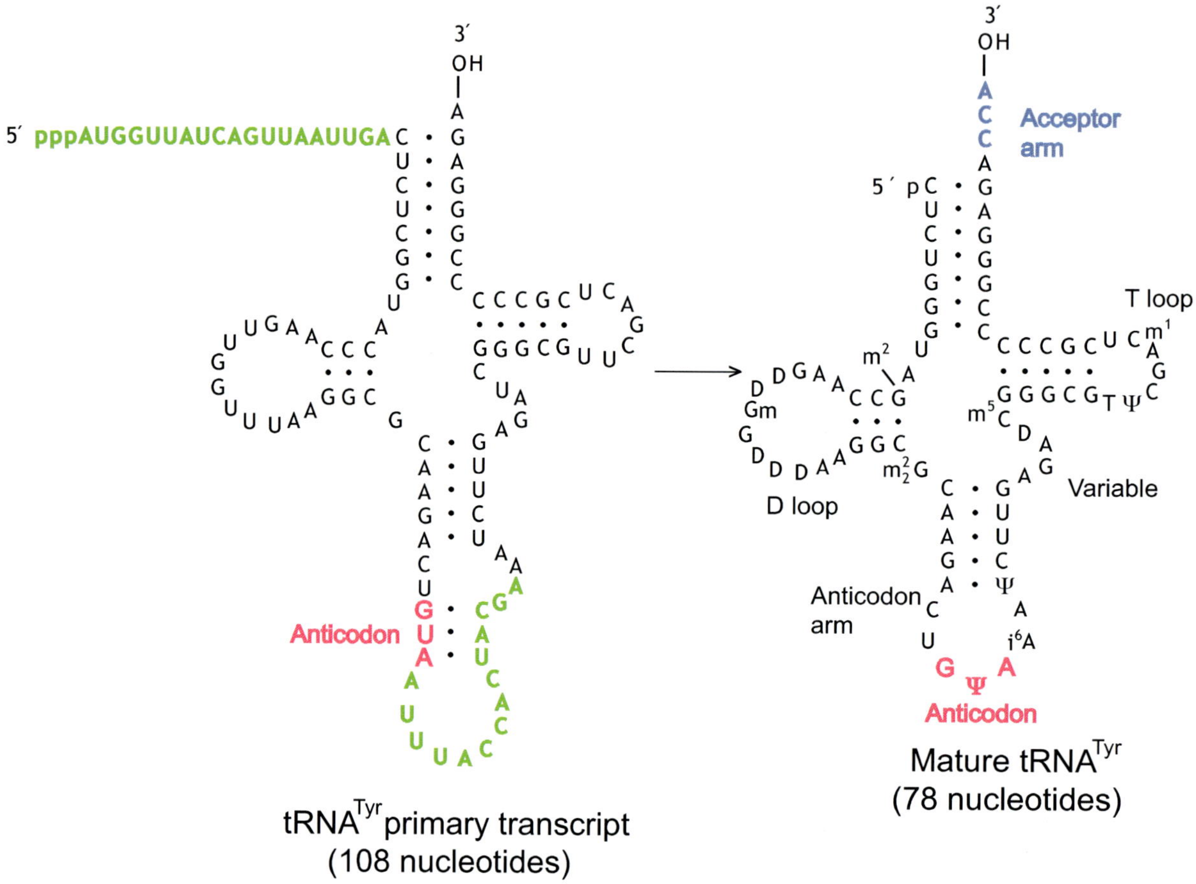

Figure 4-16. The post-transcriptional processing of yeast tRNA^{Tyr}. A 14-nucleotide intron next to the anticodon (*red*) is removed, as is the 19-nucleotide 5′-terminal sequence (green). The sequence –CCA (*blue*) is added at the 3′-end, and several of the bases are modified (Figure 4-3).

Each tRNA has an L-shaped 3-D structure comprising two domains, the acceptor branch (produced by stacking of the acceptor arm and T-arm) and the anticodon branch (the anticodon arm and the D-arm) (Figure 4-17). The angle between both branches is about 90° and the distance between the extremities of the two branches, i.e. between the acceptor end and the anticodon triplet, is 75–80 Å. This 3-D structure is generated by a number of interactions between conserved or semiconserved nucleotides that are distant in the covalent structure but close in the folded conformation. Especially important are those in the D- and T-loops and within the core of the molecule. Some of the nine tertiary interactions present in each tRNA involve triple interactions between three nucleotides simultaneously: 8–14–21, 9–12–23, 10–25–45, 13–22–46.

Transfer RNAs are classified according to the lengths of their variable regions, which can be either four or five nucleotides (class I tRNA) or 10–24 (class II tRNA). The only class II tRNAs are tRNA^{Leu}, tRNA^{Ser} and tRNA^{Tyr} from eubacteria and some organelles, plus that for selenocysteine, the '21st amino acid' (see Figure 4-22). Some tRNAs from mitochondria possess peculiar structural features, such as the absence of one or more stems and loops within the secondary structure and/or the absence of conserved nucleotides. Nonetheless, these tRNAs are thought to fold into the usual

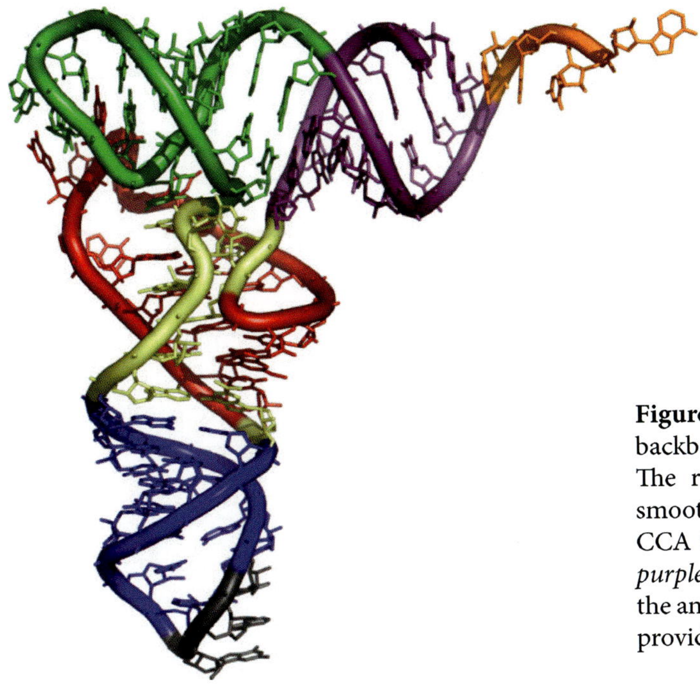

Figure 4-17. The course of the polynucleotide backbone in the 3-D structure of a typical tRNA. The ribose–phosphate backbone is depicted as a smooth rod, with the bases protruding from it. The CCA tail is *orange*, the amino acid acceptor site is *purple*, the D-arm is *red*, the anticodon arm is *blue*, the anticodon is *black*, and the T loop is *green*. Kindly provided by Vossman.

L-shaped 3-D structure (Figure 4-17), which is believed to be necessary to fulfill their functions. Over 150 mutations with documented pathogenicity have been identified in the genes for tRNAs within the human mitochondrial genome. The physiological effects of mutationally altering the tRNA sequences are multi-faceted and complex but are ultimately due to the destabilization of certain structural features, perturbing the native tRNA fold that is required for all aspects of its function.

The crystal structure of leucyl-tRNA synthetase complexed with $tRNA^{Leu}$ in the post-transfer-editing conformation. M. Tukalo *et al.* (2005) *Nature Struct. Mol. Biol.* **12**, 923–930.

The crystal structure of tRNA. B. F. Clark (2006) *J. Biosci.* **31**, 453–457.

Crystal structure of a 70S ribosome–tRNA complex reveals functional interactions and rearrangements. A. Korostelev *et al.* (2006) *Cell* **126**, 1065–1077.

Structure of the 70S ribosome complexed with mRNA and tRNA. M. Selmer *et al.* (2006) *Science* **313**, 1935–1942.

4.2.C. Ribozyme Structures

A ribozyme is an RNA molecule capable of catalyzing a chemical reaction by itself in the absence of proteins (Chapter 8). Such catalytic RNAs fold into compact 3-D structures that efficiently activate a single phosphate linkage for a chemical reaction. There are several major types of ribozymes. Some have been identified in **satellite RNAs** from plants and animals, where they participate in the replication of these double-stranded RNAs. A more common category includes the group I and group II **self-splicing introns**. Another category comprises **ribonuclease P**, which is associated with a protein and participates in the processing of tRNAs.

Most of the naturally occurring catalytic RNAs require divalent metal cations. While the metal specificity varies substantially between ribozymes, Mg^{2+} or Mn^{2+} will usually suffice. These metals often have both a structural and catalytic role. The P5abc subdomain of the *Tetrahymena* group I intron dramatically illustrates the roles of divalent metal ions in RNA folding. Three Mg^{2+} ions interact with disparate phosphate groups within the subdomain, which allows the RNA to fold inside-out, in that the phosphates point into the structure and the nucleotide bases point out to the solvent. Some structural metals in RNA can often be substituted with polycations such as spermidine (Equation 2.4) or cobalt hexamine, which emphasizes the importance of charge neutralization for RNA folding.

One of the simplest structures is that of the **hammerhead ribozyme**, which requires only 42 nucleotides for a folded and catalytically active minimal structure. Yet quite dramatically different structures for this minimal ribozyme have been determined crystallographically and by NMR. Each consists of three double-helical stems, but in remarkably different orientations: the first and second helical stems differ in orientation by about 150°. Apparently the crystal lattice induces or selects a more compact structure than predominates in solution, which vividly illustrates the intrinsic flexibility of such RNA molecules. The crystal structures did not explain how catalysis takes place, and it seemed that very substantial changes in conformation would have to occur during the reaction (Section 8.2). The structure of a full-length hammerhead ribozyme, with 63 nucleotides (Figure 4-18), however, resolved the inconsistencies. The 50% additional nucleotides produce significant changes throughout the structure, further illustrating the plasticity of RNA structures. This full-length molecule is 1000 times more active than the minimal molecule, and the structure can explain how catalysis takes place (Section 8.2).

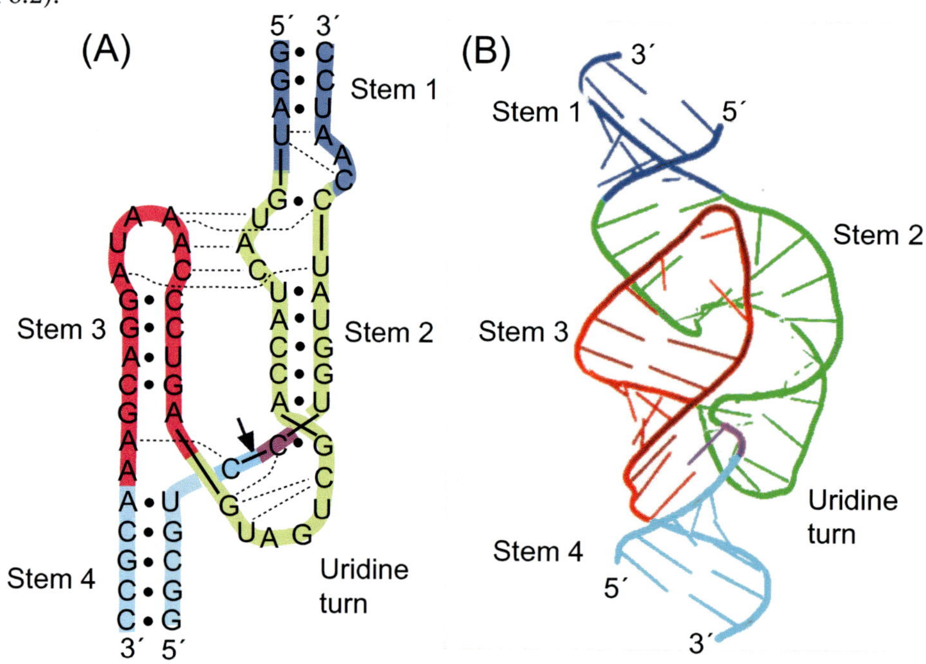

Figure 4-18. The structure of a full-length hammerhead ribozyme consisting of 63 nucleotides in two polynucleotide chains. (A) Schematic diagram of the sequences and secondary structures of the two strands. The colored ribbons and the solid black lines indicate the linear sequences. The dots indicate the base pairs that define the secondary structure. The long-range interactions that determine the tertiary structure are indicated by the dashed lines. The phosphodiester bond that is cleaved by the ribozyme is indicated by the arrow. (B) The 3-D structure of the ribozyme determined crystallographically. The polynucleotide backbone is depicted as the thin cylinder, with the bases depicted as lines off it. Data from M. Martick and W.G. Scott (2006) *Cell* **126**, 309-320.

The **hairpin ribozyme** is a small catalytic RNA comprising two internal loops carried on two adjacent arms of a four-way helical junction. To achieve catalytic activity, the ribozyme folds into a compact conformation that facilitates the formation of tertiary interactions between the two loops (Figure 4-19-A).

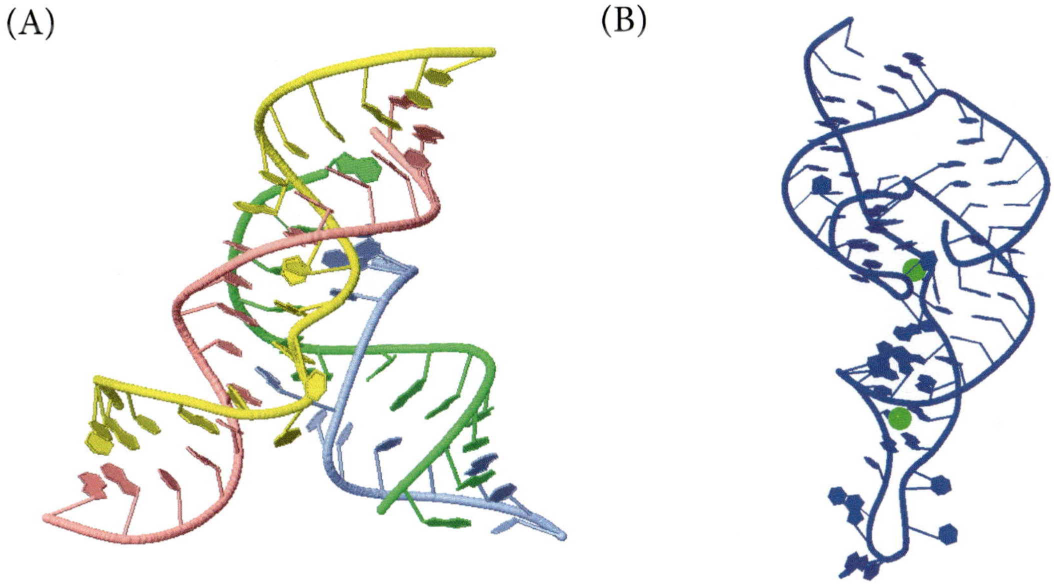

Figure 4-19. Tertiary structures of two types of ribozymes. (A) The crystal structure of a hairpin ribozyme model, comprising four RNA chains. (B) The crystal structure of the precursor of the hepatitis delta virus ribozyme. The *green spheres* are magnesium ions. The bottom of the RNA molecule is distorted because it was bound to a protein in the crystal. Figure generated from PDB files *1x9k* and *1sj3* using the program Jmol.

The **hepatitis delta virus ribozyme** is only somewhat more complex (Figure 4-19-B). It is folded into a double pseudoknot containing five helical stems (P1, P1.1, P2, P3 and P4). The P1 helix is coaxially stacked upon P1.1 and P4, while P2 is stacked on P3. The two stacks are positioned side-by-side, linked by five-strand cross-overs and constrained by the P1.1 pairing.

The **group I and group II introns** are larger and fold into distinct tertiary structures comprising many conserved secondary structural elements. The hallmarks of a group I intron include a common secondary structure of 10 paired segments and several single-stranded joiner (J) segments between the double helices (Figure 4-20). Group II introns also have distinct structures (Figure 4-21). The intron is organized into six double-helical domains (1–6) that originate from a central wheel-shaped loop. Each of the domains has a particular function in the activity of the intron, although D1, D5 and D6 form the intron core. Domain 1 is an extended multi-helical element that includes two looped regions (termed EBS1 and EBS2) that are complementary to two sites in the 5′-exon (termed IBS1 and IBS2). There are also interdomain interactions that allow domain I to serve as the scaffold upon which the intron active site is built. Many of the phylogenetically conserved nucleotides are concentrated in domain 5, which constitutes the catalytic center of the group II intron. Domain 6 includes the branch-point adenosine whose 2′-OH nucleophilically attacks the 5′-splice site during the catalytic action (Figure 8-2). In isolation, domains 5 and 6 together comprise a single double-stranded hairpin (Figure 4-21-B).

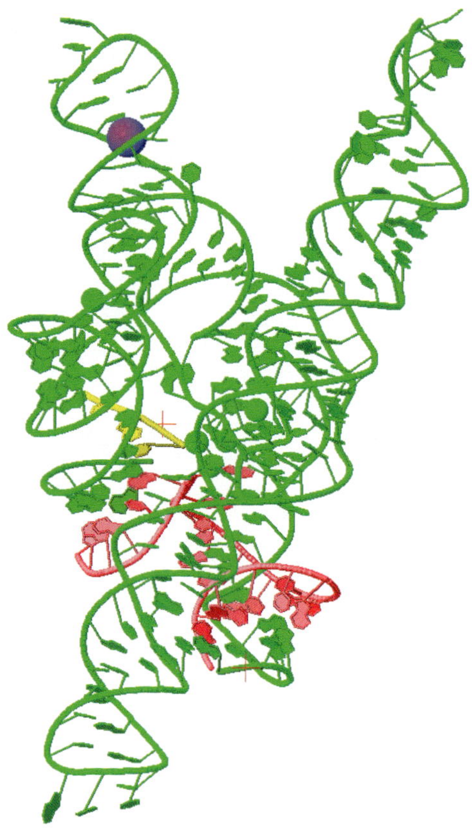

Figure 4-20. The 3-D structure of the *Tetrahymena* group I self-splicing intron. The *blue sphere* is a K^+ ion, while the *green spheres* are Mg^{2+} ions. Figure generated from PDB file *1zzn* using the program Jmol.

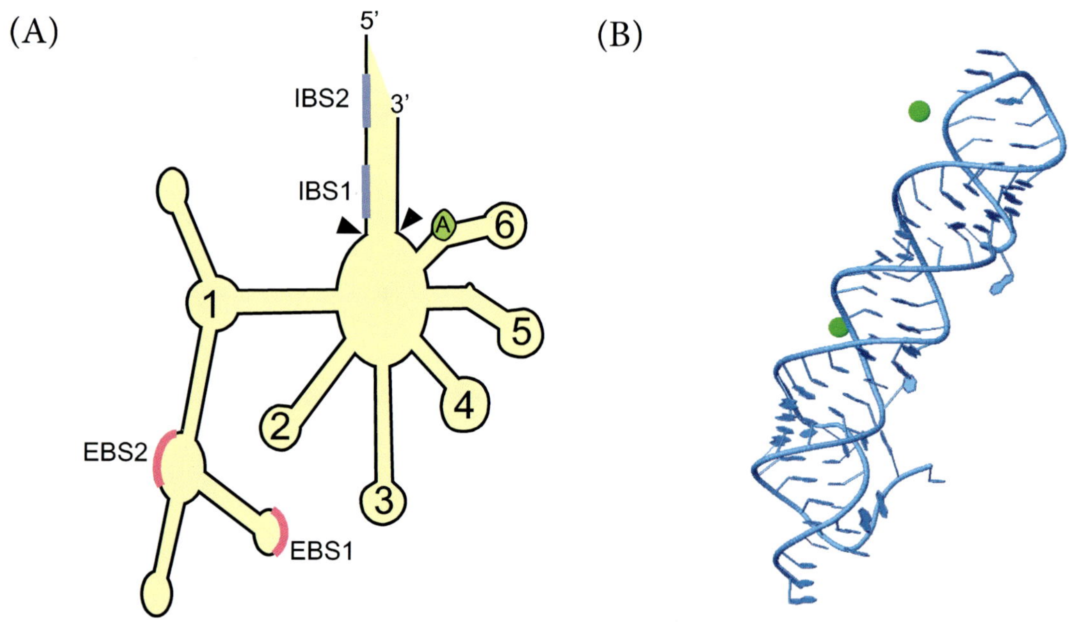

Figure 4-21. Structures of group II self-splicing introns. (A) The predicted secondary structure, with the stem structures being antiparallel base-paired helices. The EBS sequences are complementary in sequence to the IBS sequences; both are indicated as *thick lines*. The circled A in domain 6 is a bulge in the double helix, causing it to bend, and acts as the nucleophile in the first step in splicing. The two *arrowheads* indicate the splice sites. (B) The 3-D structure of domains 5 and 6 in isolation, which comprise a single double-stranded hairpin. Figure generated from PDB data file *1kxk* using the program Jmol.

Ribozymes with new folds and biochemical activities have been found after only limited alterations of the nucleotide sequence (Section 8.3). The probability of finding such altered structures increases considerably as the mutational distance from the parental ribozyme increases, suggesting that it must first escape from the 3-D fold of the parent.

The structure–function dilemma of the hammerhead ribozyme. K. F. Blount & O. C. Ulenbeck (2005) *Ann. Rev. Biophys. Biomol. Struct.* **34**, 415–440.

The tertiary structure of the hairpin ribozyme is formed through a slow conformational search. G. Pljevaljcic *et al.* (2005) *Biochemistry* **44**, 4870–4876.

New catalytic structures from an existing ribozyme. E. A. Curtis & D. P. Bartel (2005) *Nature Struct. Mol. Biol.* **12**, 994–1000.

Tertiary contacts distant from the active site prime a ribozyme for catalysis. M. Martick & W. R. Scott (2006) *Cell* **126**, 309–320.

4.3. QUATERNARY STRUCTURE OF RNA

Relatively few examples of the association of RNA molecules to form supramolecular quaternary structures have been well-characterized, but they are relatively important. During splicing of messenger RNA (mRNA) precursors, to remove the introns, for example, the precursors associate with five major ribonucleoprotein (RNP) particles called **small nuclear RNPs** that interact with each other and the mRNA. The dynamic disruption and formation of these RNA–RNA quaternary interactions are essential for RNA splicing. Oligomerization usually depends on the concentrations of cations or polyamines.

The quaternary association of RNA molecules generally occurs by conventional Watson–Crick base pairing. For example, small regulatory RNAs with sequences complementary to other RNA molecules (known as **antisense RNAs**) form intermolecular duplexes during the control of gene expression in both prokaryotes and eukaryotes. Similarly, **guide RNA**s recognize complementary sequences to identify sites where mRNAs are edited after transcription. More complex RNA quaternary structures do not rely exclusively on Watson–Crick pairing but base pairs are still very important. So-called 'kissing-hairpins' (Figure 4-12) form between loop nucleotides in two stem–loop structures with complementary sequences. These structures provide protein recognition sites during the regulation of the copy number of prokaryotic plasmids.

Supramolecular association of RNA molecules by means of nonWatson–Crick interactions is best-characterized in the so-called 'G-quartet structures' (Section 2.2.C.4). These structures form readily *in vitro* for both RNA and DNA sequences that contain stretches of guanidines or uracils, but it is not clear whether they occur *in vivo*.

Regulation of the structure of RNA assemblies through RNA tertiary interactions. R. Ohmori *et al.* (2005) *Nucleic Acids. Symp. Ser.* **49**, 193–194.

4.4. RNA STRUCTURE PREDICTION

The folded structures of RNA molecules are dictated by their nucleotide sequences and are adopted spontaneously, so it should be possible to predict the folded structure from just the nucleotide sequence; the problem is comparable to that of predicting the structures of proteins. A fundamental difference between RNA and protein structures is that the RNA secondary structure is often stable on its own, whereas that of proteins is at best only marginally stable in the absence of the tertiary structure. Formation of the secondary structure dominates RNA folding, and the tertiary structure forms through relatively weak interactions between pre-formed secondary structure motifs. Consequently, secondary structure is especially easy to be predicted successfully from just the RNA nucleotide sequence, because complementary base pairing in double helices is the most important consideration. Even in the simplest case, however, it is not a trivial exercise because the base pairing in RNA double helices is not restricted to Watson–Crick base pairs and because a sequence of N nucleotides can adopt approximately 1.8^N different secondary structures.

4.4.A. Prediction of Secondary Structure

The most probable secondary structure for a given RNA sequence can usually be identified using computer programs that find the maximum amount of complementary base pairing. The **thermodynamic approach** attempts to find the secondary structure with the lowest free energy, whereas the **phylogenetic approach** uses information available in homologous sequences that have arisen from a common evolutionary ancestor. Comparison of secondary structures of 16S and 16S-like ribosomal RNA predicted solely on thermodynamic calculations with those derived by the phylogenetic approach demonstrated that the quality of the thermodynamic predictions is variable: the percentages of correctly predicted base pairs ranged between 10% and 90%. Often, the programs produce several solutions that seem equally likely, so identifying the correct one requires additional chemical or biological information.

Revolutions in RNA secondary structure prediction. D. H. Mathews (2006) *J. Mol. Biol.* **359**, 526–532.

Statistical and Bayesian approaches to RNA secondary structure prediction. Y. Ding (2006) *RNA* **12**, 323–331.

Computational RNA secondary structure design: empirical complexity and improved methods. R. Aguirre-Hernandez *et al.* (2007) *BMC Bioinformatics* **8**, 34.

---

1. Thermodynamic Approach

The thermodynamic approach is based on a set of energy parameters for the stabilities of structures formed by short oligonucleotides, plus the observation that the secondary structure of an RNA molecule can be decomposed into elementary motifs that recur in other structures. The sum of the free energies of the secondary elements is assumed to be a reasonable approximation of the total free energy of the RNA, because tertiary interactions are weaker than secondary interactions. The free energy of a given base pair is assumed to be due primarily to hydrogen bonding and base stacking, so it depends only on the immediate neighbors (**nearest-neighbor approximation**). The base pairs considered are the standard Watson–Crick pairs, G•C and A•U, as well as the frequent G•U (Equation 4.1). Stereochemical considerations are also taken into account, such as the requirement that at least three nucleotides occur in a loop between two paired strands (Figure 4-5). Unfortunately, the thermodynamic parameters used are imprecise and incomplete. On the other hand, such calculations have confirmed the expectation that the RNA sequences from organisms that live at higher temperatures have evolved RNA sequences with higher melting temperatures.

Pseudoknots are difficult to predict, because most algorithms are based on decomposition into substructures, which is not possible mathematically with pseudoknots (Figure 4-8). With luck, however, one of the two helices of a pseudoknot might be predicted on the basis of its stability, and inspection of the remaining single-stranded regions might uncover the second helix.

Many computer programs use recursive algorithms based on **dynamic programming** principles, which solve and combine solutions corresponding to subproblems. Each subproblem is solved just once and then saved in a table, avoiding the need to compute the same answer every time the subproblem is encountered. Such algorithms are guaranteed to find the structure that is optimal for a given set of rules. The number of nucleotides, N, can be as large as 10,000. The required computer time, however, is proportional to N^3, and the computer memory required is proportional to N^2. Decomposition into elementary structures makes it possible to consider more complete thermodynamic data, at the price of increased computational time and storage requirements, depending on the way in which energy is assigned to the loops. The free energies of the base pairs and any additional experimental data are included in the energy function, permitting fast computation of the optimal 2-D structure.

Prediction of RNA secondary structure by free energy minimization. D. H. Mathews & D. H. Turner (2006) *Curr. Opinion Struct. Biol.* **16**, 270–278.

A method for finding optimal RNA secondary structures using a new entropy model (vsfold). W. Dawson *et al.* (2006) *Nucleosides Nucleotides Nucleic Acids* **25**, 171–189.

A set of nearest neighbor parameters for predicting the enthalpy change of RNA secondary structure formation. Z. J. Lu *et al.* (2006) *Nucleic Acids Res.* **34**, 4912–4924.

2. Phylogenetic Approach

The phylogenetic approach assumes that the function and folding architecture have been conserved during evolution while the nucleotide sequence has changed, producing a family of homologous sequences; consequently, different versions of an RNA sequence should be compatible with the same secondary structure. The power of this comparative approach stems from the general observation that molecular 3-D architectures change during evolution much more slowly than do sequences. A set of homologous sequences with common ancestry and function can pinpoint secondary structure elements that are common to all the sequences.

The first step is to align the sequences, arranging them horizontally so that the lengths of the paired regions juxtapose vertically (Figure 4-22), with inclusion of blanks or gaps in a fashion similar to the alignment of protein sequences. The next step consists of searching for nucleotide interactions by measuring correlations between pairs of RNA positions in the alignment, such as compensatory base changes (e.g. an A•U pair changed into a C•G pair). The more compensatory base changes in the sequences, the more probable the secondary structure prediction is correct. The level of ambiguity can always be reduced by using additional sequences. Aligning the sequences and assessing pairs of

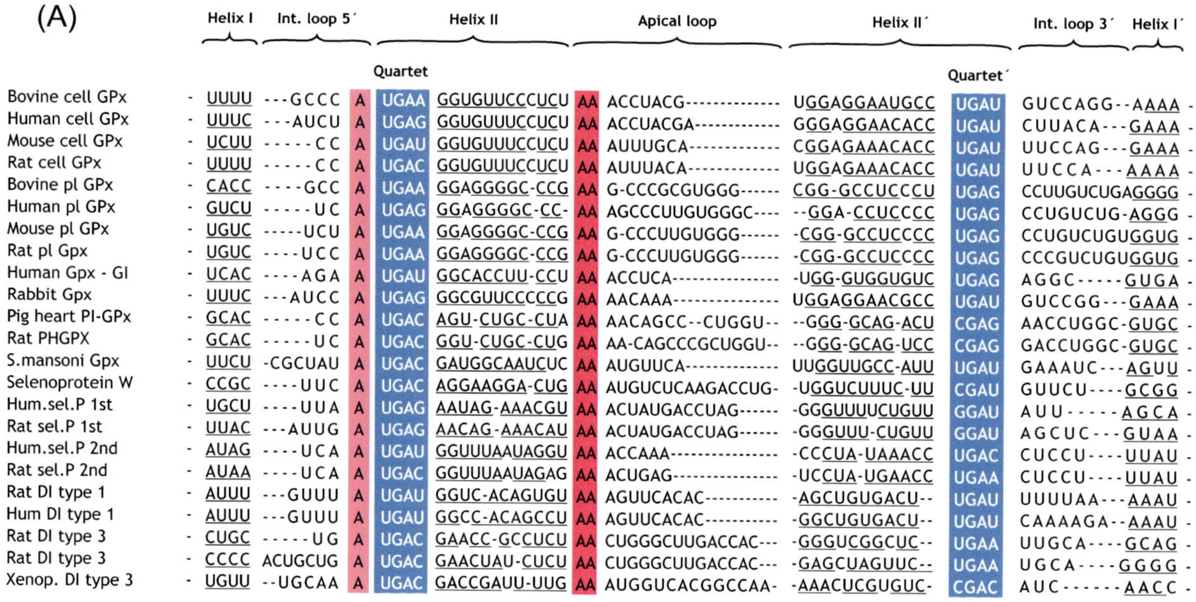

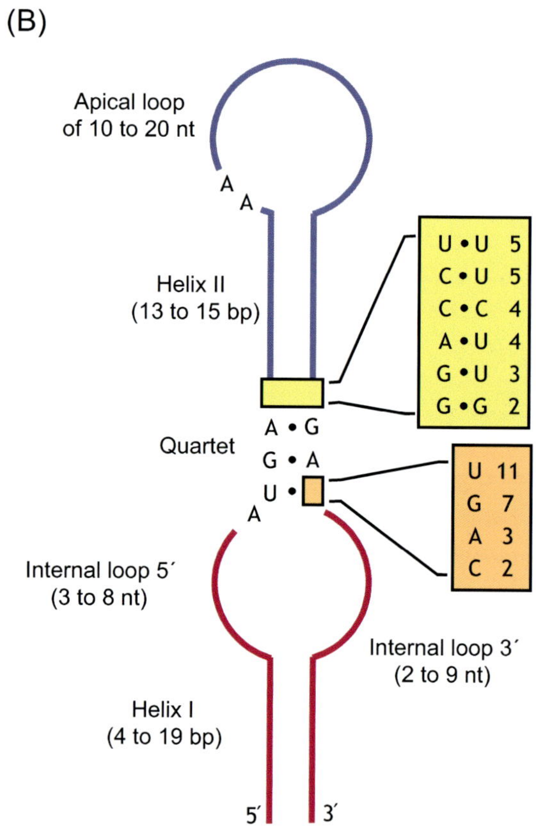

Figure 4-22. Identifying the signal for inserting selenocysteine residues (Section 7.2.O) from the sequences of mRNAs. (A) Sequence alignments of mRNAs coding for various selenocysteine-containing proteins (Gpx, glutathione peroxidases; DI, iodothyronine deiodinase). The base-paired segments, denoted Helix I/I′ and Helix II/II′ corresponding to the 5′- and 3′-strands, are *underlined*; note the significant number of noncanonical base pairs. The invariant residues in the internal and apical loops are highlighted (*black on red*). At the base of helix II, the quartet made of four nonWatson–Crick pairs that constitute a signature of the SelenoCysteine Insertion Sequence (SECIS element) is boxed and highlighted (*white on blue*). Gaps have been introduced to maximize superpositions between secondary structure elements and sequence similarities. (B) Secondary structure of the SECIS element predicted from the sequence alignment. The central 'quartet' of nonstandard base pairs consists primarily of tandem sheared G•A base pairs involving hydrogen bonds between N2(G) and N7(A), as well as between N3(G) and N6(A). The number of occurrences of the various bases or base pairs at the boxed sites are given in the two boxes (bp, base pairs; nt, nucleotides). Data from R. Walczak *et al.* (1996) *RNA* **2**, 367–379.

them should identify the conserved core of the secondary structure. All such methods, however, are plagued with problems of statistical relevance: with only four bases, sequences may be similar purely by coincidence.

The phylogenetic approach is the best method when a set of homologous sequences is available for RNAs with the same biological function. The sequences should be arranged in groups according to the phylogenetic classification or on the basis of their functional properties. The overall power of the approach increases with the diversity of the sequences and the evolutionary distances separating them, while the accuracy of each prediction depends on the number of alterations in each group and subgroup that are related, such as changing both bases of a base pair. A conserved core can often be found by using an iterative stepwise convergent cycle that combines the alignment search and secondary structure assessment, usually by aligning each sequence to a tentative model of the secondary structure.

Use of the information that is present in a large number of homologous sequences is illustrated in Figure 4-22.

Identification and classification of conserved RNA secondary structures in the human genome. J. S. Pedersen *et al.* (2006) *PLoS Comput. Biol.* **2**, e33.

Efficient pairwise RNA structure prediction and alignment using sequence alignment constraints. R. D. Dowell & S. R. Eddy (2006) *BMC Bioinformatics* **7**, 400.

Efficient pairwise RNA structure prediction using probabilistic alignment constraints in Dynalign. A. O. Harmanci *et al.* (2007) *BMC Bioinformatics* **8**, 130.

4.4.B. Prediction of Tertiary Structure

It should be possible, in principle, to predict the 3-D structure of an RNA from just its nucleotide sequence, but it is a formidable problem, comparable to that of predicting the tertiary structures of proteins. The prediction methods range from very mathematical, relying solely on computer algorithms, to the most pragmatic interpretations of experimental data. A predicted secondary structure is always necessary for construction of a tertiary structure of an RNA molecule.

Tertiary contacts between groups in the tertiary structure can often be inferred through sequence comparisons and chemical modification experiments (Section 2.6), which might indicate the importance of specific atomic positions or uncover some groups that are inaccessible or protected and cannot be explained solely by the secondary structure. Cross-linking experiments can indicate directly which groups are in close proximity in the structure. Such information can be used with the distance-geometry method of generating structures from a set of constraints from NMR data, although it is difficult to compile a sufficient number of constraints and there are problems in choosing the correct chiralities and avoiding knots in the structures. Other computer methods try to find the minimum free energy by simplifying the structure using a **pseudoatom approach**, with either one pseudoatom per helix or one pseudoatom per nucleotide; however, this ignores the asymmetry of the RNA fragments and, more importantly, all the detailed interactions that control RNA folding. A manual approach uses known RNA structures to extract the structure of a fragment of interest. The fragments obtained are then assembled manually on a computer graphics screen, using interactive modeling procedures. Any

3-D structure that results can be refined by restrained energy minimization, molecular mechanics or molecular dynamics procedures. The surface of the final model that is accessible to the solvent can be compared to experimental data on the reactivities of specific groups to chemical reagents.

Modeling should not be considered as a method for generating the correct structure, but primarily as a heuristic tool to help in planning and interpreting experiments. It can suggest new relationships between the various components of a modeled molecule; for example, varying the sequence of a macromolecule in the absence of a 3-D model will be somewhat random and rarely informative. At best, such experiments will confirm the secondary structure of an RNA molecule. The data can still be useful, however, for gradually generating a 3-D structure and suggesting new experiments.

The case of tRNA (Section 4.2.B) is the best example of how such approaches have clarified understanding of the secondary and tertiary structures of RNA molecules: new genome sequences can be searched using precise rules for identifying the sequences of tRNA molecules.

De novo synthesis and development of an RNA enzyme. Y. Ikawa *et al.* (2004) *Proc. Natl. Acad. Sci. USA* **101**, 13750–13755.

Bridging the gap in RNA structure prediction. B. A. Shapiro *et al.* (2007) *Curr. Opinion Struct. Biol.* **17**, 157–165.

The RNAz web server: prediction of thermodynamically stable and evolutionarily conserved RNA structures. A. R. Gruber *et al.* (2007) *Nucleic Acids Res.* **35**, W335–338.

~ CHAPTER 5 ~

DENATURATION, RENATURATION AND HYBRIDIZATION OF NUCLEIC ACIDS

The folded structures of nucleic acids, especially the helices of double-stranded DNA and the folded structures of single-stranded RNA molecules, are disrupted upon exposure to extreme conditions, such as high temperatures, chemical denaturants and extremes of pH. The conditions required to denature a nucleic acid are a quantitative measure of the stability of its structure. In most cases, denaturation of a nucleic acid involves the separation of base-paired double strands, whether two separate strands in double-stranded DNA or intramolecular stems and loops in a single RNA molecule. One of the most important aspects of such structures is that base pairs between guanine and cytosine (G•C and C•G) are generally more stable than those between adenine and thymine (A•T and T•A) or uridine (A•U and U•A) and much more than any other nonstandard (mismatched) base pairs.

In many cases, the original structure can be regained by restoring the conditions. Alternatively, new structures may be generated by adding new polynucleotides with complementary sequences, which can hybridize to the single strands. The feasibility of renaturing nucleic acids depends primarily upon how many independent molecules are generated; the greater the number of different polynucleotide chains generated, the less likely that the original strands can find each other to regenerate complementary structures by base pairing. Because there are only four types of bases in DNA or RNA, regions of some nucleotide sequence complementarity occur even in random sequences, so a variety of base-pairing interactions is possible. The denaturation and renaturation of nucleic acids are central to most techniques in molecular biology.

In many cases, one of the hybridizing strands can be a nucleic acid mimic, such as a protein nucleic acid (PNA; Section 5.4). They can be designed to have a variety of properties to make them useful.

Principles of Nucleic Acid Structure. W. Saenger (1984) Springer, NY.

5.1. DENATURATION OF DOUBLE-STRANDED NUCLEIC ACIDS

The three-dimensional (3-D) structures adopted by nucleic acids, the antiparallel double helices and more complex structures (Chapters 2–4), are only marginally stable and may be disrupted in a variety of ways comparable to those used with proteins. An exception is high pressures, which can unfold proteins but do not induce large-scale structural changes in the helical forms of DNA or RNA. Under most conditions, the latter are stabilized by high pressures, although only slightly. These observations indicate that the volume occupied by the double helix in solution is very slightly less than that of the individual strands (Table 2-9).

Effect of hydrostatic pressure on nucleic acids. R. B. Macgregor (1998) *Biopolymers* **48**, 253–263.

Thermodynamics of nucleic acids and their interactions with ligands. A. N. Lane & T. C. Jenkins (2000) *Quart. Revs. Biophys.* **33**, 255–306.

Molecular dynamics applied to nucleic acids. J. Norberg & L. Nilsson (2002) *Acc. Chem. Res.* **35**, 465–472.

Factors regulating thermodynamic stability of DNA structures under molecular crowding conditions. D. Miyoshi *et al.* (2006) *Nucleic Acids. Symp. Ser.* **50**, 203–204.

5.1.A. Methods for Monitoring Denaturation

Different techniques can monitor different aspects of denaturation, whether it be breaking of base pairs or separation of strands. The disruption of base pairing in double-stranded structures is most readily monitored by ultraviolet (UV) absorbance, using the 38% increase in absorbance (**hyperchromism**) at 260 nm (Figure 5-1). Double-stranded nucleic acids exhibit **hypochromicity**, in that the absorbance of the duplex is less than that of the constituent single strands and nucleotides. The shape of the absorbance spectrum does not change, only its intensity, as a result of the absorbing bases being stacked very closely in the double helix so that the electronic systems of neighboring bases interact with each other. Other spectral methods are also useful, especially circular dichroism with relatively small nucleic acids, although the interpretation of the results is not always straightforward. Large changes occur in the hydrodynamic volumes and viscosities of nucleic acids upon their denaturation, especially when the strands separate, but such measurements are not very convenient. Alternatively, unfolding dramatically decreases the electrophoretic mobilities of nucleic acids through gels, and the unfolding and dissociation steps may be visualized directly and easily.

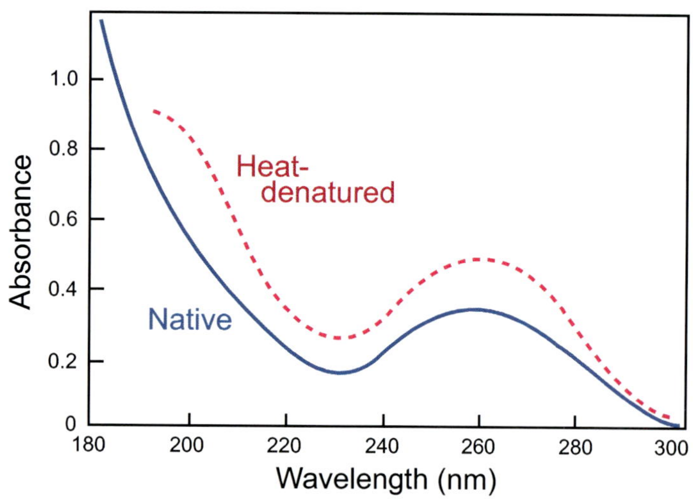

Figure 5-1. Typical UV absorbance of DNA when native (25°C) and when heat-denatured (82°C).

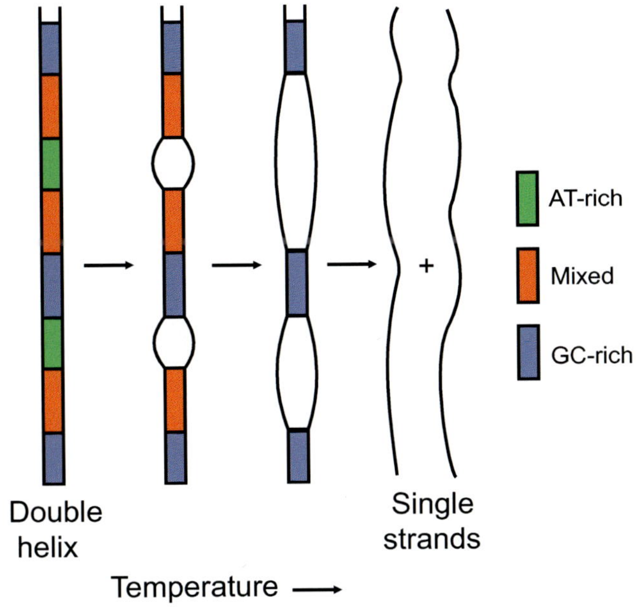

Figure 5-2. The progressive denaturation of a nucleic acid with a nonuniform sequence. The sequence of the double helix (*left*) is divided into regions of about 50 base pairs with three types of sequence, depending upon whether it is AT- or GC-rich; in reality, there is a continuum of sequence variation, from the most AT-rich to the most GC-rich. The final result at high temperatures on the *right* is the two separated and independent strands.

With long DNA molecules, the process of denaturation can be complex (Figure 5-2), but it can be observed almost directly using the technique of **denaturation mapping**. The partly opened intermediate forms are trapped by reacting them with a chemical reagent, such as **formaldehyde** ($H_2C{=}O$), or its bifunctional version, **glyoxal** (O=CH–HC=O); they react chemically with the amino groups of bases and block their participation in base pairing; the open segments cannot then regenerate a double helix. The single- and double-stranded segments can be visualized, after appropriate staining, with an electron microscope (Figure 5-3). The DNA molecules are spread on the surface of the electron microscope grid and made visible by staining with basic proteins, such as cytochrome *c*, and heavy metals. In this way, single- and double-stranded regions can usually be distinguished: the cytochrome *c* molecules bound to double-stranded DNA make it appear somewhat thicker than single-stranded regions, and the double-stranded segment is not quite as kinky. The various single- and double-stranded segments of the DNA molecule are identified simply by their positions relative to the ends of the DNA molecule. This technique of denaturation mapping is useful in physical and genetic studies of very large DNA molecules, such as the genomes of viruses, plasmids, etc.

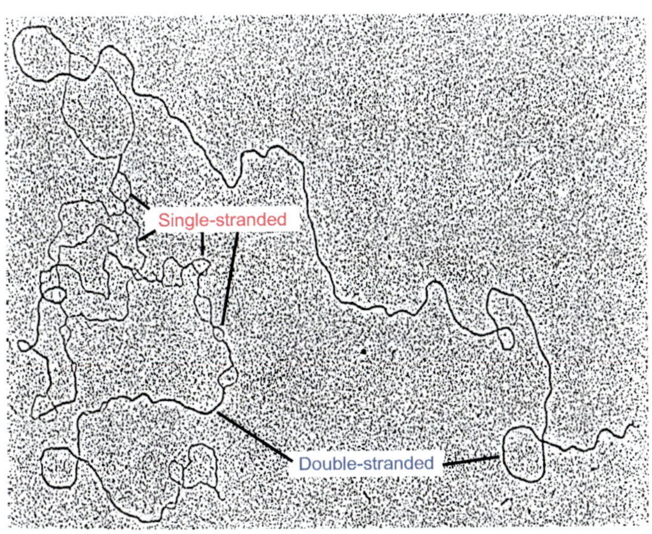

Figure 5-3. Electron micrograph of a DNA molecule that was partially denatured by alkaline pH. Separation of the two strands locally is apparent by a decrease in the thickness of the strand. The sites that are locally denatured coincide with the segments that are relatively deficient in C•G base pairs.

Single- and double-stranded DNAs or RNAs can be distinguished using the enzyme **S1 nuclease**, which hydrolyzes the backbones of only single-stranded nucleic acids. In the case of RNA, the enzyme **RNase V_1** hydrolyzes only double-stranded forms. With all such reagents, the possibility that the structure of the nucleic acid molecule will change after being hydrolyzed must be considered, and it is advisable to use these modifications only sparingly, so that no molecules are modified more than once.

Single- and double-stranded molecules can be separated using chromatography on **hydroxyapatite**. Under conditions of low sodium phosphate concentration, both single- and double-stranded DNA bind to hydroxyapatite. At 0.1–0.2 M phosphate buffer, however, the column retains only double-stranded DNA in which at least some 50 adjacent residues are base paired. The double-stranded DNA may be eluted with 0.5 M sodium phosphate. The time–course for dissociation or reassociation may be measured in this way, although it must be kept in mind that this assay does not measure directly base-pair formation, only association or dissociation of complete single strands.

Thermal difference spectra: a specific signature for nucleic acid structures. J. L. Mergny *et al.* (2005) *Nucleic Acids Res.* **33**, e138.

5.1.B. DENATURATION OF DOUBLE-STRANDED DNA

The stability of a double-stranded nucleic acid with a nonuniform base sequence varies by region, and the base-pair composition, especially the fraction of G•C base pairs, determines the stability of each region (Figure 5-2). Under most conditions, regions of the duplex that are composed predominantly of A•T (for DNA) or A•U (for RNA) base pairs are less stable than GC-rich regions (Figure 5-4). As the conditions become increasingly denaturing, the path from fully base-paired duplex to fully-denatured single strands passes through a continuum of intermediate states, where first the AT/AU-rich regions 'melt', forming internal loops, then the regions with increasing GC content, and then finally the most GC-rich regions. The melting transitions of such random-sequence duplexes can be quite broad and the various stages of melting and strand dissociation occur at distinct temperatures.

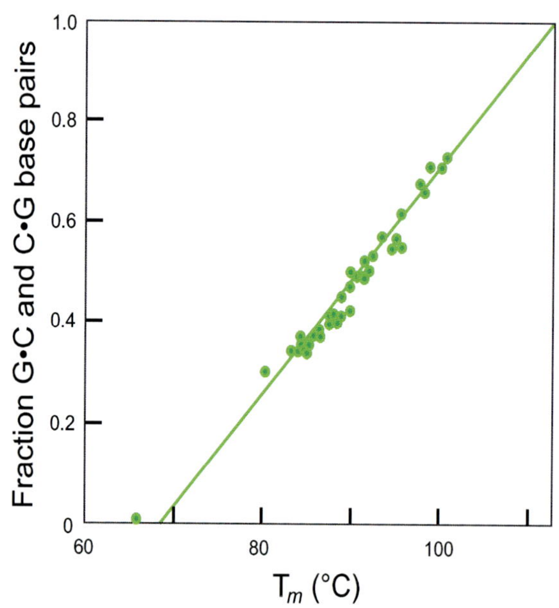

Figure 5-4. Typical variation of the melting temperature, T_m, of various DNAs with their content of G+C nucleotides. The solvent was 0.15 M NaCl and 15 mM sodium citrate (to chelate metal ions).

As the last region melts, the two strands separate (Figure 5-2). Up until that point, the two strands will very quickly regenerate their original structure if the conditions are restored to normal, because this is an intramolecular process. Once the two strands have separated, however, their reassociation is a bimolecular reaction and subject to the concentrations of the two strands and the presence of any other molecules that might compete (Section 5.3).

Thermodynamic and kinetic characterization of the dissociation and assembly of quadruplex nucleic acids. C. C. Hardin *et al.* (2000) *Biopolymers* **56**, 147–194.

Bubbles and denaturation in DNA. T. S. van Erp *et al.* (2006) *Eur. Phys. J. E* **20**, 421–434.

---

1. Thermal Melting

Heating is the most common method of denaturing nucleic acids, and their stabilities are often given as their melting temperature, T_m, the temperature at the mid-point of the thermal unfolding transition measured by the breakage of base pairs. **The T_m is largely independent of the length of the polynucleotide strands and their concentration,** even though melting involves dissociation to two single strands, because its value is determined primarily by the pseudo-intramolecular local melting processes that precede melting of the final segment and dissociation of the two strands (Figure 5-2). Only with short oligonucleotides of fewer than 50 base pairs is the T_m dependent upon the length and the concentration. Because strand dissociation occurs only at the last step, different temperatures correspond to the loss of base pairing and to the separation of the strands; the T_m usually reflects the loss of half of the base pairs, and strand separation will take place only at a somewhat higher temperature.

Under typical conditions, the value of T_m for DNA depends primarily upon the salt concentration (see below) and the base composition, because G•C base pairs melt less readily than do A•T. The T_m value is observed to be linearly dependent upon the mole fraction of G•C base pairs, f_{GC} (Figure 5-4), and the T_m in 0.15 M salt may be predicted on just this basis:

$$T_m = (69.3 + 41\,f_{GC})\ °\text{C} \tag{5.1}$$

The width of the melting transition of a synthetic DNA with a unique regular nucleotide sequence is quite sharp: the fraction of double helix measured by the hyperchromicity can decrease from 0.75 to 0.25 over a temperature range of 1°C or less. In contrast, the widths of the melting transitions of natural DNAs, with nonregular sequences, measured in the same way are several times broader, because segments approximately 50 base pairs in length with different sequences melt at somewhat different temperatures. Even in the simplest case, the melting process is not readily characterized in terms of thermodynamic parameters, because it is a multi-state reaction, even at equilibrium (Figure 5-2).

Thermal denaturation of double-stranded DNA: effect of base stacking. M. Kohandel & B. Y. Ha (2006) *Phys. Rev. E* **73**, 011905.

Loop dynamics in DNA denaturation. A. Bar *et al.* (2007) *Phys. Rev. Lett.* **98**, 038103.

---

2. Denaturants

The bases of nucleic acids are relatively hydrophobic (Table 2-3) whereas the sugar–phosphate backbone is very polar. The bases are removed from contact with water within the double helix, and the hydrophobic effect is one of the forces stabilizing the double helix. Consequently, denaturants such as urea and guanidinium chloride that diminish the hydrophobic effect, as well as form hydrogen bonds with the bases, tend to destabilize the double helix. The destabilizing effects of such reagents are independent of their ability to compete in hydrogen bonding but reasonably correlated with their actions in solubilizing the bases (Figure 5-5). **Urea** and **ethylene glycol** both bind preferentially to single-stranded DNA relative to double-stranded DNA and the bulk solution. Urea exhibits a stronger affinity for adenine and thymine bases, while that of ethylene glycol is practically independent of base composition. **Glycine, sarcosine** (*N*-methyl glycine) and **glycine betaine**:

$$CH_3\text{–}NH\text{–}CH_2\text{–}CO_2^- \qquad\qquad (CH_3)_3{}^+N\text{–}CH_2\text{–}CO_2^-$$

$$\text{Sarcosine} \qquad\qquad \text{Glycine betaine} \qquad\qquad (5.2)$$

do not bind preferentially to single strands relative to bulk solution but are excluded from the negatively charged surface of double-stranded DNA. Their preferential binding to single strands relative to double-stranded DNA increases with GC content, demonstrating a stronger affinity for guanine and cytosine bases. Increasing ionic strength diminishes the preferential binding of urea, glycine, sarcosine and glycine betaine, as a result of counterion release upon melting, which is dependent on the water activity and, hence, on co-solute concentration.

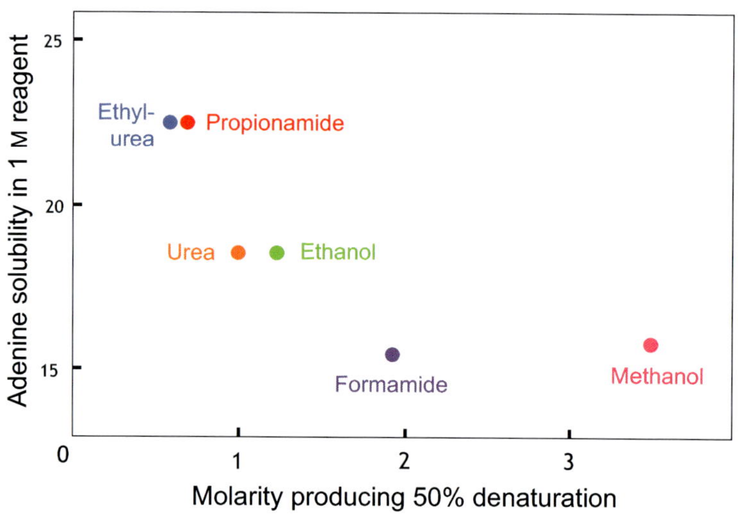

Figure 5-5. Correlation between the potency of six reagents in promoting DNA denaturation and their effects on the solubility of adenine. Data from L. Levine *et al.* (1963) *Biochemistry* **2**, 168.

Formamide ($H\text{–}CO\text{–}NH_2$) is used most frequently as a denaturant of nucleic acids, because it reduces the T_m of duplexes significantly at moderate concentrations, is readily miscible with water, has low optical absorbance in the UV region and does not react chemically with the DNA.

Some denaturants of nucleic acids, such as **formaldehyde** (H–CO–H), react chemically with the amino groups of the bases of the single-stranded form and prevent it regenerating the double helix. Formaldehyde can also produce single-strand breaks and cross-link nucleic acids to proteins (Figure 2-34).

Effect of ethylene glycol, urea, and N-methylated glycines on DNA thermal stability: the role of DNA base pair composition and hydration. L. J. Nordstrom *et al.* (2006) *Biochemistry* **45**, 9604–9614.

The denaturation transition of DNA in mixed solvents. B. Hammouda & D. Worcester (2006) *Biophys. J.* **91**, 2237–2242.

Effect of diethylsulfoxide on the thermal denaturation of DNA. S. A. Markarian *et al.* (2006) *Biopolymers* **82**, 1–5.

Enthalpies of DNA melting in the presence of osmolytes. C. H. Spink *et al.* (2007) *Biophys. Chem.* **126**, 176–185.

3. *pH*

The stability of standard nucleic acid duplexes is relatively insensitive to pH in the neutral pH region of 5–9 because none of the functional groups present in typical nucleic acids titrate in this region. Modified bases might, however, and the stability would then change. Below pH 5, or above pH 9, the standard duplex is destabilized because of titration of the polar groups on the bases (Figure 2-7). These groups are involved in the hydrogen bonding between pairs of bases, which is altered by their ionization, and the bases acquire a net charge in spite of being buried in the double helix. Low pH decreases the solubility, can cause depurination and strand breakage, and stabilizes triple helices consisting of pyrimidine–purine–pyrimidine and containing cytosine residues (Section 2.2.C.2). High pH, up to about 13, is more benign and is often used to denature nucleic acids, although depurination and strand breakage are still a hazard.

4. *Salt Effects*

The double helices of DNA and RNA are highly charged due to the ionized phosphate groups of the backbone. Consequently, these polyelectrolytes tend to bind cations to neutralize the charge density, and this condensation of counterions contributes to the conformation and stability (Section 2.3). Nonspecific 'territorial binding' or 'counterion condensation' occurs when the linear charge density of a polymer exceeds a critical value, and it produces an elevated cation concentration in the vicinity of the nucleic acid over that in the bulk solution. The local concentration of Na^+ ions on double-stranded DNA can be greater than 1 M, even in the absence of salt in the bulk solvent, and it reduces the effective charge on each phosphate group at pH 7 from –1 to –0.24. Both single- and double-stranded nucleic acids have such condensed counterions, although the linear charge density of the isolated single strands is less than that of the duplex, because the single strands are more free to elongate and diminish the charge density. Consequently, a duplex binds more cations that do the isolated strands. When the strands separate, the excess cations are released into the solution. Greater bulk concentrations of cations make this cation release increasingly unfavorable, so more energy is required to produce dissociation of the duplex, and it is stabilized by high salt concentrations. This effect depends upon the valence of the counterion, with higher valence counterions promoting greater stabilization than monovalent ions at the same concentration. Consequently **the stability of the double-stranded conformation depends markedly upon the ionic strength, and its melting temperature is increased as the ionic strength is increased**.

Anions tend not to interact directly with nucleic acids, due to charge repulsions, and they affect duplex stability only indirectly, according to the Hofmeister series.

The salt effects on the stabilities of natural nucleic acids are independent of their nucleotide sequence and content. The T_m is observed to vary linearly with the logarithm of the total salt concentration, C_s, and the slope is independent of base composition. Consequently, a wide variety of melting data on natural DNAs can be described by the empirical equation:

$$T_m = (16.6 \log[C_s/(1 + 0.7\ C_s)] + 41 f_{GC} + 81.5)\ °C \tag{5.3}$$

where f_{GC} is the fraction of bases that are guanine or cytosine.

Thermodynamic analysis of ion effects on the binding and conformational equilibria of proteins and nucleic acids: the roles of ion association or release, screening and ion effects on water activity. M. T. Record Jr *et al.* (1978) *Quart. Rev. Biophys.* **11**, 103–178.

The molecular theory of polyelectrolyte solutions with applications to the electrolyte properties of polynucleotides. G. S. Manning (1978) *Quart. Rev. Biophys.* **11**, 179–246.

5. *Prediction of the T_m*

The melting temperature of double-stranded DNA is directly proportional to its G+C content, i.e. the fraction of G•C and C•G base pairs (Figure 5-4). The melting temperature may be estimated at neutral pH in the presence of varying salt concentrations (indicated here as the concentration of sodium ion, $[Na^+]$) and with varying concentrations of formamide (indicated as the volume %, $\%_f$) for any normal DNA molecule from its length in nucleotides, L, and the mole fraction of G•C and C•G base pairs (f_{GC}):

$$T_m = (81.5 + 16.6 \log([Na^+]/(1 + 0.7[Na^+])) + 41 f_{GC} - 500/L - 0.63\ \%_f)\ °C \tag{5.4}$$

Such equations are valid only for L > 50 base pairs; with shorter strands, the strands separate more readily, so T_m will depend upon the concentrations of the hybridizing strands. Furthermore, the salt effects depend upon the exact length of a short oligonucleotide and its base composition. Equations such as 5.4 will also not be valid if any other components are present that interact differently with the single strands and the duplex.

The stabilities of DNA double helices are independent of G+C content in 3.2 M $(CH_3)_4N^+\ Cl^-$ or 2.4 M $(CH_3\text{-}CH_2)_4N^+\ Cl^-$, where the melting temperatures are 94°C and 63°C, respectively. In 5.6 M glycine betaine (Equation 5.2), which has no net charge, the melting temperature is independent of G+C and can be varied from 75°C down to 45°C by varying the KCl concentration from 1 M down to 0.01 M.

Reliable data are available to make reasonably accurate predictions of the T_ms of duplexes involving only short oligonucleotides. They involve primarily the experimentally determined parameters for the various pairs of neighboring base pairs in stabilizing the double helix (Table 5-1), including the enthalpy and entropy changes that permit calculation of the free energy at any temperature, and the effects of base mismatches (Table 5-2). The stabilizing effects of the base pairs must overcome the opposing contribution of the entropy loss on annealing, estimated to be +1.96 kcal/mol at 37°C. A self-complementary oligonucleotide will form intramolecular double helices, and consequently will

Table 5-1. Thermodynamic parameters for DNA nearest-neighbor Watson–Crick base pairs in 1 M NaCl at 37°C

Sequence[a]	ΔH° (kcal/mol)	ΔS° (cal/mol/°C)	ΔG° (kcal/mol)
AA/TT	−7.6	−21.3	−1.00
AT/TA	−7.2	−20.4	−0.88
TA/AT	−7.2	−21.3	−0.58
CA/GT	−8.5	−22.7	−1.45
GT/CA	−8.4	−22.4	−1.44
CT/GA	−7.8	−21.0	−1.28
GA/CT	−8.2	−22.2	−1.30
CG/GC	−10.6	−27.2	−2.17
GC/CG	−9.8	−24.4	−2.24
GG/CC	−8.0	−19.9	−1.84

[a] The slash indicates that the sequences are given in antiparallel orientations. For example, AC/TG means 5′–AC–3′ is base-paired with 3′–TG–5′.

Data from J. SantaLucia & D. Hicks (2004) *Ann. Rev. Biophys. Biomol. Structure* **33**, 415–440.

bind less tightly to another strand, by +0.43 kcal/mol at 37°C. The data in Table 5-2 demonstrate that the order of decreasing stability of the base pairs is: G•C > A•T > G•G > G•T > G•A > T•T > A•A > T•C > A•C > C•C. It is frequently assumed that all mismatches have the same effect on the stability of a heteroduplex, and a common 'rule of thumb' is that 1% of base-pair mismatches decrease the melting temperature by 1°C. The experimental data of Tables 5-1 and 5-2 demonstrate, however, that the different mismatches have effects that vary as much as 2700 in the equilibrium constant for duplex formation and 4.9 kcal/mol in free energy. The two extremes are GGC/CGG and ACT/TCA. Guanine is the most promiscuous base, as it forms the strongest normal pair (G•C) and the strongest mismatches. In contrast, cytosine is the most discriminating base, as it forms the strongest normal pair and the three weakest mismatches.

The thermodynamics of DNA structural motifs. J. SantaLucia & D. Hicks (2004) *Ann. Rev. Biophys. Biomol. Structure* **33**, 415–440.

Table 5-2. Increments in ΔG° (kcal/mol) for single internal matches and mismatches adjacent to Watson–Crick base pairs in 1 M NaCl at 37°C

		Y			
Sequence[a]	**X**	**A**	**C**	**G**	**T**
GX/CY	A	0.17	0.81	–0.25	**–1.30**
	C	0.47	0.79	**–2.24**	0.62
	G	–0.52	**–1.84**	–1.11	0.08
	T	**–1.44**	0.98	–0.59	0.45
CX/GY	A	0.43	0.75	0.03	**–1.45**
	C	0.79	0.70	**–1.84**	0.62
	G	0.11	**–2.19**	–0.11	–0.47
	T	**–1.30**	0.40	–0.32	–0.12
AX/TY	A	0.61	0.88	0.14	**–0.88**
	C	0.77	1.33	**–1.30**	0.64
	G	0.02	**–1.30**	–0.13	0.71
	T	**–0.88**	0.73	0.07	0.69
TX/AY	A	0.69	0.92	0.42	**–0.58**
	C	1.33	1.05	**–1.30**	0.97
	G	0.74	**–1.44**	0.44	0.43
	T	**–1.00**	0.75	0.34	0.68

[a] The slash indicates that the sequences are given in antiparallel orientations. For example, AC/TG means 5′–AC–3′ is base-paired with 3′–TG–5′. Watson–Crick pairs are in *bold*.

Data from J. SantaLucia & D. Hicks (2004) *Ann. Rev. Biophys. Biomol. Structure* **33**, 415–440

5.1.C. Double-stranded RNA

Double-stranded RNA is rare in nature but its stability can actually be somewhat greater than that of DNA of the same base composition (but thymine replacing uracil). It adopts an A-form helix rather than the B-form of DNA (Section 2.2.B). The dependence on base composition and formamide concentration is somewhat different, and the equation corresponding to Equation 5.4 is:

$$T_m = (78 + 16.6 \log([Na^+]/(1 + 0.7[Na^+])) + 70 f_{GC} - 500/L - 0.35\ \%_f)\ °C \qquad (5.5)$$

5.1.D. DNA•RNA Heteroduplexes

DNA•RNA heteroduplexes also adopt the A-type double helix, like RNA•RNA double helices. They are usually less stable than either homoduplex, except if they have high G+C content; their T_m may be estimated from the equation:

$$T_m = (67 + 16.6 \log[[Na^+]/(1 + 0.7[Na^+])] + 80\,f_{GC} - 500/L - 1.0\,\%_{f1} - 0.3\,\%_{f2})\ °C \qquad (5.6)$$

where $\%_{f1}$ is the volume percentage of formamide up to 20%, and $\%_{f2}$ is any additional volume percentage. DNA•RNA heteroduplexes are less sensitive to formamide than are the others.

Discovery of the hybrid helix and the first DNA–RNA hybridization. A. Rich (2006) *J. Biol. Chem.* **281**, 7693–7696.

5.1.E. Single-stranded Nucleic Acids

The structures of single-stranded nucleic acids depend crucially upon their nucleotide sequences, in particular their potential for complementary base pairing. Sequences with the potential for substantial base pairing will adopt compact folded structures, like those of transfer RNA and ribozymes (Section 4.2). Other single-stranded RNA or DNA molecules with no significant ability for base pairing between complementary segments will maintain a more unfolded, random coil-like structure. Long nucleic acid polymers with nonuniform sequences, however, invariably have segments with complementary sequences, simply by chance. These single-stranded molecules will consequently form double-stranded hairpins and loops, like those depicted in Figure 5-6. These structures will also unfold at high temperatures, in the presence of denaturants, or at extremes of pH.

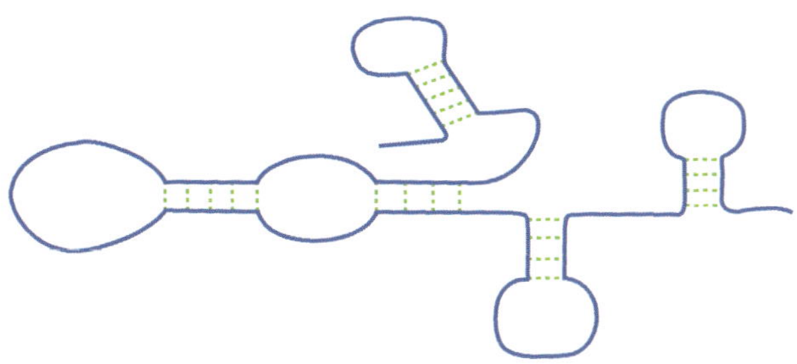

Figure 5-6. Intramolecular base-paired structures formed by nucleic acid single strands. The *long continuous line* is the backbone of the single strand, whereas the *dashed lines* represent complementary base pairs.

Even in the absence of sequence complementarity and base pairing, a wide variety of data indicate that single polynucleotide chains are not random coils (Section 1.3.A) at room temperature. Only at very high temperatures, approaching 100°C, do model polynucleotides approach the properties expected for a polynucleotide random coil. At lower temperatures, the adjacent bases tend to stack and the polynucleotide backbone tends to adopt a helical conformation like it would have in a double helix, but without the other strand. The most vivid illustration of such a structure is that determined by fiber diffraction of polycytosine (poly C). Although superficially similar to one strand of a double helix, the conformation of poly C has only six bases per turn, instead of the 10 or 11 for double helices. Similar structures in other single-stranded nucleic acids are apparent from their circular dichroism spectra.

In solution, each nucleotide of a single polypeptide chain appears to be in equilibrium between a random coil conformation and one in which its base is stacked with adjacent bases (Table 2-5). Such stacking is only slightly cooperative, in contrast to the helix–coil transition in polypeptides; the equivalent of the cooperativity parameter σ in the Zimm–Bragg model for α-helices is only slightly less than 1 with polynucleotides, whereas it is 10^{-2}–10^{-4} with polypeptide chains. The degree of stacking increases with length of the polynucleotide chain, but only up to about 10 nucleotides; beyond that, the stacking is essentially independent of length. The equilibrium between stacked and flexible conformations shifts with temperature, with flexible conformations predominating near 100°C and stacked conformations predominating only at temperatures lower than 0°C.

RNA denaturation: excluded volume, pseudoknots, and transition scenarios. M. Baliesi *et al.* (2003) *Phys. Rev. Lett.* **91**, 198102.

Random coil phosphorus chemical shift of deoxyribonucleic acids. C. N. Cho & S. L. Lam (2004) *J. Magn. Res.* **171**, 193–200.

Heat capacity changes in RNA folding: application of perturbation theory to hammerhead ribozyme cold denaturation. P. J. Mikulecky & A. L. Feig (2004) *Nucleic Acids Res.* **32**, 3967–3976.

---

1. Physical Stretching

If DNA is stretched physically, it undergoes a large, cooperative structural change at forces greater that 70 picoNewtons (pN), when it stretches to about 1.7 times its crystallographic length. If only the 3′-ends are pulled, the double helix is believed to unwind to a structure resembling a ladder. If only the 5′-ends are pulled, the helix persists but with a large inclination of the base pairs, a narrowing of the minor groove and a thinning of the double helix by roughly 30%. When the molecule is stretched to double its original length, the base pairs rupture and the two strands separate.

5.2. Unfolding and Refolding of Single-stranded RNA Molecules

The unfolding of folded single-stranded RNA molecules, such as transfer RNA and ribozymes (Section 4.2), differs from the denaturation of double helices described above in that it is a solely unimolecular process and does not result in separation of individual polynucleotide chains. It is similar, on the other hand, in that antiparallel double strands, formed by hairpins in this case, are major parts of the structure and can be unfolded in the same way. RNA molecules typically fold across a highly rugged energy landscape, with many energy peaks and minima, and correspondingly there are long-lived folding intermediates, multiple folding pathways and heterogeneous conformational dynamics. Stable secondary structures are believed to be at least partly responsible for the rugged energy landscape of RNA molecules.

Folded RNA molecules can be unfolded and will then refold spontaneously, so the information for their structures resides in their nucleotide sequences. Unfolding can be induced by a gradual change of temperature or a gradual change in the concentrations of the stabilizing Mg^{2+} ions or a denaturant like urea. Unfolding is not very cooperative in the case of RNAs, and intermediates with various structures predominate throughout the transition region between the fully folded and unfolded states. These are equilibrium intermediates, which should have the structures with the lowest free energies under the intermediate conditions. They are readily interconverted and so cannot be assumed to be the intermediates that occur during the kinetic process of unfolding or refolding. In general, the least denaturing conditions cause the tertiary structure to unfold, followed by the various elements of the secondary structure in order of their intrinsic stabilities. Unfolding of tertiary and secondary structures can usually be distinguished because the stability of only the tertiary structure is dependent upon the presence of Mg^{2+} ions. The intermediates usually correspond to isolated double helical hairpins, as they have significant stabilities in isolation (Table 4-2) without being incorporated into higher order structures. A single base pair can stabilize a hairpin by 1–3 kcal/mol, and the secondary structure of a group I self-splicing intron, with 129 base pairs, has been estimated to be 84.7 kcal/mol more stable than the fully unfolded molecule. This is the reason for the much lower cooperativity of unfolding transitions of RNA relative to proteins, where the secondary structure is not stable by itself.

The equilibrium thermal unfolding of a pseudoknot is illustrated in Figure 5-7. The folded pseudoknot contains an unpaired base, A15, between the two helices and a bulge, A35, in the lower helix. Because of the destabilizing effect of this bulge, the lower helix melts first, and the upper helix is then extended in the intermediate helix by pairing A15 with a previously inaccessible uracil base. At higher temperatures, the more stable helix unfolds.

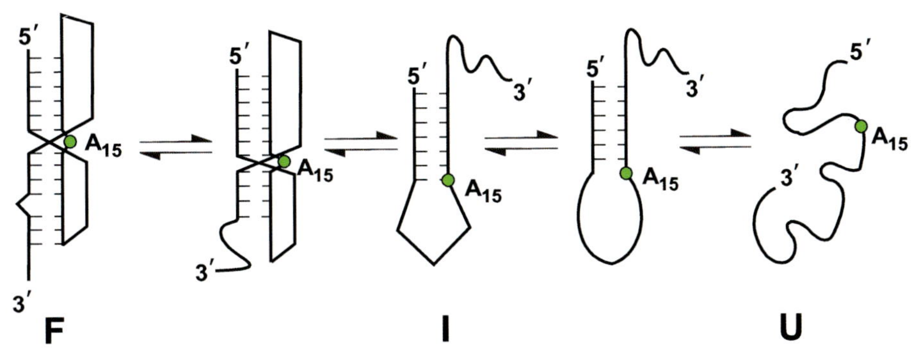

Figure 5-7. Thermal unfolding pathway at equilibrium of a pseudoknot. The *thick line* is the polynucleotide backbone, while the *short thin lines* are bases that are base-paired in the helices. F is the fully folded molecule at low temperatures, I the predominant intermediate, and U the unfolded form at high temperatures. The three intermediates were populated to detectable extents at intermediate temperatures. Nucleotide A15 is unpaired in the folded structure but becomes base-paired in the intermediate after unfolding of the lower helix. Data from C. A. Theimer & D. P. Giedroc (1999) *J. Mol. Biol.* **289**, 1283–1299.

In kinetic studies, the change in structure is caused by rapidly changing the conditions, and the course of the change is followed as a function of time in the new conditions. In this case, the kinetic pathway need not be reversible, and intermediates do not necessarily correspond to low free energy states. RNA molecules are observed to get trapped in metastable states from which it is difficult to escape, reflecting the ruggedness of the energy landscape for a large RNA molecule, with many local energy minima. The folding process occurs over a wide range of time scales: adjacent bases can stack in nanoseconds, individual helices can form in milliseconds, larger secondary structural domains that involve reorganization of certain helices can require seconds, while formation of the fully folded structure can take minutes. The local secondary structure, such as a hairpin loop, forms first, as the two halves of a helix are in close proximity in the unfolded polynucleotide chain. Longer range helices will form more slowly and may only form after previous folding of intermediary short-range helices brings together the two distant halves of the long-range helix. Progressively larger secondary structure domains are generated, and larger domains can form by rearranging and combining some of the smaller elements, but these involve substantial energy barriers between structures, as some helices have to be broken up before other more stable ones can be added (Figure 5-8). As the sizes of domains increase, the height of the energy barrier increases, as does the time required. In general, the transition state occurs early in forming the secondary structure, i.e. the transition state is close to the unfolded polynucleotide, whereas it occurs late in forming the tertiary structure: this latter transition state is close to the fully folded state. Finally, the structure can become frozen in one that does not have the lowest possible free energy but has resulted from the kinetic pathway. Many folded RNA molecules exist in alternative folded conformations. The larger the structure, the greater the discrepancy with the predicted structure of minimum free energy.

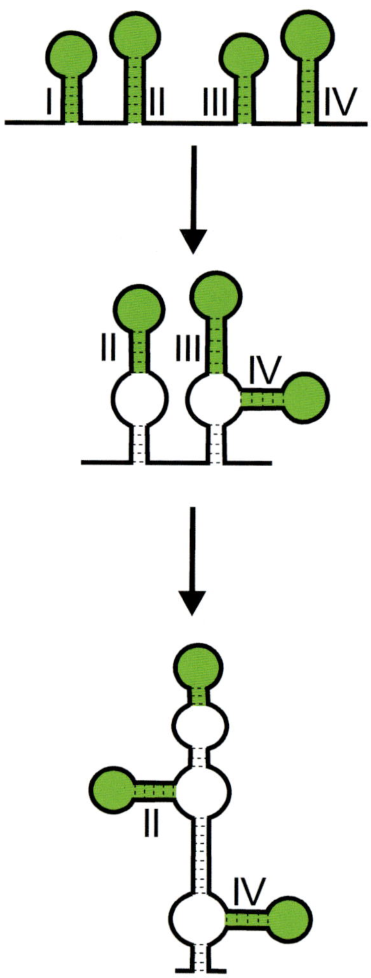

Figure 5-8. Schematic representation of the reorganization of the secondary structure during RNA folding that leads to the formation of progressively larger domains, with base pairing between segments more distant in the sequence. The *thick line* is the polynucleotide backbone, while the *dashed lines* between them represent base pairs in the stem. The *colored* stems have the helices made of segments adjacent in the sequence.

Strategies for RNA folding and assembly. R. Schroeder *et al.* (2004) *Nature Rev. Mol. Cell. Biol.* **5**, 908–919.

RNA folding and unfolding. B. Onoa & I. Tinoco (2004) *Curr. Opinion Struct. Biol.* **14**, 374–379.

Mg^{2+}–RNA interaction free energies and their relationship to the folding of RNA tertiary structures. D. Grilley *et al.* (2006) *Proc. Natl. Acad. Sci. USA* **103**, 14003–14008.

Concordant exploration of the kinetics of RNA folding from global and local perspectives. L. W. Kwok *et al.* (2006) *J. Mol. Biol.* **355**, 282–293.

RNA folding during transcription. T. Pan & T. Sosnick (2006) *Ann. Rev. Biophys. Biomol. Struct.* **35**, 161–175.

5.2.A. Transfer RNA Unfolding/Refolding

The unfolding and refolding of transfer RNA molecules has been characterized extensively. The process depends critically upon the salt concentration, particularly the presence of Mg^{2+} ions, which are bound at a few specific locations in the folded tertiary structure; free Mg^{2+} ions stabilize the structure. At high Mg^{2+} concentrations, the tertiary structure is disrupted only at high temperatures, and the remainder of the structure, the secondary structure of double-helical stems, unfolds rapidly thereafter; consequently, **unfolding at high Mg^{2+} concentrations appears to be a single-step, all-or-nothing, cooperative process**. At lower Mg^{2+} concentrations, however, the tertiary structure unfolds much more readily; under such less severe conditions, the various elements of the secondary structure, which include the double-helical stems, but with mismatches, loops and bulges, are sufficiently stable to remain folded, at least transiently. These elements of secondary structure then unfold relatively independently of each other.

Refolding occurs by the reverse of the unfolding process. The individual elements of secondary structure form, and then they interact via unpaired bases to generate the folded tertiary structure. The D-helix is formed in the overall rate-limiting step, so it serves as a nucleus for further folding, which is much faster. It also provides a scaffold for strong Mg^{2+} binding sites. The D-helix is involved in several tertiary interactions with the remainder of the molecule (Figure 4-17). The loop conformations and their interactions can change upon forming the tertiary structure, and a single base pair may change, but the helices do not. In these cases, the secondary structure in the tertiary structure is that which is most intrinsically stable in isolation, and the interactions stabilizing the tertiary structure are significantly weaker. In other cases, however, particularly ribozymes (Section 4.2.C), the secondary structure formed initially is not the same as that in the final tertiary structure, so **rearrangements of the secondary structure can take place during the tertiary folding process**. These can be substantial: in one case, six base pairs are broken and six new ones are formed, a tetraloop is disrupted, and a single-nucleotide bulge is introduced. The rearranged secondary structure is less intrinsically stable than that formed initially, but this is compensated for by the interactions stabilizing the tertiary structure.

Internally mismatched RNA: pH and solvent dependence of the thermal unfolding of $tRNA^{Ala}$ acceptor stem microhairpins. E. Biala & P. Strazewski (2002) *J. Am. Chem. Soc.* **124**, 3540–3545.

Study of the influence of metal ions on $tRNA^{Phe}$ thermal unfolding equilibria by UV spectroscopy and multivariate curve resolution. M. Vives *et al.* (2002) *J. Inorg. Biochem.* **89**, 115–122.

Highly conserved modified nucleosides influence Mg^{2+}-dependent tRNA folding. K. N. Nobles *et al.* (2002) *Nucleic Acids Res.* **30**, 4751–4760.

5.2.B. Ribozyme Unfolding/Refolding

The kinetic folding pathway of the *Tetrahymena* ribozyme (Figure 5-9) includes both Mg^{2+}-dependent and Mg^{2+}-independent steps. The structures of the intermediates on the folding pathway were determined by rapidly hybridizing their single-stranded segments with complementary oligodeoxyribonucleotides, then digesting them with **ribonuclease H**, which is specific for DNA•RNA heteroduplexes. Only segments that were not base paired survive this treatment intact. Both the hybridization and cleavage steps were more rapid than the rates of the folding steps being analyzed. One of the two main structural domains, P4–P6, forms first, and the overall rate-limiting step is a Mg^{2+}-dependent rearrangement preceding stable formation of the second main structural domain, P3–P7. The P3 and P7 helices form in an interdependent manner, and the two domains can be considered kinetic folding units. The rate-limiting step is formation of the triple-helical scaffold, involving strands entering and leaving the shallow groove of P6 and the deep groove of P4. **The intermediates that accumulate kinetically during folding possess some but not all of the fully folded structure**. In particular they have most of the secondary structure and at least some local cooperativity.

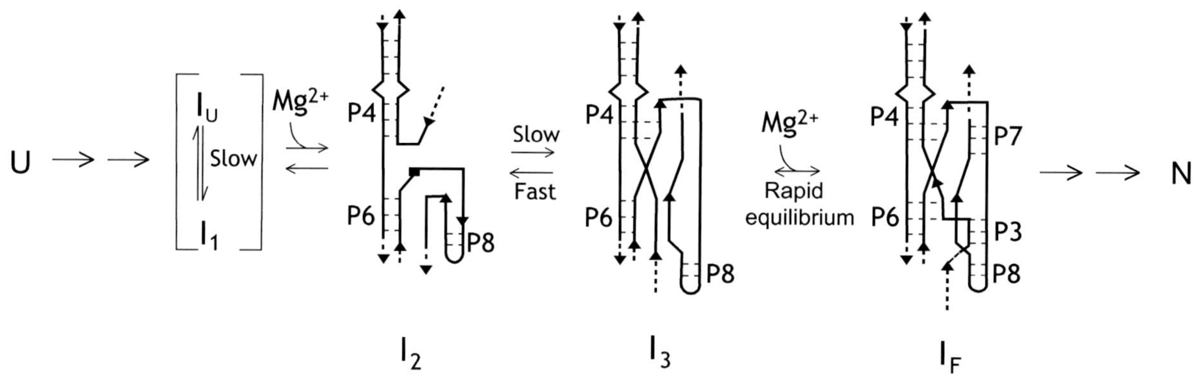

Figure 5-9. Kinetic folding pathway elucidated for the *Tetrahymena* ribozyme. The rate-limiting step in folding is between intermediates I_2 and I_3 and corresponds to the formation of the triple-helical scaffold at the junction between the stacked helices P4 and P6. The two main structural domains of these ribozymes are P4–P6 and P8–P3–P7. Only some of the steps involve binding of magnesium ions. Data from P. P. Zarrinkar & J. R. Williamson (1996) *Nature Struct. Biol.* **3**, 432–438.

Distinct parallel pathways lead the RNA to its native form. The structures of the intermediates populating the pathways are not affected by variation of the concentration and type of background monovalent ions, but they are altered by a mutation that destabilizes one domain of the ribozyme. Starting from different conformational ensembles, but folding under identical conditions, indicated that **the electrostatic environment modulates the molecular flux through different pathways, but the initial conformational ensemble determines the partitioning of the flux**. Misfolding occurs frequently, and kinetically trapped nonproductive intermediates occur. One quasi-native intermediate is only slowly converted to the correctly folded structure.

The folding process of a 115-nucleotide sequence known as SV-11 demonstrates the occurrence of a metastable intermediate (Figure 5-10). The sequence of this RNA is almost a palindrome, and the most stable structure is a single, almost perfect, hairpin. Folding proceeds, however, through a metastable structure involving only local hairpins. The local hairpins form much more readily and rapidly than the single double helix of the hairpin, which would require bringing together the opposite ends of the molecule. The nucleotides that make up the loop in the final hairpin are not involved in such a loop in the metastable intermediate.

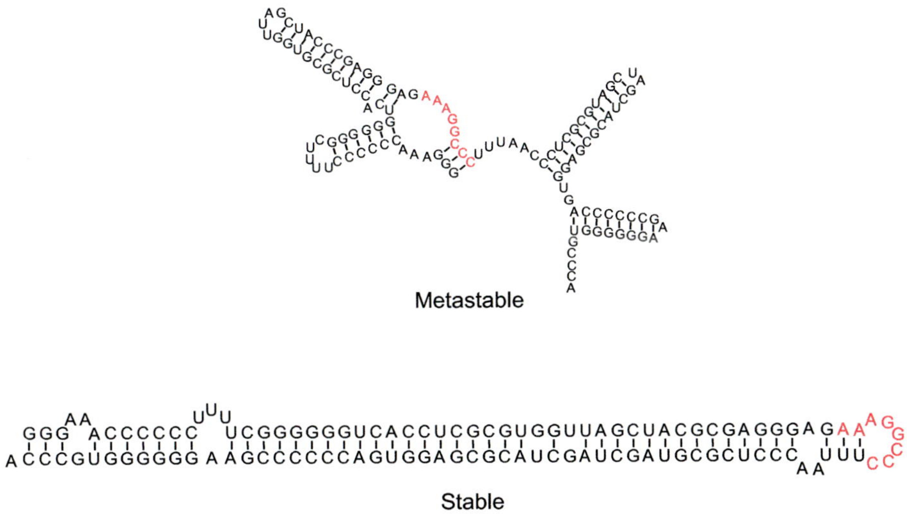

Figure 5-10. The stable and kinetic metastable structures of one strand of the double-stranded RNA molecule SV-11. Its sequence is a nearly perfect palindrome, and the most stable structure is a single, almost perfect, hairpin. The nucleotides in the loop of this structure, –AAAGGCCC– (*red*), have a different structure in the metastable structure. Data from C. K. Biebricher & R. Luce (1992) *EMBO J.* **11**, 5129–5139.

Perturbation of the hierarchical folding of a large RNA by the destabilization of its scaffold's tertiary structure. I. Shcherbadova & M. Brenowitz (2005) *J. Mol. Biol.* **354**, 483–496.

The paradoxical behavior of a highly structured misfolded intermediate in RNA folding. R. Russell *et al.* (2006) *J. Mol. Biol.* **363**, 530–544.

Distinct contribution of electrostatics, initial conformational ensemble, and macromolecular stability in RNA folding. A. Laederach *et al.* (2007) *Proc. Natl. Acad. Sci. USA* **104**, 7045–7050.

5.2.C. Unfolding Using Mechanical Force

Unfolding by pulling the two ends of an RNA molecule apart is simpler than thermal unfolding, because the helices tend to unfold by breaking the base pairs at the end furthest from the loop and closest to the ends being pulled. The folding-energy landscape can be manipulated to control the fate of an RNA molecule: individual RNA molecules can be induced into either native or misfolding pathways by modulating the relaxation rate of the applied force, and they can even be redirected during the folding process to switch from misfolding to native folding pathways.

Mechanical unfolding and refolding of single RNA molecules were originally studied using optical traps attached to the ends of the polynucleotide chain to change suddenly the extension of the molecule. Two methods were traditionally used: 'force-ramp', with the applied force changing continuously, and 'hopping', when the force is held constant and the molecule jumps spontaneously between two different states. The rates of unfolding and refolding are measured directly, but only over very narrow ranges of forces. With a third, 'force-jump', method, the applied force is rapidly stepped to a new value and either the unfolding or refolding event is monitored through changes in the molecular extension. This makes it possible to measure the unfolding and refolding rates independently over a wider range of forces. A 52-nucleotide RNA hairpin with a three-nucleotide bulge exhibits very similar unfolding kinetics and estimated Gibbs free energies when studied using all three methods. The RNA unfolds and refolds without detectable intermediates in constant-force conditions (hopping and force-jump) but it exhibits partially folded intermediates in force-ramp experiments at higher unloading rates. Consequently, folding of even simple RNA hairpins can be more complex than a simple single-step reaction.

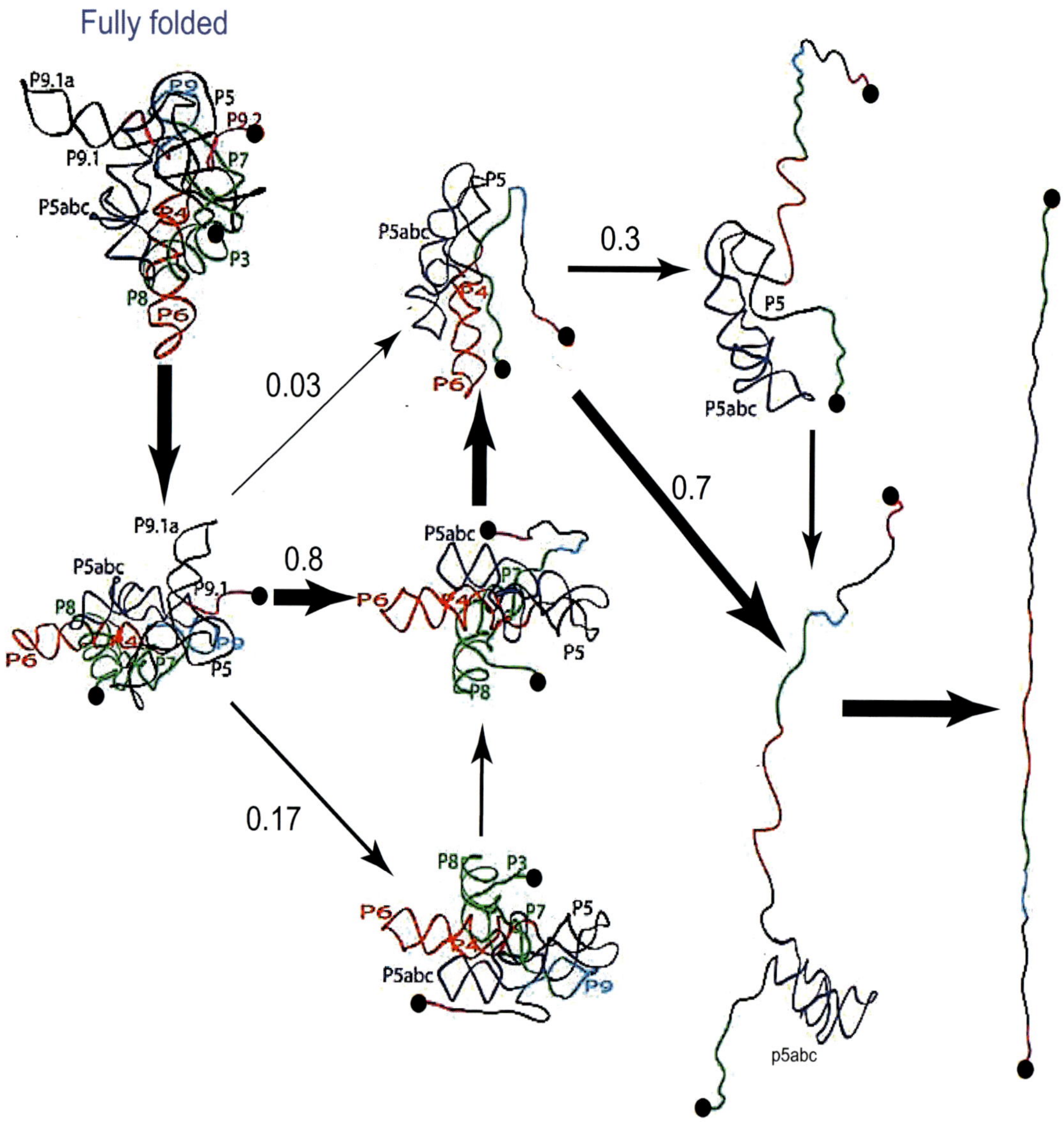

Figure 5-11. Pathway of unfolding of the *Tetrahymena* ribozyme when the two ends are pulled apart mechanically. The flux through the pathway is indicated by the thicknesses of the arrows; the numbers associated with some arrows indicate the probability of that step being followed. Data from C. A. Bustamante (2005) *Quart. Rev. Biophys.* **38**, 291–301.

The *Tetrahymena* ribozyme, with about 400 nucleotides, unfolds through eight intermediates (Figure 5-11). The pathway of unfolding differs for each measurement, presumably because the thermal fluctuations responsible for crossing the kinetic barriers are stochastic. The intermediates were identified by studying mutant forms and subdomains of the RNA, and by adding complementary oligonucleotides that compete with secondary and tertiary interactions by hybridizing to single-stranded segments. The kinetic barriers are caused primarily by tertiary interactions stabilized by Mg^{2+} ions. In the absence of Mg^{2+}, the kinetic barriers disappear and no intermediates are detected.

Single-molecule RNA folding. G. Bokinsky & X. Zhuang (2005) *Acc. Chem. Res.* **38**, 566–573.

Determination of thermodynamics and kinetics of RNA reactions by force. I. Tinoco *et al.* (2006) *Quart. Rev. Biophys.* **39**, 325–360.

Probing the mechanical folding kinetics of TAR RNA by hopping, force-jump, and force-ramp methods. P. T. Li *et al.* (2006) *Biophys. J.* **90**, 250–260.

Modelling RNA folding under mechanical tension. J. R. Vieregg & I. Tinoco (2006) *Mol. Phys.* **104**, 1343–1352.

Force unfolding kinetics of RNA using optical tweezers. I. Effects of experimental variables on measured results. J. D. Wen *et al.* (2007) *Biophys. J.* **92**, 2996–3009.

Real-time control of the energy landscape by force directs the folding of RNA molecules. P. T. Li *et al.* (2007) *Proc. Natl. Acad. Sci. USA* **104**, 7039–7044.

5.3. RENATURATION, ANNEALING AND HYBRIDIZATION

Two nucleic acid strands with complementary base sequences can associate to generate a double helix under the appropriate conditions. This can occur after denaturation of a sample of double-stranded nucleic acid, known as renaturation, or upon mixing two single strands, hybridization. Another variation is adding a single-stranded molecule, often a short oligonucleotide, with a sequence complementary to one part of the target molecule to a mixture of denatured nucleic acid, to compete with reassociation of the original strands.

Base pairing between complementary strands can be either a relatively straightforward or a very complex process, depending upon the length and sequences of the polynucleotide chains involved. For short oligonucleotides with totally complementary sequences, the reassociation process is fast and complete. When a short double-stranded oligonucleotide is denatured and then re-annealed under nondenaturing conditions, the optical and hydrodynamic properties return quickly to their original values. The rate is determined by the concentrations and frequency of encounter of the complementary strands; once the first base pairs are formed, the pairing of the remaining bases is rapid. For longer chains, however, the process is much more complex, slower and does not necessarily go to completion, because a variety of structures and complexes can be generated initially (Figure 5-6) and hinder the search for the correct complementary partner.

The denaturation of synthetic polymers with repeating sequences is fully reversible when measured by the change in UV absorbance, which indicates that most bases are in intact base pairs in the regenerated double helix. There is hysteresis, however, in the loss and regain of the hydrodynamic properties, in that renaturation does not follow the reverse of the pathway of denaturation; this is due to the formation of branched structures comprising hairpin loops and concatenated aggregates during renaturation. Very long natural DNAs display hysteresis in both their optical and hydrodynamic properties, indicating a failure of a significant fraction of the bases to find their correct partner base and the presence of nonlinear structures.

Before they can base pair, two complementary strands must encounter each other by diffusion, but the presence of other segments with partially complementary sequences will compete. When two long nucleic acid polymers encounter each other, there is only a very small probability of an exact match of many of the base pairs in the initial complex. Instead, small stretches of complementary duplex are formed initially:

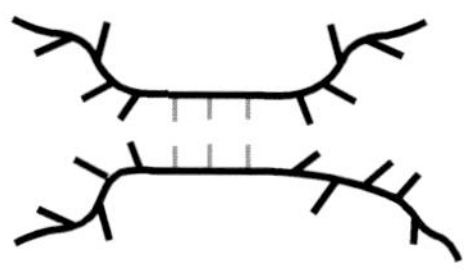

(5.7)

The two single strands encounter each other in a bimolecular reaction, so the rate is proportional to their concentrations, and just a few adjacent bases form base pairs. These complexes may dissociate, as they are relatively unstable, or they may grow if the neighboring residues are also complementary:

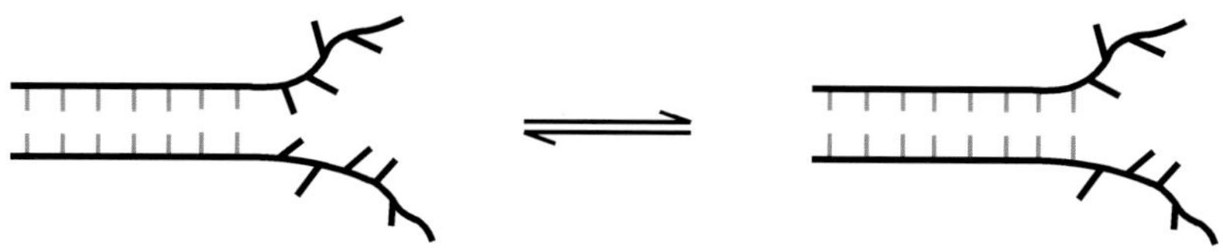

(5.8)

Zippering of the remainder of the double helix can be very rapid, as it is essentially a unimolecular process. If the remainder of the sequences are not complementary, the initial duplexes formed must dissociate and wait for more productive encounters; this depends upon the temperature being sufficiently high to melt short sequences, but low enough to permit duplex formation. Otherwise, at low temperatures, the unproductive complexes will persist and the most stable base pairing will not be found.

Intramolecular hairpins and internal loops present in the single-stranded nucleic acids (Figure 5-6) present a significant impediment to annealing to another strand. These structures are formed intramolecularly, and thus can be generated much more rapidly than can duplex structures. The first-order rate constant for hairpin formation is 10^5–10^7 s^{-1}, so such structures are formed on the microsecond time scale. The second-order rate constant for oligonucleotide duplex formation is 10^5–10^7 M^{-1} s^{-1}; at typical concentrations used in such experiments, duplexes form generally no faster than in seconds, and often much slower. The competing structures must be unfolded for the annealing process to proceed towards formation of the complementary duplex. The activation energy for such unfolding is large, so they do not dissociate rapidly at temperatures much below the T_m.

Within the double-stranded segments formed in the initial duplexes, defects can occur in the alignment of the base sequences (Figure 5-12), including hairpins and internal loops, as well as mismatches and bulges. Here the activation energy to change them is even greater than for single-stranded structures. These regions may form nucleation complexes that cannot be extended, resulting in large numbers of nonproductive intermediate states. In highly repetitive sequences, large numbers of such structures may be formed initially, but in this case they are readily interconvertible and the proper sequence alignment can be achieved.

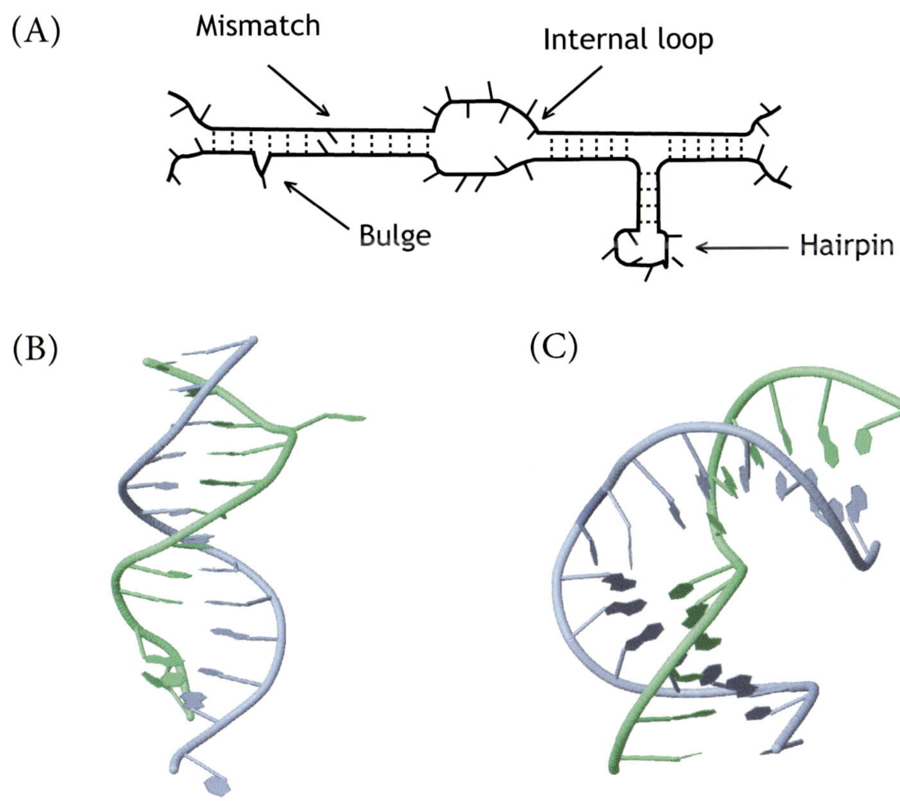

Figure 5-12. Structures of misaligned sequences in nucleic acid duplexes. (A) Schematic illustration of the types of structures formed. The *long continuous line* is the polynucleotide backbone, whereas the *short lines* indicate the bases, most of which are paired in complementary base pairs. (B, C) Crystallographic structures of (B) a single-nucleotide bulge at the *top* and a mismatch at the *bottom*, and (C) a four-nucleotide bulge. (B) and (C) generated from PDB files *1d31* and *1jst* using the program Jmol.

A common strategy to renature nucleic acids efficiently is to incubate them at the T_m of the desired duplex, followed by slow cooling to permit the facile reorganization of the duplex regions, to optimize their sequence alignment. The T_m of the desired duplex may be estimated from the empirical Equations 5.3–5.6. The T_m can be adjusted by manipulating the salt concentration and the amount of formamide; a given T_m can be obtained by many different combinations of the two.

With a short oligonucleotide that is complementary to one segment of a much longer single-stranded nucleic acid, a duplex can form without appreciable competing structures. The process is still hindered, however, by any hairpin structures in the long polynucleotide (Figure 5-6). The two processes differ in that duplex formation is bimolecular, while hairpin formation is unimolecular, so they can be distinguished experimentally by the concentration dependence. The hairpin structure will be favored at low concentrations of the strands, the duplex at high concentrations. The salt concentration also affects these equilibria; the hairpin is favored at low salt concentrations due to its lower charge density. Therefore a combination of salt and oligonucleotide concentrations can be used to generate the desired structure.

Formation of a duplex involves two nucleic acid strands, so their concentrations and lengths also influence the stability of the duplex. There is a complex interplay between length and concentration in determining the stability. The initial nucleation event (Equation 5.7), in which the strands make initial productive contacts, is a bimolecular process that depends upon the strand concentrations. Subsequent base pairing is propagated (Equation 5.8) in a zippering process, in which the nucleated base pairing is extended, which is a pseudo-unimolecular process and occurs with a rate constant of 10^7 s^{-1}. Formally, more than one strand is involved, but the formation of a closed base pair adjacent to a preformed base pair is effectively a monomolecular process analogous to extension of a hairpin.

For short oligonucleotides, the nucleation event dominates complex formation. The concentration dependence of the stability is simple and described adequately by conventional mass action for a biomolecular reaction. Initial encounters probably involve only one or two adjacent base pairs, but this *encounter complex* will be very unstable and dissociate rapidly, unless the adjacent nucleotides can form base pairs as well. If the two sequences are sufficiently complementary, the duplex will extend by a zippering reaction. The rate-limiting step is believed to be formation of the second or third base pair. The apparent rate constant is approximated by:

$$k_2 = k'_N (L_S)^{1/2} / N \tag{5.9}$$

where L_s is the length of the shortest strand participating in duplex formation, N is the total number of base pairs present in nonrepeating sequences, and k'_N is the rate constant for forming the initial nucleus. The value of k'_N is approximately 4×10^5 M^{-1} s^{-1} under optimal conditions. The dependence of k_2 on the square root of L_s rather than L_s is due to the nucleation sites becoming less available as the length of the single strand increases. The value of k_2 is zero at temperatures at or above the melting temperature, because the nucleation complex is not stable at high temperatures but increases at lower temperatures, until it reaches a broad maximum at a temperature about 25°C lower than the melting temperature (Figure 5-13). The rate decreases at lower temperatures, gradually for homopolymers with simple sequences but drastically for normal nucleic acids, due to intramolecular base pairing and the consequent decrease in availability of nucleation sites. The rate of hybridization is maximal at high salt concentrations and gradually decreases at monovalent cation concentrations less than 1 M. Mismatched base pairs affect the stability of the final duplex, but not the rate of forming it, so long as fewer than 10% of the base pairs are mismatched, consistent with the rate-limiting nucleation event involving only about two standard base pairs. It must be remembered that a unique sequence in a large genome (such as the 3 billion base pairs in the human genome) will be present at an extremely low concentration, so hybridization reactions with nucleic acids from an entire genome may reach equilibrium only extremely slowly.

At optimal temperatures, k'_N is about 20–50% smaller for DNA•RNA hybrid formation than for DNA duplexes, depending upon the degree of RNA secondary structure. The rates of forming RNA•RNA duplexes are similarly reduced.

For very long complexes, nucleation accounts for a relatively small part of the total stabilizing free energy, and the concentration dependence of complex formation disappears for long chains. For complexes of intermediate length, both nucleation and propagation make significant contributions to complex formation and stability; the concentration dependence of their stability is reduced relative to that for short complexes of short oligonucleotides, but it does not vanish.

The stability of a long duplex depends upon the fraction of G•C and C•G base pairs (Figure 5-4) but appears to be independent of the actual sequence (Equations 5.3–5.6). This is an illusion, however, that applies only to long duplexes where sequence effects are averaged out over the polymer. For short oligonucleotides, the actual sequence must be taken into account.

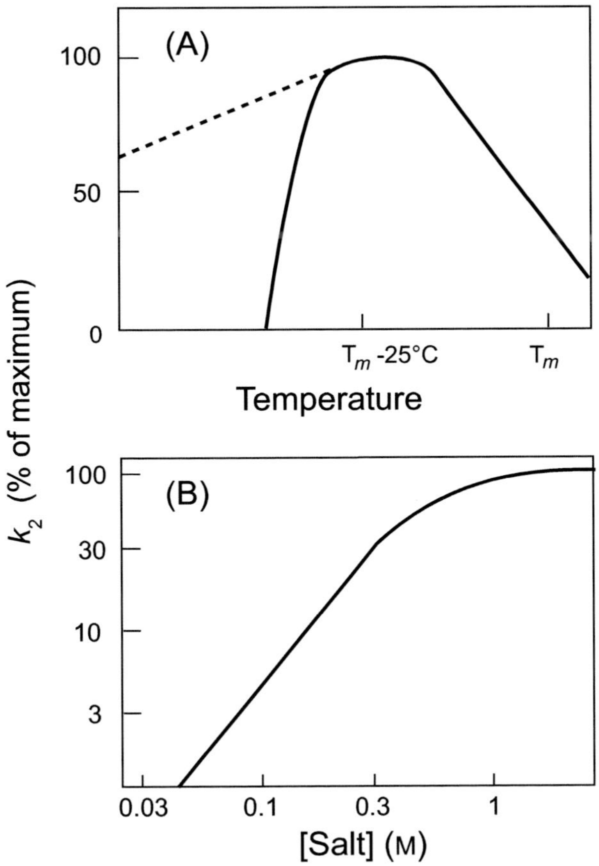

Figure 5-13. Dependence of the rate of hybridization of two complementary nucleic acid strands, k_2, on the temperature (A) and concentration of a monovalent cation (B). The melting temperature of the resulting duplex is T_m. The *solid curves* are those observed with complex natural nucleic acids with very diverse nucleotide sequences, whereas the *dashed curve* in (A) is observed with homopolymers with simple sequences that do not form intramolecular base pairs.

The power of hybridization techniques comes from the dependence of the stability of the hybrid duplex on the degree of complementarity of the constituent strands. Hybridization of nucleic acid strands of different origins produces heteroduplexes *in vitro*, and their stability is a measure of their degree of sequence similarity. The greater the degree of sequence complementarity, the greater the stability of the heteroduplex, so measurement of its stability, by varying the stringency, can give an indication of the degree of sequence identity. When 1% of the bases are mismatched in long strands, the T_m is generally decreased by approximately 1°C, although in short strands this depends upon the nature of the mismatches and their neighboring bases (Table 5-2). There are occasions when the exact sequence of the target DNA or RNA is not known, for example when the sequence of a gene is inferred from the amino acid sequence of its protein product. In this case, the base inosine can be used at the uncertain positions in the probe, as it has less stringent base-pairing criteria than adenine, cytosine, guanine and thymine.

Two single-stranded nucleic acid polymers with at least partial complementarity will form duplexes with stretches of double helix, corresponding to the regions of nucleotide sequence complementarity, interleaved with unpaired stretches, corresponding to noncomplementary regions. Two types of singe-stranded loops are introduced by noncomplementarity in the heteroduplex:

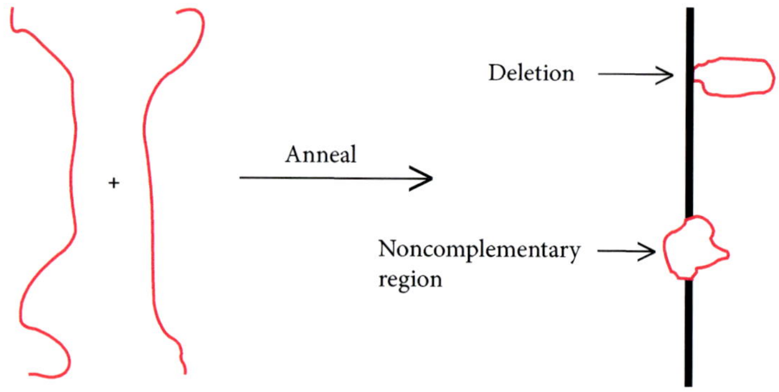

(5.10)

Regions where the bases do not match will produce a loop of nonbase-paired segments in both strands. The smallest loops and those with the same number of nucleotides on each strand are the most stable. On the other hand, the presence of extra nucleotides in only one of the strands, or a deletion in the other, will produce a loop in only that strand.

The thermodynamics of DNA structural motifs. J. SantaLucia (2004) *Ann. Rev. Biophys. Biomolec. Structure* **33**, 415–440.

Statistical thermodynamics and kinetics of DNA multiplex hybridization reactions. M. T. Horne *et al.* (2006) **91**, 4133–4153.

Secondary structure effects on DNA hybridization kinetics: a solution versus surface comparison. Y. Gao *et al.* (2006) *Nucleic Acids Res.* **34**, 3370–3377.

Electrostatic free energy landscapes for nucleic acid helix assembly. Z. J. Tan & S. J. Chen (2006) *Nucleic Acids Res.* **34**, 6629–6639.

Influence of secondary structure on kinetics and reaction mechanism of DNA hybridization. C. Chen *et al.* (2007) *Nucleic Acids Res.* **35**, 2875–2884.

5.3.A. Competing Intramolecular Structures in Individual Single Strands

Formation of a duplex between two single-stranded DNA molecules will be in competition with formation of any intramolecular structures. The most likely intramolecular structures are hairpins, which will be formed whenever the nucleotide sequence has segments with inverted complementarity. They can be formed very rapidly, much more rapidly than the two-stranded duplex under most conditions, so the intramolecular structure will form first and must then be undone in order to generate the duplex, which will be more stable if the two strands are truly complementary. For example, the single-stranded 13-mer 5′-d(GC*T*ACGTAGT*C*GC) folds to form a hairpin structure with a highly stabilizing central cGNAg loop (*bold*) that accommodates a CT mismatched base pair within the double-stranded stem region (*italics*):

```
  / A— G—T— C—G—C
T       •  •     •  •
  \ G—C —A—T— C—G
```

(5.11)

This loop is especially stable because the guanine and adenine bases form a 'sheared' base-pair conformation stabilized by two hydrogen bonds, and it is stabilized by the base-paired stem as well, so the hairpin of loop and stem is a cooperative structure. The hairpin is in equilibrium with the double-stranded duplex form, which is slow on the nuclear magnetic resonance (NMR) time scale (1–2 s^{-1} at 35°C). The hairpin folds very rapidly (>500 s^{-1}), much more rapidly than the duplex is formed, so it produces a 'kinetic overshoot' that is followed by a much slower equilibration to the equilibrium mixture of hairpin and double-helical duplex, with a rate constant of approximately 0.13 s^{-1} at 25°C at the concentrations used.

Melting of a DNA hairpin is not a cooperative two-state process but involves numerous intermediate structures that are compact, with bases stacked but not base-paired. The melting of such stacking interactions occurs on a time scale of 700 ps to 2 ns, while dissociation of base pairs in the helical segment requires microseconds or longer.

The initial step of DNA hairpin folding: a kinetic analysis using fluorescence correlation spectroscopy. J. Kim *et al.* (2006) *Nucleic Acids Res.* **34**, 2516–2527.

Direct measurement of the full, sequence-dependent folding landscape of a nucleic acid. M. T. Woodside *et al.* (2006) *Science* **314**, 1001–1004.

Structure and folding dynamics of a DNA hairpin with a stabilising d(GNA) trinucleotide loop: influence of base pair mis-matches and point mutations on conformational equilibria. G. D. Balkwill *et al.* (2007) *Org. Biomol. Chem.***5**, 832–839.

DNA folding and melting observed in real time redefine the energy landscape. H. Ma *et al.* (2007) *Proc. Natl. Acad. Sci. USA* **104**, 712–716.

5.3.B. C_0t and R_0t Curves

Monitoring the rate of reassociation of double-stranded DNA is a classic method for evaluating the size and relative sequence complexity of genomic DNA. The duplex DNA of the entire genome is fragmented to moderate-sized pieces, approximately 400 nucleotides long, by physical shearing or cleavage by endonucleases, and the fragments are denatured thermally to produce single-stranded fragments. The reassociation of these fragments is then monitored as a function of time. Being a second-order reaction, the extent to which two complementary DNA strands reassociate should be proportional to C_0t, where C_0 is the initial concentration of nucleotides (not of strands) and t is the time; the degree of reassociation is plotted versus C_0t (Figure 5-14).

If the sequences of all the initial DNA fragments were identical, complementary fragments would encounter each other readily and the mixture would reanneal rapidly, at a low value of C_0t. At the other extreme, reannealing would be very slow and occur only at high C_0t values if the original DNA contained many different sequences; in this case, the concentrations of complementary strands are very low and they encounter each other only infrequently. For a given quantity of DNA, a small genome will have greater concentrations of the same segment than a large genome. The value of C_0t at which the fractional extent of reassociation is 0.5 is used as a measure of the complexity of the genome. One complication with randomly fragmented DNA, with different breakage points, is that complementary segments may hybridize but have sequences at one or both ends that are not present in the other.

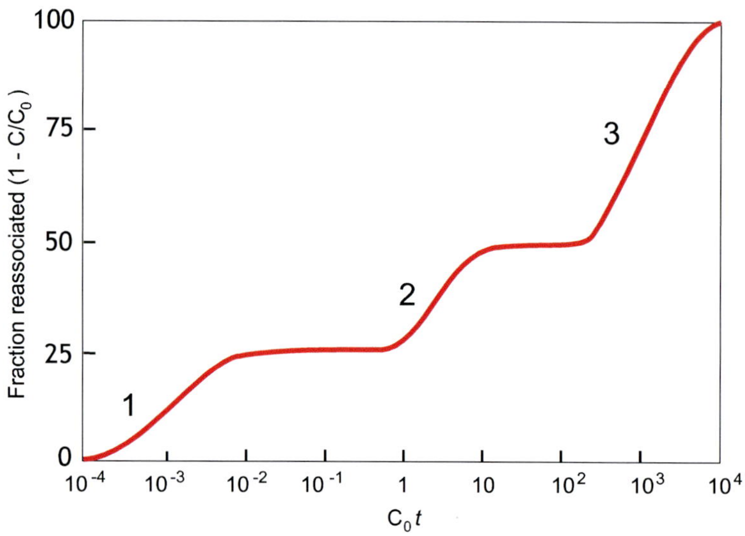

Figure 5-14. Idealized C_0t plot for a DNA that has three distinct components. The first consists of 25% of the DNA that contains about a million copies of repeated DNA and thus reassociates rapidly. The second consists of another 25% that contains 100–1000 copies of repeated DNA. The remaining 50% of DNA consists of single copies of unique sequences; it reassociates most slowly, because the concentrations of the complementary fragments are the lowest.

Such studies on genomic DNA provided the first indications that genomes often contain repetitive DNA, in which the same sequence is present in many copies in the genome. In the fragmented genomic DNA, such DNA segments will be present at higher concentrations than those segments that are present in only one copy per genome, and this fraction of the DNA reanneals at a correspondingly lower C_0t value. The results indicate that different sequences are present in widely varying frequencies throughout the genome, with only a fraction occurring only once per genome.

Hybridization of DNAs from different sources is a ready measure of the average degree of their sequence identity and their genetic and evolutionary relatedness.

A similar approach characterizes the quantity of a specific messenger (m) RNA in a sample by its rate of hybridization to its DNA gene. The DNA gene is radiolabeled and denatured to single strands. An excess of mRNA is then added. The excess mRNA minimizes reannealing of the DNA strands, causes its concentration to remain constant, and causes the hybridization reaction to follow pseudofirst-order kinetics. The amount of DNA•RNA hybrid is measured by separating it from the free single-stranded DNA by hydroxyapatite chromatography or by digesting the single-stranded DNA using S1 nuclease, which does not degrade DNA•RNA hybrids. The time when half of the DNA strands have been converted to DNA•RNA hybrids, $t_{1/2}$, will be inversely proportional to the second-order rate constant for the hybridization, k, and the original concentration of the mRNA of interest, R_0:

$$t_{1/2} = \frac{\log_e 2}{R_0 k} \tag{5.12}$$

The rate constant k is proportional to the size of the mRNA, and its value can be obtained from comparison with the rates of association of comparable pure RNAs and DNAs. Knowing the value of k, the value of R_0 for any sample of mRNA can be determined using Equation 5.12 simply by measuring $t_{1/2}$.

Distortion of quantitative genomic and expression hybridization by Ct-1 DNA: mitigation of this effect. H. L. Newkirk *et al.* (2005) *Nucleic Acids Res.* **33**, e191.

Use of competitive DNA hybridization to identify differences in the genomes of bacteria. O. C. Shanks *et al.* (2006) **66**, 321–330.

DNA–DNA hybridization values and their relationship to whole-genome sequence similarities. J. Goris *et al.* (2007) *Int. J. Syst. Evol. Microbiol.* **57**, 81–91.

5.3.C. Probe Hybridization

A large number of powerful techniques for identification and manipulation of the genetic information present in natural DNA and RNA molecules exploit the ability of complementary nucleic acids to form hybrid duplexes. Most of these hybridization techniques rely on the use of a labeled (usually radioactive) nucleic acid probe to hybridize with a target nucleic acid. If the target is one strand of a double-stranded nucleic acid, it must be denatured first. After removal of any unreacted probe, detection of the remaining probe identifies and quantifies the hybrid duplex; its stability can also provide a measure of the extent of the sequence complementarity between the probe and the target. The probe must have sufficient affinity for its target sequences to resist being removed during the washing procedures.

Locked nucleic acids have been designed to have less conformational flexibility when single stranded and to be restricted to a conformation like that they will adopt when hybridized. Consequently, they have increased affinities for complementary strands. The locking is accomplished by a covalent bridge between the 2′ and 4′ C atoms of the ribose ring. This 'locks' the ribose in the 3′-*endo* conformation, that present in A-form RNA and DNA. Other modifications increase the stabilities of the probes by decreasing their susceptibility to nucleases that would hydrolyze normal nucleic acids. Still others diminish their tendencies to adopt intramolecular secondary structures, which interfere with their hybridization with another molecule; the sequence of a probe should be designed with this in mind.

The length of a specific hybridization probe can be as small as about 15 bases and as long as is necessary, depending upon the complexity of the target sequence and the source of the probe. The minimum length depends upon the size of the target, being the number of bases required to locate the specific sequence uniquely. Long probes have increased sequence specificity, and they can incorporate more label, thereby increasing the signal intensity. They are generally synthesized enzymatically off a template strand of DNA (Section 6.1). Shorter probes of 15–30 nucleotides can be synthesized chemically (Sections 6.6 and 6.7), and the incorporation of reporter groups as label is straightforward. Short probes bind less tightly to the target than do long probes, but weaker binding is often a practical advantage. The stability of the hybrid duplex can then be measured, and it will be sensitive to even single defects of sequence complementarity, including mismatches, deletions, bulges and small internal loops. The binding of a probe is stabilized if it is immediately adjacent to another molecule binding to the same strand, so that there is an additional stacking interaction:

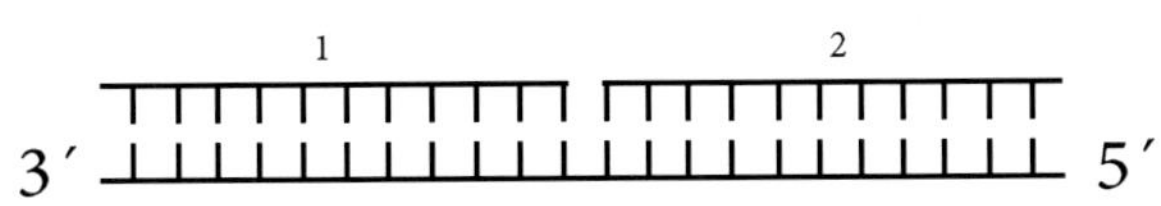

(5.13)

If the probe is shorter than the target, **dangling ends** are produced:

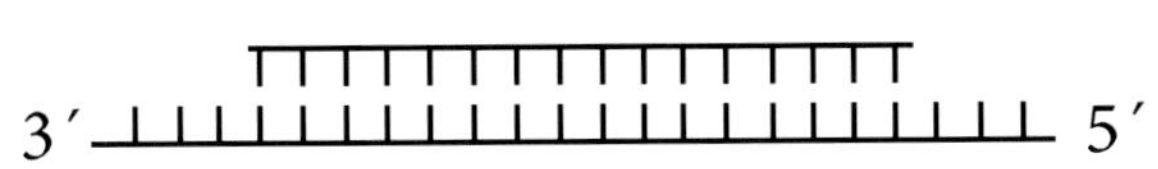

(5.14)

Noncanonical base pairs (mismatches) destabilize the duplex, as do extra or missing nucleotides, which produce a **bulge** in one strand (Figure 5-12):

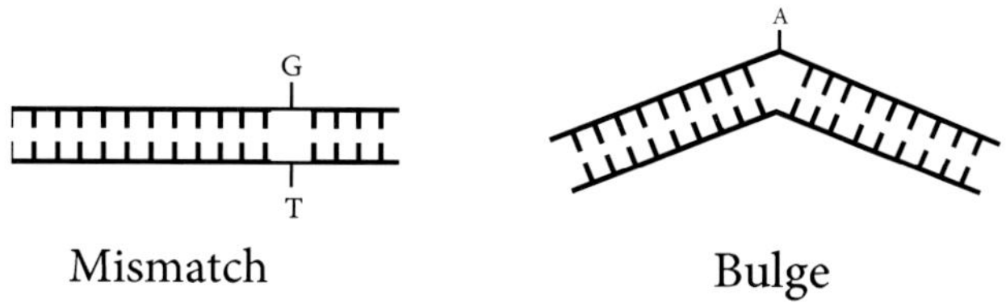

(5.15)

If the probe overlaps another oligonucleotide binding to the same strand, this generates a mixture of structures in which base pairing by one or the other in the overlap region is interconverted by **branch migration**:

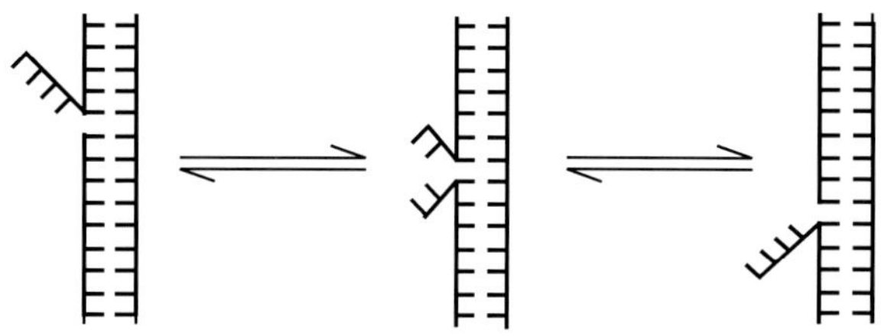

(5.16)

In many cases, the target nucleic acid is immobilized on a surface, most frequently on nitrocellulose or nylon filters. Immobilization can avoid many competitive binding interactions; for example, if the target was double-stranded, immobilization of the denatured single strands prevents the original duplex from re-forming. In **dot blots** or **slot blots**, the target nucleic acid is deposited on a small area of a blotting matrix; a series of targets can be deposited in different places on the same filter and hybridized simultaneously to a single probe. In **Southern blots**, **Northern blots** and related techniques, target nucleic acids are separated into discrete bands on a gel by electrophoresis. They are then transferred to a filter membrane, which is subjected to a conventional hybridization procedure. Those bands that bind the probe are detected by its label.

Many different probes are attached to solid surfaces in **micro-chip arrays**. After hybridization with complementary nucleic acids in solution, those molecules that did not hybridize are removed by washing and any hybridized molecules detected. The procedures are like those used in ELISA assays.

Padlock probes form especially stable complexes. They are designed so that the sequences at the two ends of the probe are homologous to the two halves of the target sequence and the annealed probe has its termini juxtaposed (Figure 5-15). A phosphodiester bond can then be formed by a DNA ligase between the two ends of the probe, to produce a circular probe molecule. The helical nature of the DNA molecule causes the annealed segment to be wound around the complementary strand. The probe can dissociate only if its backbone is cleaved or if it can slide off an end of the target DNA molecule. It can remain catenated to the target molecule even after denaturing washes that disrupt all base pairing.

RNA labeling and hybridization of DNA microarrays. E. P. Bottinger *et al.* (2003) *Methods Mol. Biol.* **86**, 275–284.

Modified bases in RNA reduce secondary structure and enhance hybridization. H. B. Gamper *et al.* (2004) *Biochemistry* **43**, 10224–10236.

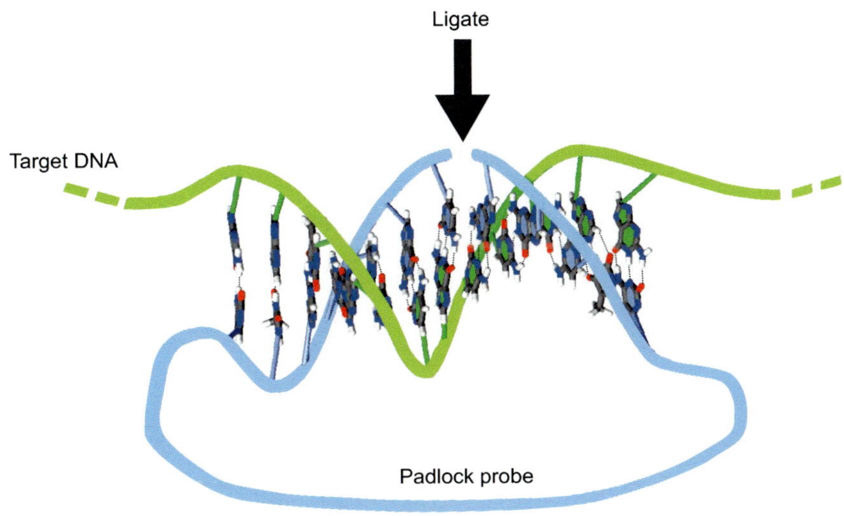

Figure 5-15. A padlock probe designed to be ligated into a circle when it hybridizes to the correct target sequence. The padlock probe is in *blue*, the target DNA sequence in *green*. Only the nucleotides involved in complementary base pairing are shown. The two ends of the linear probe hybridize in juxtaposition on the target sequence. The 5′-phosphate group on one end of the probe is in position to be ligated to the 3′-hydroxyl group at the other end, as indicated by the *arrow*. Once ligated, the padlock probe cannot dissociate, unless it can find its way to an end of the target sequence.

A general method for the synthesis of 2′-O-cyanoethylated oligoribonucleotides having promising hybridization affinity for DNA and RNA and enhanced nuclease resistance. H. Saneyoshi *et al.* (2005) *J. Org. Chem.* **70**, 10453–10460.

The forgotten variables of DNA array hybridization. E. Carletti *et al.* (2006) *Trends Biotechnol.* **24**, 443–448.

Design of LNA probes that improve mismatch discrimination. Y. You *et al.* (2006) *Nucleic Acids Res.* **34**, e60.

1. Stringency

The renaturation or hybridization of nucleic acids depends upon the solution conditions under which they are performed, in particular how they affect the stabilities of nucleic acid duplexes, which is known as the **stringency**. Low-stringency conditions favor duplex formation, while high-stringency conditions destabilize it. Consequently, many duplexes will form under low-stringency conditions, even if not totally complementary in nucleotide sequence, and base-pair mismatches, bulges and hairpins will be tolerated. In contrast, only the most complementary duplexes will form under high-stringency conditions. Different regions of a target nucleic acid can be distinguished on the basis of their sequence complementarity to the probe used.

The stringency conditions are usually adjusted in terms of the temperature, salt concentration and chemical denaturant concentration, as in Equations 5.3–5.6. They are optimized for the particular duplex that is to be formed. Adjustments in stringency can be used to discriminate among nucleic acids with different backbones and to favor a particular type of duplex. For example, use of high formamide concentrations (about 80% by volume) stabilizes RNA•DNA hybrids over DNA•DNA duplexes (Equations 5.4 and 5.6).

Most hybridization solutions contain EDTA to bind divalent cations, a polymer to enhance hybrid formation (presumably by an **excluded volume effect**) and a denaturant, in a buffered salt solution. Chemical denaturants, especially formamide, are frequently used in hybridization reactions to reduce the T_m to an experimentally convenient range. This is particularly important for experiments using RNA, which is readily degraded at high temperature.

Both the rate of formation of the duplex and its stability determine the optimal hybridization conditions. The rate of hybridization usually reaches a maximum at a temperature approximately 25°C below the T_m (Figure 5-13), which can be estimated with Equations 5.3–5.6. The same T_m can be obtained with a variety of different salt and denaturant concentrations.

Optimizing stringency for expression microarrays. J. E. Korkola *et al.* (2003) *Biotechniques* **35**, 828–835.

Specificity assessment from fractionation experiments (SAFE): a novel method to evaluate microarray probe specificity based on hybridisation stringencies. A. L. Drobyshev *et al.* (2003) *Nucleic Acids Res.* **31**, E1–1.

2. Analyzing the Extent of Complementarity

How the hybrid duplex is detected and measured depends upon what labels are incorporated into the probe molecule, and a huge number of different methods are used. Radioactive isotopes are the classic means of labeling the probe, but **chemiluminescent** and **bioluminescent** detection schemes are now in common use. These enzymes catalyze reactions that initiate a cascade of reactions that lead to the production of light. They have advantages of speed, cost and safety and are comparable to isotopic techniques in sensitivity. In many cases, a **biotin** label in the probe is detected by streptavidin coupled to alkaline phosphatase or horseradish peroxidase.

The degree of sequence complementarity between two short oligonucleotides can be evaluated by monitoring the change in UV absorbance upon heteroduplex formation (Figure 5-1), or in fluorescence with suitably labeled probes. If a pair of fluorescence donor and acceptor groups are placed in the appropriate positions, so that they come closer together when hybridized to the complementary sequence, the **fluorescence resonance energy transfer (FRET)** between them can be used to give a more specific signal. In **induced fluorescence resonance energy transfer (iFRET)**, the donor is a dye that only fluoresces while interacting with double-stranded DNA and the acceptor is a dye that is covalently linked to the oligonucleotide probe. Hybridization of the probe to its complement induces excitement of the donor dye and subsequent energy transfer to the acceptor dye. The energy transfer reaction (and concomitant hybridization status) can easily be followed by monitoring the fluorescence output of the acceptor dye. **Molecular beacons** are stem–loop hairpin oligonucleotide probes labeled with a fluorescent dye at one end and a fluorescence quencher at the other end; in the free probe with a stem–loop structure, the fluorescent and quenching moieties are in close proximity, so the fluorescence is quenched. When hybridized to a complementary nucleic acid strand, however, they are at opposite ends of the linear molecule and fluorescence is observed. Molecular beacons can differentiate between bound and unbound probes in homogeneous hybridization assays with a high signal-to-background ratio and enhanced specificity compared with linear oligonucleotide probes.

With longer nucleic acids, where some segments may be more complementary than others, the heteroduplex can be subject to digestion by enzymes that degrade only single-stranded regions and leave double-stranded regions intact. The remaining double-stranded fragments can then be isolated by hydroxyapatite chromatography and identified.

Electron microscopy can be used to visualize the regions of sequence homology in long heteroduplexes, after staining with basic proteins and heavy metals. In this way, single- and double-stranded regions can usually be distinguished and the homologous and nonhomologous regions identified (Figure 5-16).

Fluorescent PNA probes as hybridization labels for biological RNA. K. L. Robertson *et al.* (2006) *Biochemistry* **45**, 6066–6074.

Detecting RNA/DNA hybridization using double-labeled donor probes with enhanced fluorescence resonance energy transfer signals. Y. Okamura & Y. Watanabe (2006) *Methods Mol. Biol.* **335**, 43–56.

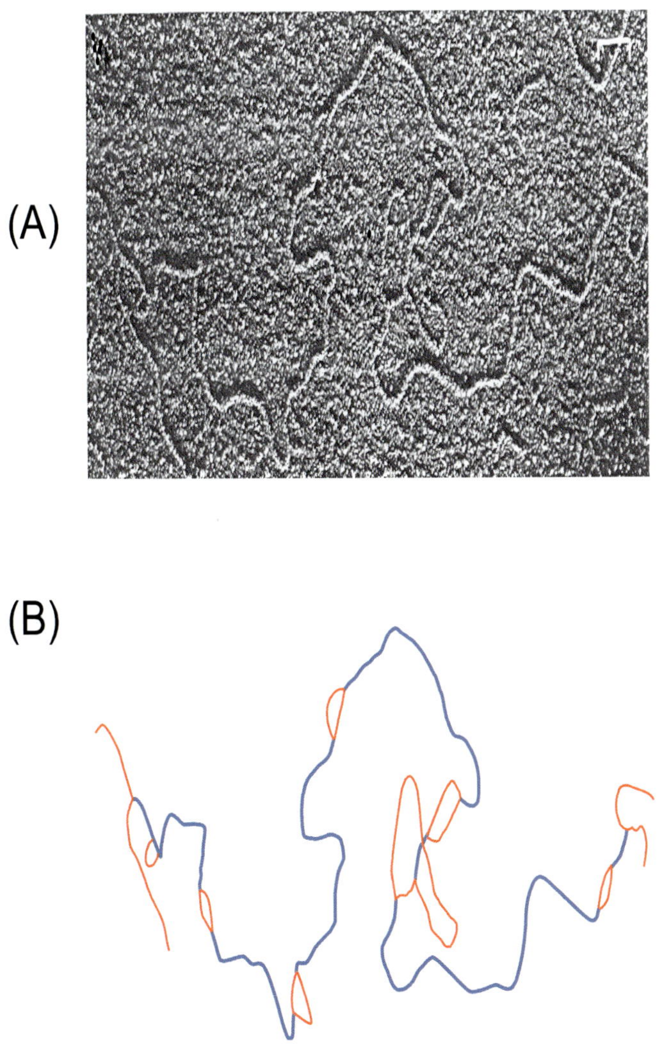

Figure 5-16. Visualization of a partly complementary heteroduplex by electron microscopy. (A) The DNA molecules have been spread on a grid and shadowed. (B) Interpretation of the micrograph. The double-stranded regions are somewhat wider and less kinky than the single-stranded and are indicated in *blue*. The single-stranded regions are in *red*.

Detection of DNA hybridization using induced fluorescence resonance energy transfer. W. M. Howell (2006) *Methods Mol. Biol.* **335**, 33–41.

Fluorescent hybridization probes for sensitive and selective DNA and RNA detection. A. A. Marti *et al.* (2007) *Acc. Chem. Res.* **40**, 402–409.

3. In situ *Hybridization*

Performing hybridization techniques *in situ* makes it possible to visualize the spatial distribution of a target nucleic acid in a tissue, cell or chromosome. There are many variations but all share a set of common steps:

(1) the sample is fixed to preserve its structure;

(2) limited proteinase digestion is carried out, to provide access of the hybridization probe to the target DNA sequence;

(3) the probe is added and hybridization is allowed to proceed under the appropriate conditions;

(4) washing removes unhybridized probe;

(5) the spatial distribution of the immobilized probe is visualized.

Small probes can diffuse into the sample more efficiently than longer probes. This is partially countered, however, by their tendency to be lost in the washing step: the stability of the hybrid duplex involving a short oligonucleotide depends upon its concentration, so the dilution that accompanies the washing step can result in release of some of the probe.

The probe can incorporate radioisotopes or fluorescent groups, but chemiluminescent techniques are now preferred. When detection is by fluorescence (or its quenching), the technique is known as **fluorescence *in situ* hybridization (FISH)**. Visualizing the immobilized probe can be difficult if only a small number of target molecules are present. This has led to the development of techniques that amplify the number of DNA sequences present, usually by the polymerase chain reaction (Section 6.1.C).

Expression analysis of murine genes using *in situ* hybridization with radioactive and nonradioactively labeled RNA probes. A. Chotteau-Lelievre *et al.* (2006) *Methods Mol. Biol.* **326**, 61–87.

Oligonucleotide probes for RNA-targeted fluorescence *in situ* hybridization. A. P. Silverman & E. T. Kool (2007) *Adv. Clin. Chem.* **43**, 79–115.

RNA *in situ* hybridizations on *Drosophila* whole mounts. C. Wulbeck & C. Helfrich-Forster (2007) *Methods Mol. Biol.* **362**, 495–511.

5.4. DNA MIMICS: PEPTIDE NUCLEIC ACIDS

Many practical uses could be imagined for oligonucleotides in medicine, such as in controlling the expression of particular genes by binding to them or their messenger RNAs (Section 6.2.D).

Oligonucleotides are not, however, ideal pharmaceuticals. They are highly charged and thus do not cross cell membranes readily. They are not as stable as would be required, and the sites where they should act are also rich in enzymes (nucleases) that will destroy them. Consequently, much effort has been expended in trying to find analogs that do not have these shortcomings and that, ideally, would bind even more strongly to their targets. The most promising are the peptide nucleic acids (PNA), DNA mimics in which the phosphodeoxyribose backbone of DNA has been substituted by a pseudopeptide backbone to which the normal nucleobases are tethered, to produce an *oligoamide* (Figure 5-17). These DNA mimics can perform many of the functions of RNAs and DNAs. **PNAs of mixed sequences containing all four DNA bases form duplexes with DNA and RNA that obey the Watson–Crick base-pair rules, and two PNAs with complementary base sequences can hybridize into a double-helical structure.**

Only minor modifications of the original amino ethyl glycine backbone can be tolerated; extending any of the three possible linkages of each monomer diminishes the stability of its complexes. The secondary amino group in the nucleobase linker restricts its flexibility and is necessary, because changing it to a tertiary amine inhibits hybridization. Moving a methylene group from the ethyl to glycine moiety has the same effect. On the other hand, substituents may be placed on the α-position of the glycine moiety of the PNA backbone in various ways, so the glycine residue can be replaced by other natural amino acids. So long as the right stereoisomer is chosen, even cyclic substituents are acceptable, such as cyclohexyl at the amino ethyl linker.

PNA is regarded as having great potential for diagnostic and pharmaceutical applications. Other promising DNA mimics are based on a pyrrolidine amide backbone but will not be described here.

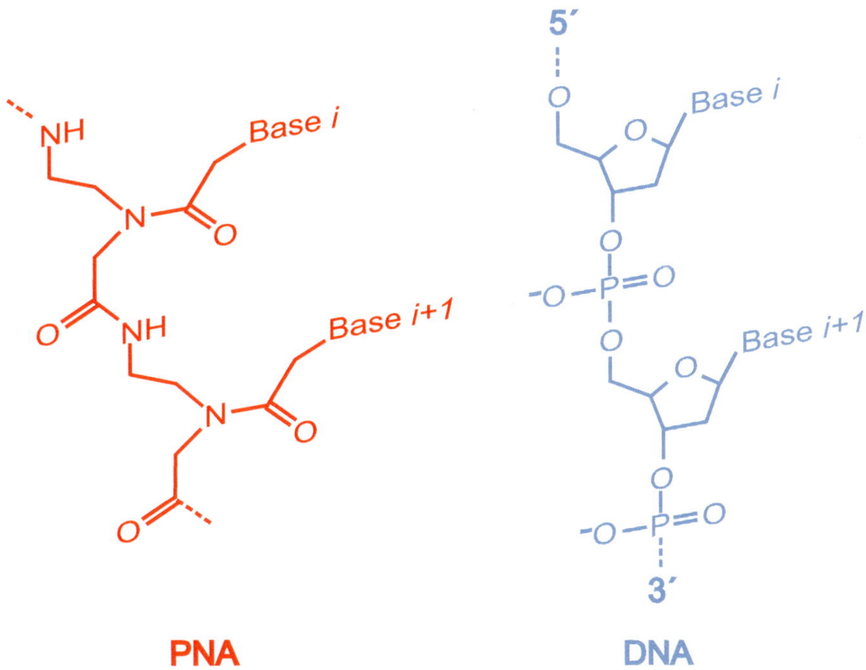

Figure 5-17. Comparison of the chemical structures of DNA and a PNA.

Gene targeting using peptide nucleic acid. P. E. Nielsen (2005) *Methods Mol. Biol.* **288**, 343–358.

Peptide nucleic acids (PNAs), a chemical overview. A. Porcheddu & G. Giacomelli (2005) *Curr. Med. Chem.* **12**, 2561–2599.

Peptide nucleic acid conjugates: synthesis, properties and applications. Z. V. Zhilina *et al.* (2005) *Curr. Top. Med. Chem.* **5**, 1119–1131.

Biological activity and biotechnological aspects of peptide nucleic acid. K. E. Lundin *et al.* (2006) *Adv. Genet.* **5**, 1–51.

Homopolymeric pyrrolidine-amide oligonucleotide mimics: Fmoc-synthesis and DNA/RNA binding properties. T. H. Tan *et al.* (2007) *Org. Biomol. Chem.* **5**, 239–248.

5.4.A. Chemistry and Synthesis

The backbone of PNA is designed to mimic the normal phosphodiester backbone of DNA and RNA and consists of 2-aminoethyl glycine linkages (Figure 5-17). The standard nucleotide bases are connected to this backbone at the amino N atoms by a methylene carbonyl linker. The properties of this polymer are of interest in themselves, but the differences also illustrate some of the important aspects of nucleic acids. In contrast to DNA and RNA, **PNA is neither ionic nor chiral**. The achirality avoids diastereomeric structures and the need to develop synthesis reactions that are stereoselective (Section 1.1.A). Reactive groups on the PNA monomers are protected using either the Boc or Fmoc chemistry that is used in peptide synthesis. Oligomers can be synthesized on a polystyrene resin using established solid-phase peptide synthesis procedures; they can be released from the support using either anhydrous hydrogen fluoride or trifluoromethanesulfonic acid, then purified chromatographically. Modified forms of PNA can be synthesized readily.

A simple γ-backbone modification preorganizes peptide nucleic acid into a helical structure. A. Dragulescu-Andrasi *et al.* (2006) *J. Am. Chem. Soc.* **128**, 10258–10267.

Peptide nucleic acids with a structurally biased backbone: effects of conformational constraints and stereochemistry. R. Corradini *et al.* (2007) *Curr. Top. Med. Chem.* **7**, 681–694.

Nucleobase modifications in peptide nucleic acids. F. Wojciechowski & R. H. Hudson (2007) *Curr. Top. Med. Chem.* **7**, 667–679.

5.4.B. Hybridization Properties

PNA is a very potent DNA mimic capable of forming both duplex and triplex structures, but with some important differences. **PNA binds equally to DNA and RNA** molecules with the same sequence.

RNA targeting using peptide nucleic acid. E. Nielsen (2006) *Handb. Exp. Pharmacol.* **173**, 395–403.

1. PNA•DNA and PNA•RNA Duplexes

The mixed-sequence PNA pentadecamer H–TGTACGTCACAACTA–NH_2 has been studied most extensively. It forms a duplex with complementary DNA that is **antiparallel, with the amino-terminus of PNA at the 3′ end of DNA**. The duplex has a T_m of 70°C, greater than the 53°C of the corresponding DNA•DNA complex. PNA is also able to bind in a parallel orientation, although with a lower affinity (T_m = 56°C). **Mixed-sequence PNA•DNA duplexes generally adopt the antiparallel orientation.** Forming antiparallel PNA•DNA duplexes is faster than forming a corresponding DNA•DNA duplex, whereas forming parallel PNA•DNA complexes is considerably slower. The rates of formation and thermal stabilities of PNA•RNA duplexes are generally greater than for PNA•DNA duplexes. **PNA in complex with DNA or RNA adopts right-handed helical structures** that are probably dictated by the DNA or RNA.

The greater thermal stabilities of PNA•DNA (or PNA•RNA) duplexes than the corresponding DNA•DNA (or RNA•RNA) duplexes are probably due to the absence of electrostatic repulsions between the phosphate groups of the two strands. The stabilities of PNA•DNA hybrids are little affected by changes in ionic strength, except at low ionic strength, where the stability increases; this is believed to be due to a more favorable entropy change upon PNA binding caused by the release of counterions from the DNA polyanion (Section 2.3).

The sequence specificities of PNA oligomers are at least as good as those of DNA: mismatch of a single base pair in the middle of a 15-mer PNA•DNA or PNA•RNA duplex decreases the T_m by 8–20°C.

Cyclohexanyl peptide nucleic acids (chPNAs) for preferential RNA binding: effective tuning of dihedral angle β in PNAs for DNA/RNA discrimination. T. Govindaraju *et al.* (2006) *J. Org. Chem.* **71**, 14–21.

2. $(PNA)_2$•DNA Triplexes

PNA oligomers with a high pyrimidine content bind complementary purine DNA by forming predominantly $(PNA)_2$•DNA triple helices, with virtually no duplex. These complexes are very stable, with a T_m > 70°C for decamer PNAs, and have well-defined melting curves. Binding is probably governed by standard Watson–Crick and Hoogsteen base pairing (Section 2.2.C.1) and, as expected, triplex formation involving cytosine in the homopyrimidine PNA is more stable at lower pH (Section 2.2.C.2). The same triplex stoichiometry is observed even at pH 9, however, where cytosine is not expected to be protonated.

3. Strand Invasion: Binding to Double-stranded DNA

PNA is able to dissociate double-stranded DNA, due to the high stability of the $(PNA)_2$•DNA triplexes. For example, homopyrimidine PNA oligomers displace the pyrimidine strands of double-stranded DNA and form $(PNA)_2$•DNA triplexes with the DNA homopurine strand. The DNA pyrimidine strand that was expelled forms a single-stranded loop (Figure 5-18). PNA•DNA complexes formed by strand invasion are both very stable thermodynamically and highly specific. The strand invasion appears to be controlled kinetically, which permits the PNA to search for the target sequence without becoming trapped in incorrect complexes that are thermodynamically stable. Forming a $PNA(pyrimidine)_2$•DNA(purine) strand displacement complex follows pseudofirst-order kinetics, and a single mismatch in a 10-residue PNA reduces the rate of binding by more than a factor of 100.

In the case of PNA•thymine trimers, binding follows somewhat more complex reaction kinetics that suggest the presence of at least one intermediate species. The rate-limiting step appears to be transient opening of a few base pairs in the DNA duplex. The reaction is second-order in PNA concentration, indicating that both PNA strands are involved in the rate-limiting step: the two PNA molecules may be associated loosely with the DNA duplex prior to the base-pair opening step. Intercalators (Section 2.5.B) bound to the starting duplex DNA increase the rate of PNA binding; the intercalation site may facilitate initiation of PNA invasion. The formation of strand displacement complexes is highly dependent on the ionic strength but complexes formed at low ionic strength are generally stable kinetically at salt concentrations up to 0.5 M NaCl.

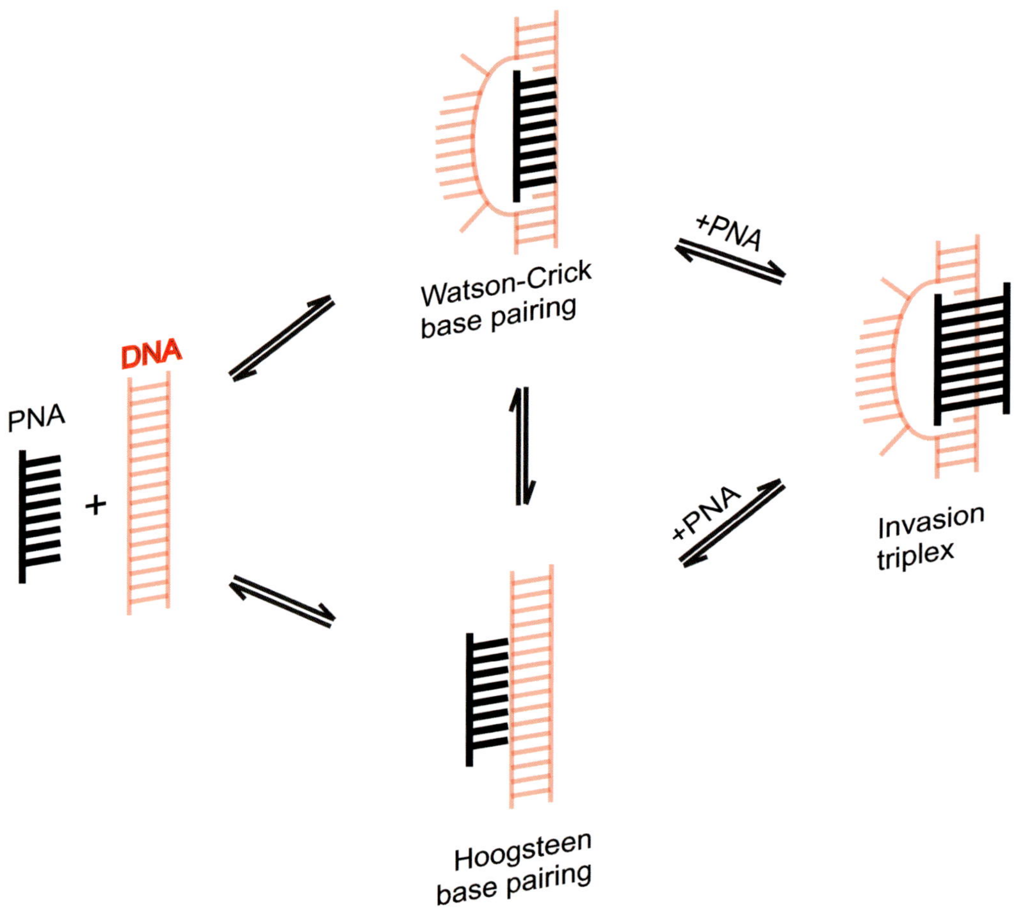

Figure 5-18. Schematic model of the possible mechanisms of PNA_2•DNA triplex formation by strand invasion of a DNA duplex.

Cytosine-rich and alternating thymine–cytosine PNA oligomers bind to their complementary sequence of intact double-stranded DNA by forming PNA•DNA_2 triplexes, which probably involve Hoogsteen base pairing. With purine-rich PNA sequences, duplexes are formed by the invasion into double-stranded DNA at a complementary pyrimidine target (making a PNA•DNA duplex). PNA binds less well to DNA sequences without purine-rich stretches.

Cooperative strand invasion of double-stranded DNA by peptide nucleic acid. T. Sugiyama *et al.* (2005) *Nucleic Acids Symp. Ser.* **49**, 167–168.

End invasion of peptide nucleic acids (PNAs) with mixed-base composition into linear DNA duplexes. I. V. Smolina *et al.* (2005) *Nucleic Acids Res.* **33**, 3146.

Structural diversity of target-specific homopyrimidine peptide nucleic acid–dsDNA complexes. T. Bentin *et al.* (2006) *Nucleic Acids Res.* **34**, 5790–5799.

4. PNA Duplexes and Triplexes

PNA molecules containing complementary sequences form stable antiparallel duplexes with Watson–Crick base paring and very high thermal stability: the T_m of a 10-mer PNA•PNA is 67°C, compared with 51°C for the corresponding PNA•DNA and only 33.5°C for DNA•DNA. The parallel PNA•PNA complex is less stable, with a T_m of 47°C. A PNA_3 triplex can be formed with two pyrimidine decamers and one purine decamer and probably has a helical structure similar to triplex DNA and PNA_2•DNA triplexes (Section 2.2.C.2).

PNA•PNA duplexes have no chiral constituents and probably exist as a racemic mixture of right- and left-handed helices. A chiral amino acid at the carboxyl-terminus of one strand can, however, induce formation of one predominant helical sense over about 10 base pairs, approximately one helical turn, the 'chiral persistence length'. This effect depends on the PNA sequence at the carboxyl terminus, being greatest with a G•C base pair, and on the nature of the terminal chiral amino acid. The D- and L-forms of amino acids on PNA give rise to circular dichroism spectra that are mirror images, indicating that helices of opposite handedness are formed. Whether the side-chain is hydrophobic or hydrophilic also determines the preferred helical sense. Changes in helix chirality occur within minutes at room temperature and follow first-order kinetics. The transition state has a high entropy, as would be expected for a transition from one helical sense to another.

During formation of a PNA•PNA duplex, base pairing occurs rapidly, within less than a second under normal conditions. Some of the PNA•PNA duplexes formed initially undergo a conformational change to the final structure.

5.4.C. Structures of PNA Complexes

PNA•PNA complexes appear to be similar to DNA duplexes in their base-stacking interactions and degree of organization. When bound to RNA or DNA, PNA forms a right-handed helix with a base-pair geometry not very different from that found in B-DNA and A-DNA helices, respectively. In contrast, parallel PNA•DNA and PNA•RNA complexes appear to have different base stacking.

PNA–T_8 hybridized to poly(dA) forms a right-handed PNA_2•DNA triplex in which the base triplets have their planes approximately perpendicular to the helix axis, similar to the standard $(poly(dT))_2$•poly(dA) triplex. The PNA_2•DNA triplex is a remarkably rigid structure, which suggests efficient stacking interactions. Titration of PNA–T_8 added to poly(dA) shows that no duplex stretches are present. Instead, PNA_2•DNA triplexes are distributed evenly on the poly(dA) strand, separated by flexible single-stranded regions. The complex becomes rigid only when the poly(dA) is saturated with 2:1 PNA:DNA bases, completing formation of the triple strand and stacking between each PNA triplex segment.

Antiparallel complementary binding of PNA allows the backbone of the nucleic acid partner to retain a nearly 'normal' conformation: in the RNA•PNA duplex, the RNA strand has an approximately A conformation with C3′-*endo* sugar puckering, whereas the DNA in the PNA•DNA duplex is closer to B-form, with C2′-*endo* sugar puckering. The base-pair positions are, however, more A-like in terms of their helix displacement, although more perpendicular to the helix axis. The PNA•DNA duplex has a pitch of about 13 base pairs, greater than the 10–11 base pairs usually observed with nucleic acid duplexes.

The PNA_2•DNA triplex has hydrogen bonds between phosphate groups of the DNA backbone and amide protons in the PNA backbone of the Hoogsteen strand, which could be important for its stability. The two PNA strands are nearly identical and resemble the PNA strand in a PNA•DNA duplex. The PNA_2•DNA triplex and the PNA•PNA duplex have relatively wide helices of 26 Å and 28 Å, respectively. They also have a large pitch of 16 and 18 bases per turn, respectively, which is probably the natural conformation of helices with a PNA backbone. The base pairs in the PNA•PNA duplex have stacking overlaps between base pairs that are close to those of A-form nucleic acids. Consequently, PNA helices appear to have a unique structure that only resembles known nucleic acid helices.

The similarity in the base-pair stacking in the PNA•DNA, PNA•RNA and PNA•PNA duplexes indicates that the helical stacking is not influenced markedly by the absence of either backbone charge or chirality in one or both of the strands. The helicity can thus be regarded a result of the stacking interactions and the architecture of the backbone, rather than arising in nucleic acids from repulsions between phosphate groups striving to increase the separation of negative charges along the backbones.

Molecules that intercalate or bind to major grooves in double-stranded DNA (Section 2.5) have no apparent affinity for PNA•DNA complexes, whereas classical binders of the minor-groove bind weakly. No simple ligand able to bind to PNA•PNA duplexes has been found, illustrating perhaps the importance of electrostatic attractions for ligand binding to DNA.

Insights into peptide nucleic acid (PNA) structural features: the crystal structure of a D-lysine-based chiral PNA–DNA duplex. V. Menchise *et al.* (2003) *Proc. Natl. Acad. Sci. USA* **100**, 12021–12026.

Crystal structure of a partly self-complementary peptide nucleic acid (PNA) oligomer showing a duplex–triplex network. B. Petersson *et al.* (2005) *J. Am. Chem. Soc.* **127**, 1424–1430.

~ CHAPTER 6 ~

MANIPULATING NUCLEIC ACIDS

Manipulating nucleic acids is central to molecular biology. Specific segments of DNA or RNA can be identified, isolated, amplified, sequenced and synthesized. Any desired changes can be made to the sequence. The nucleic acid can be inserted into cells and its effect observed. These many techniques make it possible to identify genes and their protein or RNA products, to alter them at will, and to determine their physiological roles.

Nucleic acids have the very great advantage that all those of the same type and size behave almost identically. Consequently, a technique developed with one nucleic acid is very likely to work just as well with any other, in contrast to proteins, where each one is unique. The most important parameter for manipulating a nucleic acid is its length. A second, less important, parameter is its base composition, as nucleic acids with more G•C base pairs denature less readily (Section 5.1.B).

This chapter will not describe all the many techniques that have been devised to manipulate nucleic acids. Instead, it will describe the most fundamental ones that illustrate the physical and chemical principles of them all.

6.1. REPLICATING DNA

The most fundamental characteristic of living organisms is their ability to replicate themselves and multiply. Replication of their genomes is the crucial step. It is easy to envisage **the DNA or, rarely, RNA double helix of the genome simply being dissociated into the individual strands, which then serve as templates for a complementary sequence to be synthesized; the end result is two copies of the original DNA or RNA molecule**. Entire genomes are enormous molecules, however, which means that a multitude of factors are required *in vivo*, especially to prevent these very long molecules from becoming hopelessly entangled (Chapter 3). In particular, **DNA helicases** are required to unwind the duplex DNA ahead of the point of DNA replication. The three distinct stages of genome replication are initiation, elongation and termination; each requires a complex apparatus. Most complex is the initiation step, as replication of the genome must be coordinated with the division of the entire cell. Each cell usually has a single genome, and replicating the genome in the absence of cell division would be deleterious.

This discussion will be limited to much more simple *in vitro* systems, in which a relatively short piece of DNA is replicated using defined enzymes. The first requirement to replicate a double-stranded DNA molecule is to dissociate it into the individual single strands, which can then be used as templates to synthesize a new copy of the missing strand. Although a complex process *in vivo*, in the laboratory it simply requires that the DNA molecules be heated to high temperatures (Section 5.1.B.1).

6.1.A. DNA polymerase

The enzymes that catalyze formation of the complementary strand of a DNA are known as **DNA polymerases**. Each organism has at least one DNA polymerase, but the one used most commonly in molecular biology is DNA polymerase I from *Escherichia coli*. It can synthesize the complementary strand of a DNA single strand in the 5′- to 3′-direction in the presence of all four deoxynucleoside triphosphates (dNTPs) (Figure 6-1). DNA polymerases cannot initiate synthesis, however, so a primer complementary to the site at which replication is to be initiated must be supplied and annealed to the DNA template strand (Figure 6-2). After the DNA polymerase has bound to the primer, the appropriate dNTP binds to the active site of the polymerase, and the polymerase changes conformation. This positions the dNTP so that its α-phosphate group is positioned near the 3′-hydroxyl group of the primer (Figure 6-3) for nucleophilic attack, which produces a phosphodiester bond and releases pyrophosphate (Figure 6-4). The conformational change and nucleophilic attack function in concert to sense that the appropriate Watson–Crick base pairing is occurring and that the correct nucleotide is being introduced. The nascent second strand has now been elongated by one nucleotide. The polymerase then translocates to the next nucleotide of the template strand and attaches the appropriate nucleotide to the growing chain.

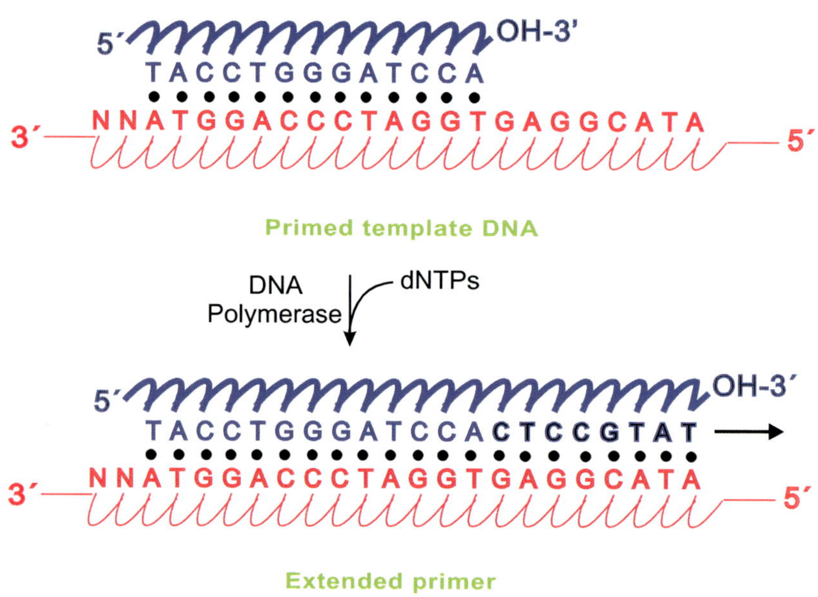

Figure 6-1. DNA chain elongation by DNA polymerase. The *jagged line* represents the DNA backbone, and the bases are indicated by the *letters*; *black circles* represent complementary pairing between the bases. Nucleotides that have been added to the 3′-end of the primer are indicated in *bold*. Addition of a dNTP to the 3′-end of the primer strand produces a new 3′-OH group, to which another dNTP can be added for continued polymerization of the growing chain.

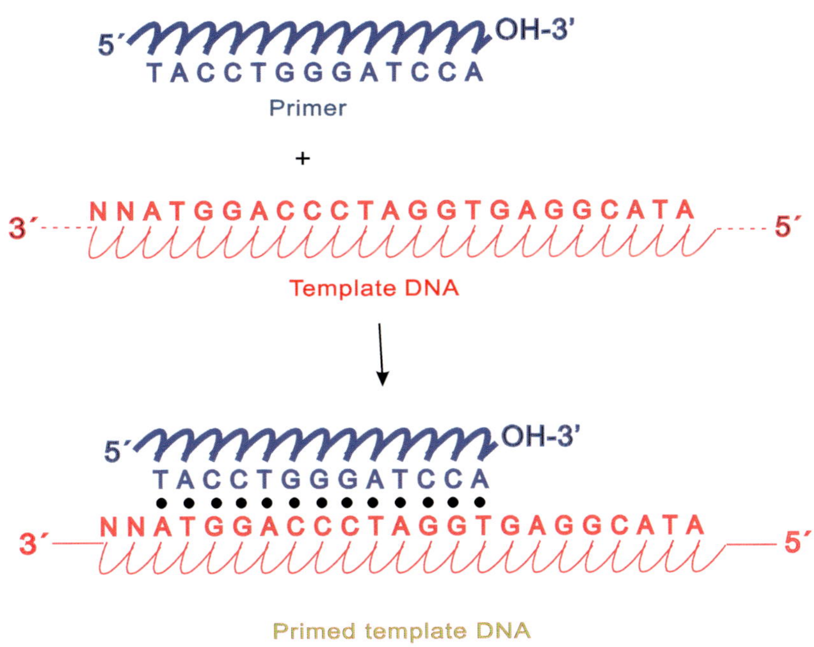

Figure 6-2. Annealing of primer to template DNA. The primer is designed to be complementary in sequence to the site where replication is to be initiated.

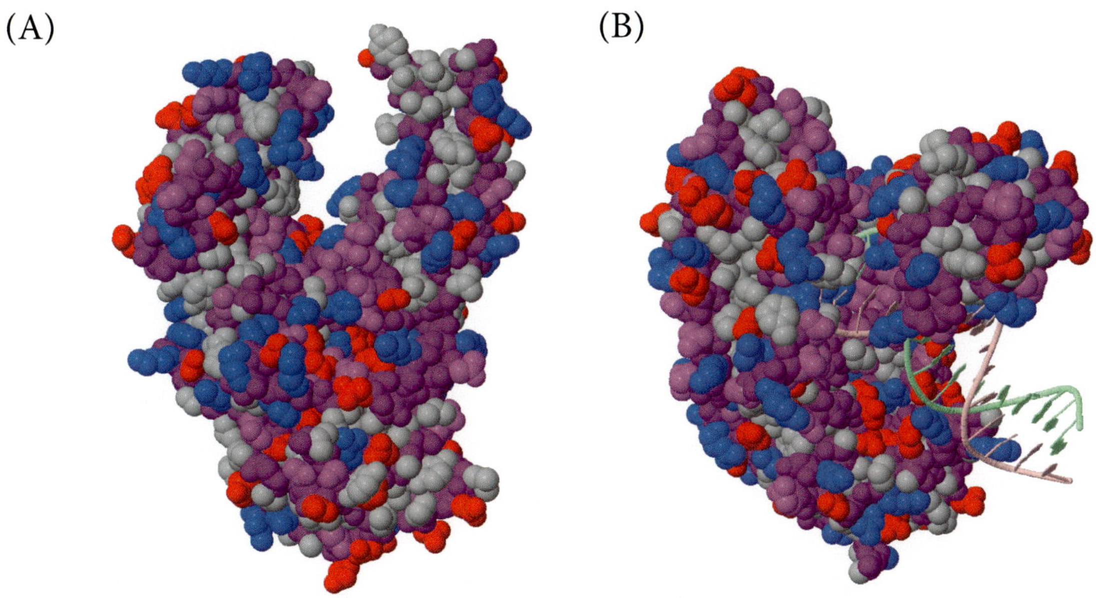

Figure 6-3. Structure of the Klenow fragment of DNA polymerase I. (A) The enzyme from *E. coli* bound to only CTP and pyrophosphate. The atoms of the protein are depicted with their van der Waals radii and colored according to their chemical properties. (B) The enzyme from *Bacillus stearothermophilus* bound to a DNA single strand and a nascent second strand. The DNA is shown in skeletal form, without the details of the backbone. The enzyme in the crystal has added four nucleotides to the original strand, but has been halted by a G•T mismatch. The two upper domains of the enzyme have closed over the DNA molecule. Figure generated from PDB files *1kfd* and *1nkc* using the program Jmol.

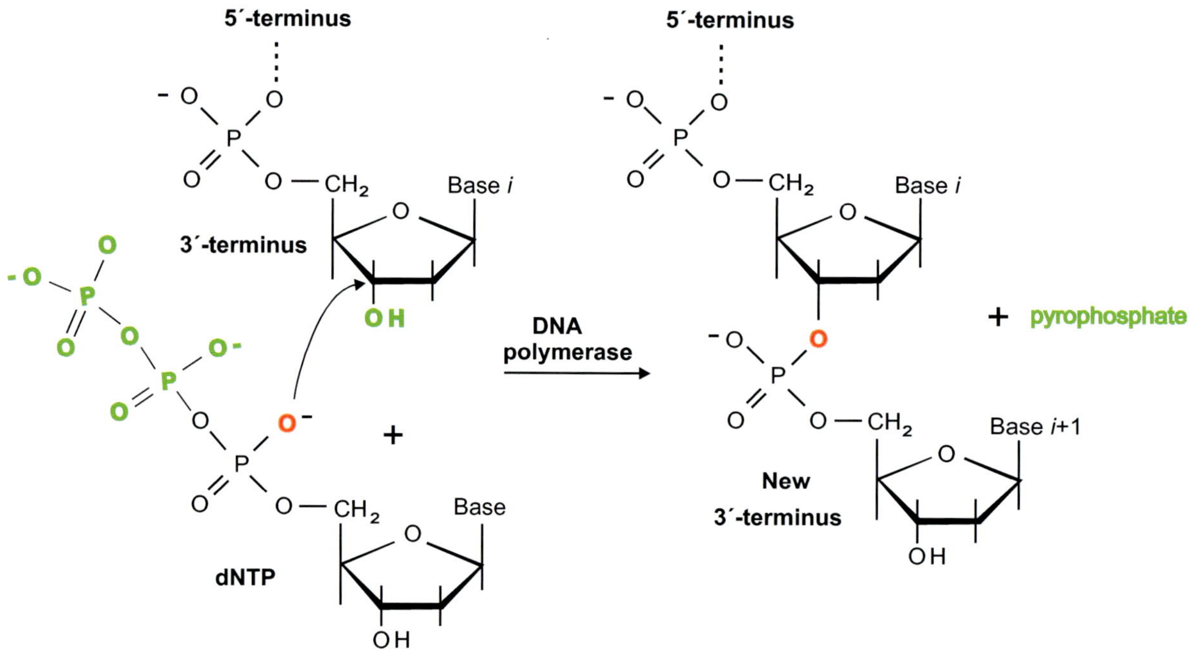

Figure 6-4. Biochemistry of nucleotide chain elongation. DNA polymerase catalyzes the condensation of the appropriate dNTP at the 3′-end of a primed template, releasing pyrophosphate.

Inserting the correct nucleotide is crucial for this process, so the process involves editing and proofreading the sequence of the assembled chain. If an improperly paired base is inserted, the elongation process is interrupted and introduction of the next nucleotide is very slow. The DNA molecule then has time to move to an adjacent site on the DNA polymerase, the exonuclease site, where the incorrect nucleotide is removed. This **3′→5′ exonuclease** activity decreases the frequency of mistakes, to about one mistake in every 50,000 nucleotides added, whereas the error rate is about one per 1000 nucleotides in its absence.

E. coli DNA polymerase I also has a **5′→3′ exonuclease activity** that it uses *in vivo* to degrade certain DNA and RNA fragments. This activity is detrimental to synthesizing DNA copies *in vitro*, and it can be removed by cleaving the DNA polymerase I into two fragments: one of 323 amino acid residues at the N-terminus that comprises the 5′→3′ exonuclease activity, and the other of about 605 residues with the polymerase and 3′→5′ exonuclease activities. The latter is generally known as the **Klenow fragment** and it is used most frequently in molecular biology. The Klenow fragment extends a DNA chain at a relatively modest rate of 50 nucleotides per second. It remains bound to the DNA for about one second on average, before dissociating. Another molecule of DNA polymerase must be bound to continue extending the polynucleotide chain.

Although DNA polymerase I is used extensively in molecular biology to replicate DNA, that is not its main function in nature, which is to fill gaps during the repair and replication of DNA. This function requires the assistance of a DNA ligase to produce a single, intact strand of DNA.

Mechanism and dynamics of translesion DNA synthesis catalyzed by the *Escherichia coli* Klenow fragment. A. Sheriff *et al.* (2008) *Biochemistry* **47**, 8527–8537.

Fingers-closing and other rapid conformational changes in DNA polymerase I (Klenow fragment) and their role in nucleotide selectivity. C. M. Joyce *et al.* (2008) *Biochemistry* **47**, 6103–6116.

Structure and function of 2:1 DNA polymerase•DNA complexes. K. H. Tang & M. D. Tsai (2008) *J. Cell. Physiol.* **216**, 315–320.

6.1.B. DNA ligase

A crucial step in assembling an intact DNA double helix is often the joining of one DNA chain to another, using a **DNA ligase** enzyme. In many cases, this is simply the formation of a phosphodiester bond between two nucleotides that are adjacent in the sequence of a double helix, being brought into proximity by base pairing to a complementary third strand (Figure 6-5). Forming a phosphodiester bond requires energy, which is usually supplied by the hydrolysis of ATP, although in some organisms the coenzyme NAD^+ is used instead.

The ligase reaction requires that the 5′-fragment have a 3′-hydroxyl group and the other strand have a 5′-phosphate group. The two fragments are best abutted without a gap, with no nucleotide missing

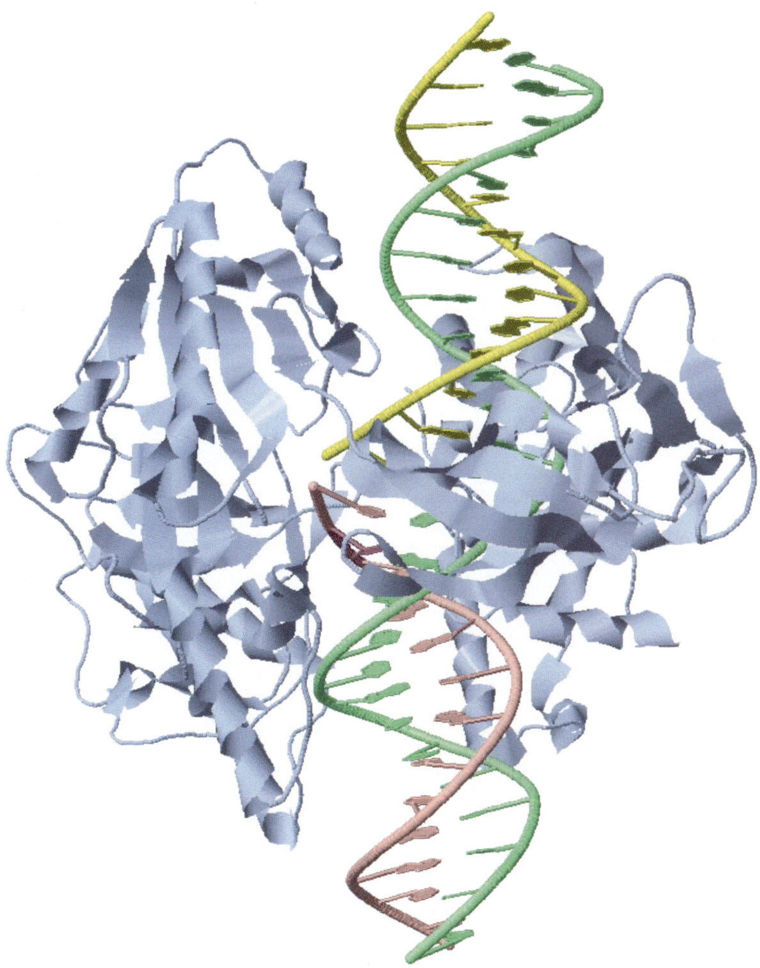

Figure 6-5. Structure of the *E. coli* DNA ligase bound to a nicked DNA double helix. The DNA substrate is indicated in skeletal form, without the details of the polynucleotide backbone. Only the polypeptide backbone of the protein is shown, and as a ribbon, so that the nicked DNA at the active site can be observed. Otherwise, the DNA is in the center of the protein molecule, protected from solvent, and invisible from the outside. Figure generated from PDB file *2owo* using the program Jmol.

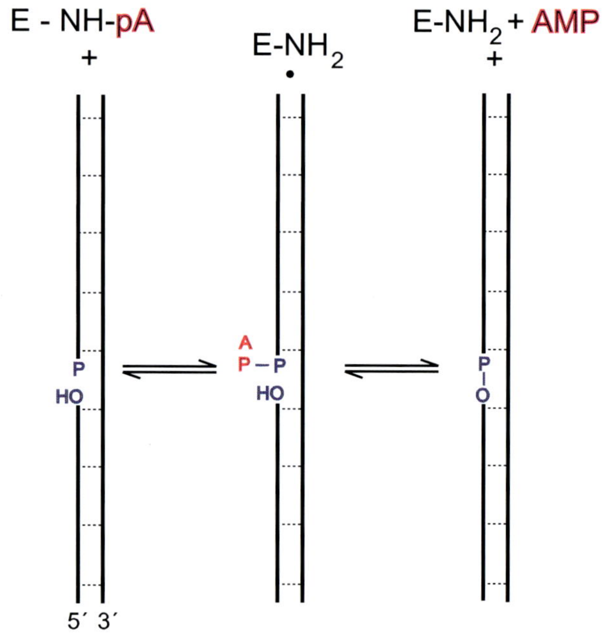

Figure 6-6. The mechanism of action of DNA ligase. On the *left* is a DNA double helix lacking one internal phosphodiester bond; OH is the 3′-hydroxyl, while P is the 5′-phosphate group. The DNA ligase (E) is adenylated on an amino group (Equation 6.1). After the DNA binds to the ligase, the adenylyl group is transferred to the 5′-phosphate. It is then expelled by nucleophilic attack by the 3′-hydroxyl group, generating the phosphodiester bond and releasing AMP and the enzyme. The DNA double helix is now intact.

from the double helix. The reaction proceeds through two intermediates (Figure 6-6). The first is formation of a covalent enzyme–AMP complex, using either ATP or NAD^+. The adenylyl group (pA) is linked to the ε-amino group of a lysine residue ($-NH_2$) of the enzyme (E), preserving the high energy of the donor, while releasing pyrophosphate (PP_i):

$$E\text{-}NH_2 + ATP \rightleftharpoons E\text{-}NH\text{-}pA + PP_i \qquad (6.1)$$

This reaction does not require DNA and is reversible, so it can be used to assay the activity of the DNA ligase. In the presence of DNA, the adenylyl group is transferred from the enzyme to the terminal 5′-phosphate of the DNA, but it is then displaced by the 3′-hydroxyl group of the adjacent DNA fragment, forming the phosphodiester bond and releasing AMP. Both strands of the double helix are now intact.

The DNA ligases used most frequently are those from *E. coli*, the thermophile *Thermus aquaticus* and bacteriophage T4.

DNA ligase: a means to an end joining. C. M. Bray *et al.* (2008) *SEB Exp. Biol. Ser.* **59**, 203–217.

Dynamics of phosphodiester synthesis by DNA ligase. A. Crut *et al.* (2008) *Proc. Natl. Acad. Sci. USA* **105**, 6894–6899.

The associative nature of adenylyl transfer catalyzed by T4 DNA ligase. A. V. Cherepanov *et al.* (2008) *Proc. Natl. Acad. Sci. USA* **105**, 8563–8568, E84 and E85.

6.1.C. Polymerase Chain Reaction (PCR)

Once a single strand has been used as a template to synthesize its complementary strand, further

DNA replication requires that the double-stranded molecule be dissociated to its individual strands; they can then be replicated in the same way, to double the number of DNA molecules present. In the laboratory, separating the strands requires only that the DNA be heated to high temperatures. Replication must be carried out promptly, however, to compete successfully with reformation of the original double strands. A primer must be present to hybridize with the single strand and provide a starting point for synthesis of the second strand. When present at high concentrations, the primer competes successfully with renaturation of the two complementary strands.

The polymerase chain reaction (PCR) is somewhat different in that it uses one primer for each of the complementary strands, which delineates the boundaries of the portion of the DNA molecule that is to be replicated. **PCR makes it possible to amplify and produce large quantities of any DNA sequence** and is one of the most useful and powerful techniques in molecular biology. In its most basic form, two oligodeoxyribonucleotides, generally 15–25 nucleotides in length, are used to select a specific target sequence, by being complementary to opposite strands at the two ends of the target sequence (Figure 6-7). Each of the two oligonucleotides serves as a primer for copying one of the two strands of the target sequence using a DNA polymerase enzyme. The two double helices that result are denatured by heating, and the process is repeated. **Each cycle virtually doubles the amount of the desired DNA**. A heat-stable DNA polymerase is used that survives the heat-denaturation of the DNA and is viable for the next cycle.

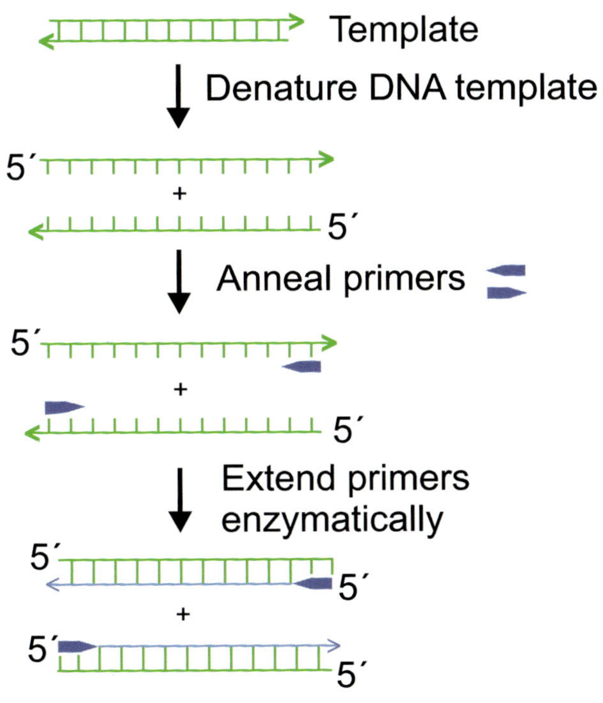

Figure 6-7. One cycle of basic PCR. The DNA of interest is denatured to separate the two strands. Primers complementary to the two ends are annealed to the single strands, and they direct copying of the template by DNA polymerase. Two identical copies of the original molecule result. Another cycle will produce four copies, and the number of molecules nearly doubles for each additional cycle. Other DNA sequences can be present in the original sample but only those that hybridize with the primers will be copied.

Operationally, each cycle consists solely of altering the temperature, from that suitable for annealing of the primers to the DNA and for synthesis by the thermostable polymerase (usually about 72°C) to that required to denature the DNA double helices (usually 94–96°C); simple PCR machines are available to carry out the entire process of several cycles automatically. The temperature for the annealing step is most crucial: a too-high temperature will inhibit annealing, whereas a too-low temperature allows nonspecific priming and the production of spurious amplification products. The optimal temperature for use with primers up to 18 nucleotides long is usually the same as the T_m of the complex and can be predicted by:

$$T = 4°C \times (\text{number of G and C nucleotides}) + 2°C \times (\text{number of A and T}) \tag{6.2}$$

The two primers used should be designed to have very similar optimal temperatures.

Each cycle doubles the amount of target DNA present: in theory, although not quite in practice. The degree of amplification accomplished depends primarily upon the number of temperature cycles, so **even extremely small amounts of target DNA can be amplified to make substantial quantities.** This exquisite sensitivity is one of the most important attributes of PCR, and makes it possible in theory to detect a single molecule of DNA; on the other hand, it also makes PCR prone to false-positive results due to, for example, exogenous contamination. Contamination ultimately limits the degree of amplification that is practicable. The primer sequences must be chosen with care, and other sequences can be added to their 5′-end to facilitate their subsequent manipulation, such as sites for restriction enzymes to facilitate cloning (Section 6.3).

In **real-time PCR**, the amount of product formed is monitored during the course of the reaction by measuring the fluorescence of dyes or probes introduced into the reaction mixture; the number of amplification cycles required to obtain a particular amount of DNA molecules can be monitored. If the amplification efficiency is known, it is possible to calculate the number of DNA molecules of the amplified sequence that were initially present in the sample. The number of DNA molecules of a particular sequence in a complex sample can be determined with unprecedented accuracy.

Nested primers are complementary to sequences 3′ to the original primers and occur within the amplified sequence. They can be used to carry out a second PCR, to increase the specificity of the reaction and produce increased amounts of product.

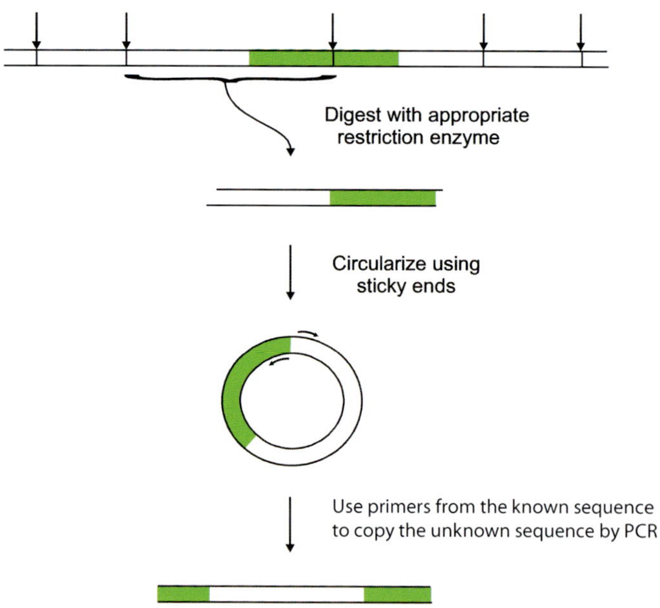

Figure 6-8. Inverse PCR. DNA with only one segment of known sequence (*colored*) is digested with an appropriate restriction enzyme at the sites indicated by *arrows*, to produce pieces that can be circularized by DNA ligase. Basic PCR is then performed using primers complementary to the segment of known sequence, designed so that the DNA polymerase extends the sequence outwards. Consequently, many copies of the restriction fragments containing the segment of unknown sequence result, and their sequences can be determined using normal methods.

Inverse PCR can be used to clone and determine the unknown DNA sequence adjacent to a segment of known sequence (Figure 6-8). The DNA is cleaved into pieces of an appropriate size, which are then converted to circles by DNA ligase (Figure 3-2). The circles containing the known sequence are subjected to PCR using primers complementary to the known sequence so that DNA synthesis occurs outwards from the known sequence. As a result, the segment with unknown sequence is amplified and ready to be sequenced in the normal manner (Section 6.4).

ImmunoPCR is based on chimeric conjugates of specific antibodies linked to nucleic acid molecules. The presence of such an antibody, for example if it is bound to a specific antigen, is detected by amplifying the attached nucleic acid molecule by PCR to generate the signal that is measured. The enormous efficiency of nucleic acid amplification typically leads to a 10^2- to 10^4-fold greater sensitivity than normal immunoassays.

An introduction to PCR techniques. G. Rumsby (2006) *Methods Mol. Biol.* **324**, 75–89.

Introduction to the polymerase chain reaction. Y. M. Lo & K. C. Chan (2006) *Methods Mol. Biol.* **336**, 1–10.

The real-time polymerase chain reaction. M. Kubista *et al.* (2006) *Mol. Aspects Med.* **27**, 95–125.

6.2. PRODUCING RNA

RNA is most commonly produced in nature using DNA as a template, but many viruses have RNA genomes that must be replicated. The enzymes responsible in both instances are known as **RNA polymerases**, being distinguished as either DNA-dependent or RNA-dependent.

6.2.A. RNA Replication: RNA Replicases

Replication of RNA occurs only with the few genomes, primarily those of viruses, that consist of RNA. The RNA-dependent RNA polymerases, or RNA replicases, that replicate viral RNAs are similar in many ways to the polymerases that replicate DNA (Section 6.1.A) but they are not so well characterized, nor are they used routinely in molecular biology. They are often complex molecules containing additional enzymatic activities, such as RNA helicase and others that modify the RNA covalently. Some of their polypeptide chains originate from their host cells. Their most remarkable features are that they (1) are specific for their viral RNA even though they function in the presence of a large excess of host RNAs, (2) amplify their RNA template many-fold during a short infection period, (3) copy the entire template RNA without the use of primers and (4) make about one mistake in every 10^3–10^4 nucleotides incorporated. The error rate is relatively high because they do not have any proofreading activity; this causes RNA viruses to mutate and evolve rapidly.

Alfalfa mosaic virus coat protein bridges RNA and RNA-dependent RNA polymerase *in vitro*. V. L. Reichert *et al.* (2007) *Virology* **364**, 214–226.

In vitro assembly of the Tomato bushy stunt virus replicase requires the host heat shock protein 70. J. Pogany *et al.* (2008) *Proc. Natl. Acad. Sci. USA* **105**, 19956–19961.

6.2.B. Transcription: DNA-dependent RNA Polymerases

The first step in expressing the genetic information in DNA genomes is to produce RNA molecules using the DNA as a template by DNA-dependent RNA polymerases, a process known as **transcription**. The RNA molecules produced may, after modification, function directly, as in the case of transfer RNAs and ribosomal RNAs, or they are messenger RNAs and translated into proteins (Section 7.5.F). The initiation of transcription tends to be where gene expression is regulated, and each gene must be subject to the appropriate control. This requires a very sophisticated apparatus, of which the DNA-dependent RNA polymerase is just a part. As might be expected, these enzymes are physically complex and consist of multiple protein subunits. They usually include a core enzyme that is fully capable of elongation of an RNA chain and its termination but is unable to initiate transcription. The other components are involved in selection of the site at which RNA synthesis will be initiated (known as the **promoter**) and the rate at which it will be transcribed. Very many different proteins are known to be involved and are referred to collectively as **transcription factors**.

Several distinct types of RNA polymerases are often present in an organism. Eukaryotes typically have three that decode the nuclear genome into RNA: DNA-dependent RNA polymerases I, II and III (Pol I, II and III). Pol I is situated exclusively in the nucleolus, where its function is to synthesize ribosomal RNAs. Pol II is responsible for transcribing genes into messenger RNA that will be translated into proteins, plus a few small nuclear RNAs. Pol III is restricted to transcribing a small set of highly expressed, nonprotein-coding (nc) RNA genes, including those for the 5S ribosomal

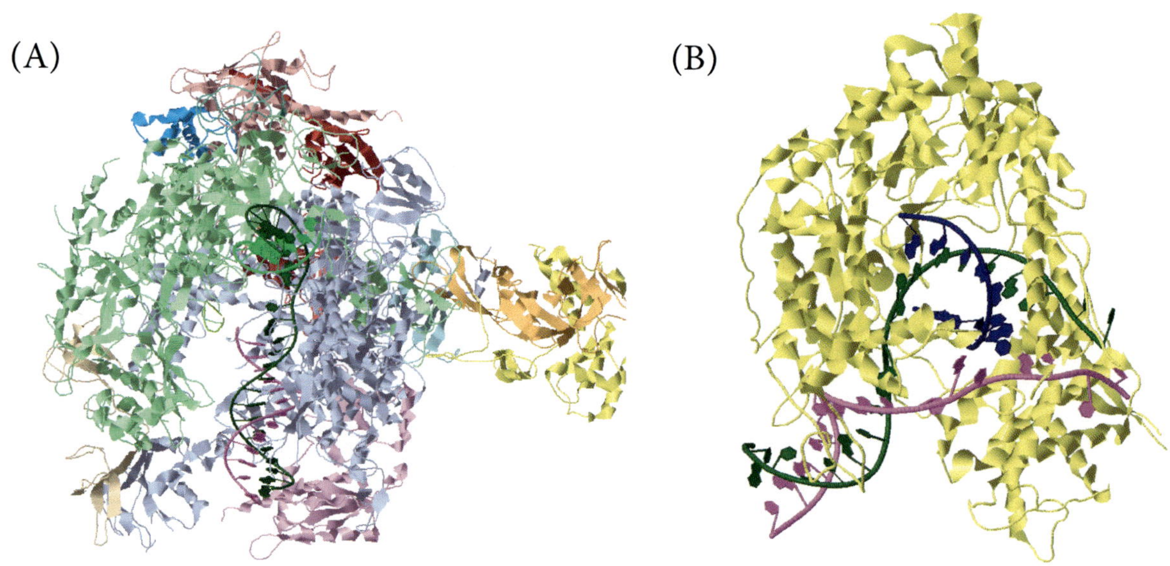

Figure 6-9. DNA-dependent RNA polymerases transcribing DNA. (A) The eukaryotic RNA polymerase complex of 12 polypeptide chains that elongates the messenger RNA. The nucleic acids are depicted in skeletal form, without the details of the polynucleotide backbone. Only the polypeptide backbones of the proteins are depicted, as ribbons, so that the nucleic acids at the active site are visible. Otherwise, they would be buried in the protein and not visible from the outside. (B) The much simpler single-polypeptide chain RNA polymerase from phage T7 transcribing a DNA duplex into RNA. The two DNA strands have been separated within the active site, and the RNA chain being synthesized off one is depicted in *purple*. Figure generated from PDB files *2owo* and *1msw* using the program Jmol.

RNA, transfer RNAs, small nuclear RNAs, microRNAs, short interspersed nuclear element-encoded or transfer RNA-derived RNAs and novel classes of ncRNA that can display significant sequence complementarity to protein-coding genes and might thus regulate their expression. Higher plants have two additional nonessential polymerases, Pol IVa and Pol IVb, that specialize in small RNA-mediated gene silencing pathways.

DNA-dependent RNA polymerases must identify the sequence where transcription should be initiated (the promoter), initiate transcription, translocate and pause along the DNA template, proofread errors, and ultimately terminate transcription at the correct location. Initiation by RNA Pol II requires the assembly of a specific group of general transcription factors at each promoter; the minimal initiation complex contains at least 30 different proteins! The process of elongation is nearly as complex (Figure 6-9-A). *In vivo*, DNA tends to be wrapped in chromatin (Section 3.7), which must be disassembled transiently; this requires the coordinated efforts of Pol II and its associated C-terminal domain kinases, elongation complexes, chromatin-modifying enzymes, chromatin-remodeling factors, histone chaperones (nucleosome assembly factors) and histone variants.

Bacteria have some of the most simple control systems, but even here transcription is complex. The core polymerase consists of four different polypeptide chains with a stoichiometry of $\alpha_2\beta'\beta\omega$; the subunits β' and β are most important catalytically and are similar to the subunits of eukaryotic enzymes. A fully functional enzyme, however, also requires a sigma factor (σ) that is essential for the RNA polymerase to bind to a specific promoter. There are various sigma factors, and bacteria exchange the sigma subunit in the RNA polymerase in response to different environmental stimuli and thereby tune their gene expression according to their current needs. Sigma factors are thought to recognize clearly distinguishable promoter DNA determinants and thereby activate distinct gene sets. The sigma factors themselves are regulated by antisigma factors, which bind and inhibit their cognate sigma factor, and 'appropriators', which deploy a particular sigma-associated RNA polymerase to a specific class of promoters. Adding to the complexity is the regulation of antisigma factors by both anti-antisigma factors, which turn on sigma factor activity, and co-antisigma factors, which act in concert with their partner antisigma factor to inhibit or redirect sigma activity. While the structures and functions of sigma factors are highly conserved, they are sufficiently diverse to enable bacteria to regulate sigma factor activities and respond to the diversity of environmental cues to which the bacterial transcription system has evolved.

The steps in transcription are similar for all DNA-dependent RNA polymerases. Initially the active form of the enzyme binds to the DNA at the appropriate promoter region and separates the two DNA strands there into a 'bubble' of individual strands, consisting usually of about 18 base pairs. Synthesis of a complementary copy of one of the strands is then initiated, but the synthesis is often aborted and the short RNA molecules produced are released. Once a transcript reaches about 8–12 nucleotides, the sigma factor dissociates from the complex and the remaining core enzyme continues extending the transcript without dissociating from the DNA. The 'bubble' of separated strands progresses along the DNA with the polymerase (Figure 6-9). The steps involved in adding each nucleotide to the complementary strands are similar to those involved in DNA synthesis using DNA polymerases (Figure 6-4). At the appropriate point, transcription is terminated and the transcript and polymerase are released.

The complexities of these DNA-dependent RNA polymerases have only been alluded to, but as a result these enzymes are not useful in the molecular biology laboratory to produce RNA from a piece of DNA. This would require all the transcription factors and other ancillary factors to be known for each piece of DNA, which would be impractical and is not feasible at the present time. Fortunately, much more simple enzymes are available from bacteriophages.

RNA polymerase: the vehicle of transcription. S. Borukhov & E. Nudler (2008) *Trends Microbiol.* **16**, 126–134.

RNA polymerase elongation factors. J. W. Roberts *et al.* (2008) *Ann. Rev. Microbiol.* **62**, 211–233.

RNA polymerase II stalling: loading at the start prepares genes for a sprint. J. Q. Wu & M. Snyder (2008) *Genome Biol.* **9**, 220.

Single-molecule studies of RNA polymerase: motoring along. K. M. Herbert *et al.* (2008) *Ann. Rev. Biochem.* **77**,149–176.

---

1. Single-subunit DNA-dependent RNA Polymerases

The enzymes from bacteriophages T7, T3, SP6 and K11 have been characterized, with that from phage T7 serving as a model. Remarkably, in comparison with the systems described above, the single polypeptide chain of this enzyme is complete and fully functional in initiating, elongating and terminating transcription of any DNA with the appropriate signals in its sequence (Figure 6-9-B). Perhaps not surprisingly, the enzyme is similar structurally to the Klenow fragment of DNA polymerase I (Figure 6-3). The enzyme binds to DNA at a region 6–17 base pairs to the 5′-side of the transcription start site (known as upstream); these base pairs are designated as being at positions –17 to –6. Transcription is initiated when the enzyme binds to sites –5 to +6. Elongation of the transcript requires a substantial change in the structure of the complex of enzyme and DNA.

The phage T7 RNA polymerase does not recognize the transcription signals of the *E. coli* host, nor do the bacterial polymerases recognize the T7 signals. A further useful property of T7 RNA polymerase is that it differs from the *E. coli* enzyme in not being inhibited by the antibiotic **rifampicin**. Consequently, transcription by the host can be blocked while not affecting that by the phage T7 RNA polymerase. The T7 transcription signals are well-known and lie upstream and downstream of the nucleotide sequences to be transcribed; consequently, they can be attached at both ends of a segment of DNA so that it will be transcribed efficiently by the T7 polymerase. A variety of plasmid vectors are available commercially for transcribing any DNA fragment that is inserted into one of them (Figure 6-12).

Mechanism for *de novo* RNA synthesis and initiating nucleotide specificity by T7 RNA polymerase. W. P. Kennedy *et al.* (2007) *J. Mol. Biol.* **370**, 256–268.

Functional architecture of T7 RNA polymerase transcription complexes. D. Nayak *et al.* (2007) *J. Mol. Biol.* **371**, 490–500.

The structure of a transcribing T7 RNA polymerase in transition from initiation to elongation. K. J. Durniak *et al.* (2008) *Science* **322**, 553–557.

T7 RNA polymerase studied by force measurements varying cofactor concentration. T. Thomen *et al.* (2008) *Biophys. J.* **95**, 2423–2433.

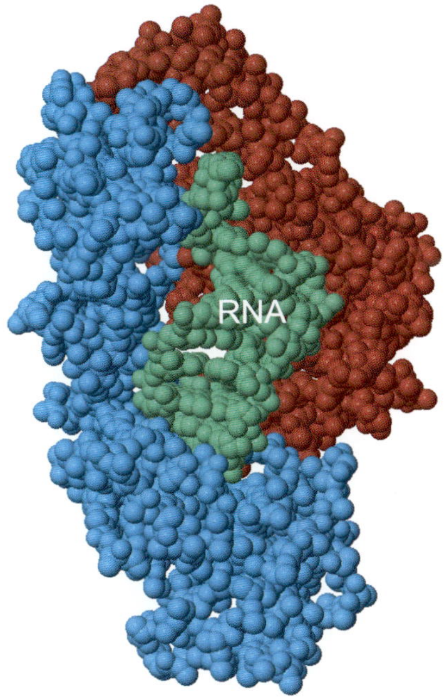

Figure 6-10. The reverse transcriptase of retrovirus HIV-1 binding an RNA pseudoknot. The van der Waals radii of the atoms of all the molecules are shown. The RNA is *green*, whereas the two polypeptide chains of the protein are *blue* and *red*. Figure generated from PDB file *1hvu* using the program Jmol.

6.2.C. Reverse Transcription: RNA into DNA

DNA is not normally produced using RNA as a template, but it does occur with retroviruses that have RNA genomes. These viruses insert their genomes into that of their host cell, which requires that the RNA genome first be converted to one of DNA. This is accomplished by an enzyme produced by the virus, **reverse transcriptase**, which is an 'RNA-dependent DNA polymerase'. It is similar in many respects to other DNA polymerases but is specific for copying RNA (Figure 6-10). The enzyme requires a primer complementary to the RNA strand to be copied. This role is played *in vivo* by a transfer RNA of the host, and the copying process is very complex but results in a double-stranded DNA copy. Reverse transcriptase has a high error rate when transcribing RNA into DNA, in the range of 1 mistake per 17,000–30,000 bases, because it has no proofreading ability. Retroviruses are responsible for human diseases such as AIDS, and currently one of the best ways to fight such a virus is to inhibit its reverse transcriptase activity; such inhibitors tend to be nucleoside analogs (Figure 2-12) and are some of the most powerful anti-AIDS medications available.

Associated with reverse transcriptase is **ribonuclease H** (H for 'hybrid') activity, which is specific for DNA•RNA hybrids and degrades the RNA strand that served as a template. This frees the DNA strand synthesized so that it can be used as a template to synthesize its complementary strand, and reverse transcriptase synthesizes both strands of the DNA copy. Other ribonucleases H can be used to target RNA molecules that have hybridized with a DNA probe.

Reverse transcriptase is used most frequently in molecular biology to make a **complementary DNA** (cDNA) copy of an RNA molecule, so that it can be cloned (Section 6.3.B). The RNA molecule is usually a messenger RNA that would be translated to make a protein. The primer in this case is poly(dT), which is complementary to the poly(A) sequence that is usually at the 3′-ends of messenger RNAs.

Reverse-transcription PCR is PCR (Section 6.1.C) performed on RNA. The first step is the production of a cDNA strand from the RNA, using the enzyme reverse transcriptase rather than DNA polymerase.

Performing quantitative reverse-transcribed polymerase chain reaction experiments. G. Lutfalla & G. Uze (2006) *Methods Enzymol.* **410**, 386–400.

RNA integrity and the effect on the real-time qRT-PCR performance. S. Fleige & M. W. Pfaffl (2006) *Mol. Aspects Med.* **27**, 126–139.

Crystal structure of type 1 ribonuclease H from hyperthermophilic archaeon *Sulfolobus tokodaii*: role of arginine 118 and C-terminal anchoring. D. J. You *et al.* (2007) *Biochemistry* **46**, 11494–11503.

Murine leukemia virus reverse transcriptase: structural comparison with HIV-1 reverse transcriptase. M. L. Coté & M. J. Roth (2008) *Virus Res.* **134**, 186–202.

6.2.D. Antisense Oligonucleotides

A complementary oligonucleotide can hybridize with a target single-stranded DNA or RNA molecule and thereby block its subsequent use. This is especially useful with a messenger RNA (mRNA), which conveys the information from the gene about the amino acid sequence of the protein product of that gene. When expression of this protein is to be blocked in this way, the complementary oligonucleotide is known as an **antisense oligonucleotide**. Formation of the antisense•mRNA heteroduplex in the cell can (1) trigger ribonuclease H activity, leading to mRNA degradation, (2) induce arrest of translation by steric hindrance of ribosomal activity, (3) interfere with mRNA maturation by inhibiting its splicing, or (4) destabilize mRNA precursors in the nucleus; all diminish expression of the target gene. Inhibiting the expression of specific genes helps to elucidate the physiological role of the gene and its product. Such an antisense oligonucleotide might also be useful therapeutically, for example to block an infection by a bacterium or virus. To be useful *in vivo*, however, an antisense oligonucleotide must be much more stable than ordinary nucleic acids, and it must be able to reach the interior of a cell (Section 5.4).

Phosphorothioate oligodeoxynucleotides have been studied extensively because they are resistant to hydrolysis by nucleases. They differ from normal DNA in that one of the nonbridging O atoms in each phosphate group is replaced by a S atom. The melting temperature of a heteroduplex between a phosphorothioate oligonucleotide and RNA is lower than for the DNA•DNA duplex by about 0.5°C for each nucleotide, and some 2.2°C lower than for an RNA•RNA duplex. Consequently, these antisense oligonucleotides must be relatively long, typically at least 17–20 nucleotides, to bind specifically to RNA, and they are not effective if the RNA sequence is involved in a double-stranded structure. Unfortunately, they have been found to bind to and inhibit enzymes involved with nucleic acids, such as DNA polymerase, be immunogenic and have a variety of unexpected side-effects. A key drawback lies in their polydiastereomerism (Section 1.1.A.3) because the various diastereomeric species with the S atoms in different positions exhibit different chemical and biological properties. Another type of molecule that can be useful as an antisense oligonucleotide is a nucleic acid mimic, such as a peptide nucleic acid (Section 5.4).

The phenomenon of antisense also seems to be used naturally, as many genes contain segments on both DNA strands of the genome that are transcribed, and the two oppositely oriented transcripts can overlap. Often, one strand codes for a protein, whereas the transcript from the other strand is nonencoding. Such **natural antisense transcripts** (NATs) can inhibit translation of the sense transcript. NATs are highly prevalent in a wide range of species; for example, around 15% of human protein-encoding genes have an associated NAT. What they do *in vivo* is a mystery.

Catalytic RNAs and DNAs (Chapter 8) can also function as antisense reagents. They cleave target RNAs after base pairing via their antisense flanking arms.

A related antisense phenomenon is **RNA interference**, in which small interfering double-stranded RNA molecules induce RNA degradation through a natural gene-silencing pathway. Being double stranded, they contain both sense and antisense regions of the target RNA. They cause destruction of the target RNA by a poorly understood natural phenomenon in which they are incorporated into the RNA interference (RNAi) targeting complex of RNA and protein that then cleaves the target RNA.

Identifying accessible sites in RNA: the first step in designing antisense reagents. W. H. Pan & G. A. Clawson (2006) *Curr. Med. Chem.* **13**, 3083–3103.

Recent advances in the stereocontrolled synthesis of antisense phosphorothioates. Y. Lu (2006) *Mini Rev. Med. Chem.* **6**, 320–330.

Genome-wide natural antisense transcription: coupling its regulation to its different regulatory mechanisms. M. Lapidot & Y. Pilpel (2006) *EMBO Rep.* **7**, 1216–1222.

Antisense oligonucleotides: target validation and development of systemically delivered therapeutic nanoparticles. C. Zhang *et al.* (2007) *Methods Mol. Biol.* **361**, 163–185.

RNA learns from antisense. D. R. Corey (2007) *Nature Chem. Biol.* **3**, 8–11.

6.3. CLONING

Manipulating DNA is at the very heart of molecular biology. For such an important technique, the physical principle is very simple: DNA is fragmented into pieces of the desired size, then the pieces are inserted into an appropriate **cloning vector** (Figure 6-11) that serves to replicate and store the DNA. The desired fragment is then identified and its cloning vector purified. This **clone** is then the source of pure DNA fragments, and cloning is one method of purifying and isolating a single homogeneous segment of DNA.

The cloning vector need have only sufficient genetic information to permit it to replicate within its host cell. Four basic types are used in the bacterium *E. coli* as a host: plasmids, bacteriophage vectors, cosmids and bacterial artificial chromosomes. All of them have been engineered to accommodate relatively large segments of foreign DNA, without impairing their ability to replicate within their host. **Plasmids** have minimal genomes, just sufficient for them to replicate within the host. They have the advantage of being smaller than the other vectors and of propagating with a high **copy number**, the number of copies of the vector per host cell. **Bacteriophage** vectors can accommodate larger inserts of DNA, up to 120,000 base pairs, and their growth and propagation are straightforward. **Cosmids** combine properties of the preceding two, in that they can replicate as plasmids or they can be packaged as bacteriophages. **Bacterial artificial chromosomes** and **yeast artificial chromosomes** can accept even larger DNA fragments, greater than 300,000 base pairs.

How the clone with the desired DNA fragment is identified depends upon many considerations, but it usually involves hybridization to a single-stranded DNA molecule with a complementary sequence (Section 5.3) or translation to produce a protein with the expected amino acid sequence, reactivity with an antibody or biological activity. Some cloning vectors permit the cloned DNA fragment to be

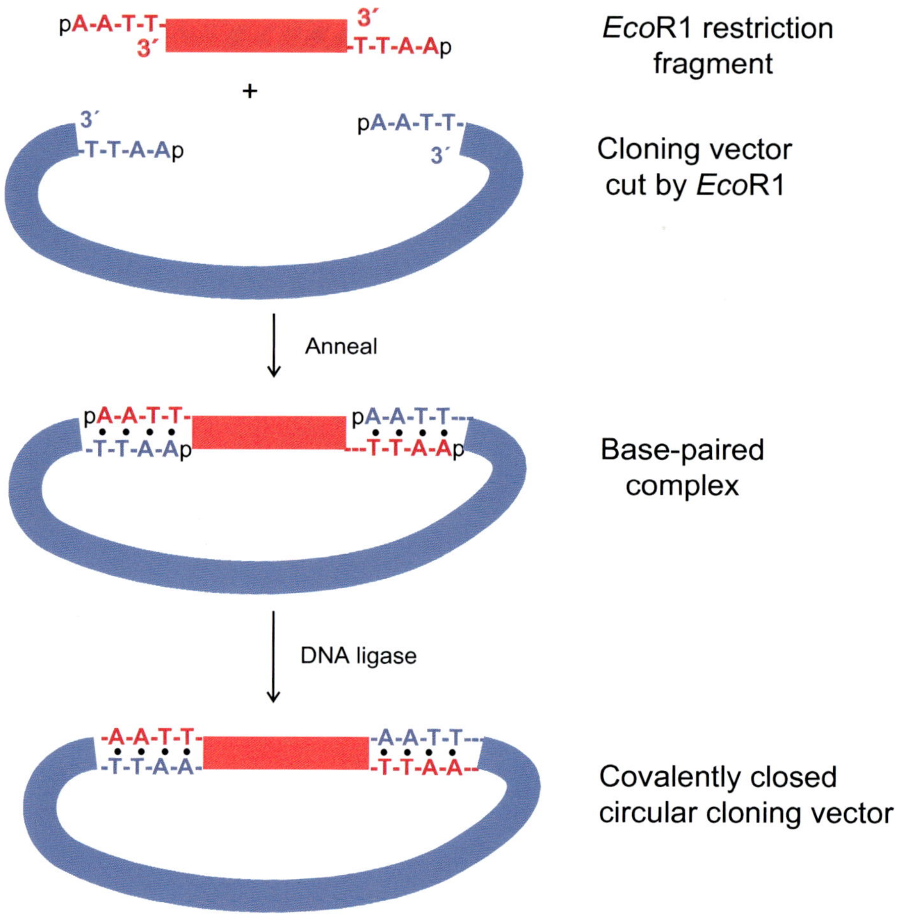

Figure 6-11. Cloning of a DNA fragment into a cloning vector using base pairing between the identical 'sticky ends' of the pair. In this case, both the DNA fragment to be cloned and the cloning vector have been generated by cleavage with the restriction enzyme *Eco*R1 (Equation 6.3). Their complementary sticky ends produce a base-paired complex. The two missing phosphodiester bonds are generated using an enzyme like DNA ligase.

transcribed, translated and expressed as a polypeptide chain (Section 6.3.A); if the product is a folded protein, its biological functions can be assayed.

Strategies for high-throughput gene cloning and expression. L. J. Dieckman *et al.* (2006) *Genet. Eng.* **27**, 179–190.

Enzyme free cloning for high throughput gene cloning and expression. R. N. de Jong *et al.* (2006) *J. Struct. Funct. Genomics* 7, 109–118.

6.3.A. Expression Vectors

Very often, the cloned DNA is to be transcribed into RNA and translated into a polypeptide chain. Special vectors for this purpose are available that contain all the sequences to signal the initiation and termination of transcription and translation. These signals are at both ends of the cloned sequence, so they are part of the vector and can be used to express any DNA fragment that is cloned in between them. The vector must be inserted into a suitable host cell for expression to occur. The host must provide all the functions necessary for expression; for example, an unprocessed gene will need the introns to be removed from the initial RNA transcript before it can be translated into protein.

A general expression vector for use in prokaryotes is described in Figure 6-12. It is a circular DNA duplex plasmid that contains a sequence serving as an **origin of replication** of the plasmid, so that it can be replicated and maintained within the host. There is usually a **selection marker gene** to ensure that the bacterial host retains the plasmid. This is usually a gene that conveys resistance to an antibiotic, so only cells containing the plasmid can grow in the presence of that antibiotic. A **promoter** sequence provides the site for initiation of transcription; it usually determines the degree to which expression will take place. Eukaryotic promoters usually have a **TATA box** about 30 nucleotides upstream of the transcription start site, with the consensus sequence T–A–T–A–A/T–A–A/T; it is recognized by one of the core transcription factors and positions the RNA polymerase on the DNA. A downstream sequence provides a site for the initiation of translation of the messenger RNA transcript, which must include a site for binding ribosomes and a **start codon**, usually AUG, for initiation of synthesis of the polypeptide chain. In prokaryotes, a **Shine–Dalgarno sequence** is positioned 4–13 nucleotides upstream of the start codon; it is complementary to the 3′-end of an RNA of the ribosome and positions the mRNA and ribosome appropriately. The DNA fragment to be cloned must be inserted just after the start codon and in such a way that the correct triplet reading frame is in phase with the start and stop codons. The coding sequence is followed by a **stop codon** to terminate polypeptide chain elongation. Finally there are appropriate signals to stop transcription and add the poly(A) tail to the messenger RNA (which is carried out by a specific enzyme). Certain sequences that have the opposite effect, **antitermination**, can be incorporated to stabilize the RNA polymerase on the DNA and increase elongation of the transcript.

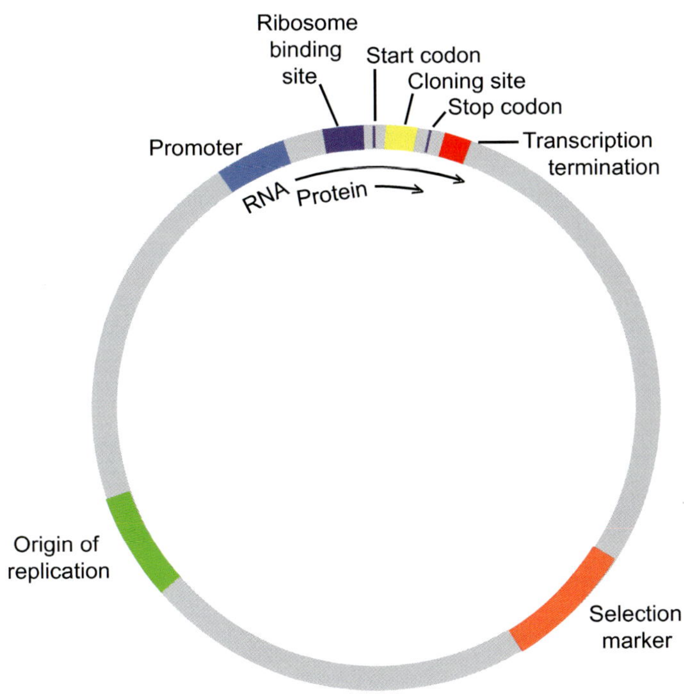

Figure 6-12. A generalized expression vector for prokaryotes. The *circle* represents the DNA duplex of the vector. The origin of replication is for replicating the vector in the host cell. The selection marker is usually a gene conferring on the host resistance to an antibiotic, so that only cells containing the plasmid can grow in the presence of that antibiotic. The promoter specifies initiation of transcription into RNA, which is terminated at the termination signal. The crucial sites on a messenger RNA are the translation site for binding a ribosome and the start and stop codons for producing the polypeptide chain. The DNA of interest is cloned between the latter two sites in such a way that its nucleotide sequence is translated into amino acid residues of the polypeptide chain.

Expression requires that the appropriate RNA polymerase be available. In prokaryotic cells it is usually that from phage T7 and its gene is present on the vector. It is often engineered to be expressed only under certain growth conditions, usually after addition of some inducer molecule. The T7 polymerase is then produced, and it can activate the expression vector. This makes it possible to control the expression of the cloned DNA, which often is deleterious to the host cell.

Baculovirus expression vector system: an emerging host for high-throughput eukaryotic protein expression. B. Shrestha *et al.* (2008) *Methods Mol. Biol.* **439**, 269–289.

pETPhos: a customized expression vector designed for further characterization of Ser/Thr/Tyr protein kinases and their substrates. M. J. Canova *et al.* (2008) *Plasmid* **60**, 149–153.

6.3.B. cDNA Libraries

Messenger RNAs contain the gene sequences of all the proteins that are produced in a cell. They can be readily converted to the amino acid sequences of those proteins because the genetic code is known and because the introns that interrupt the coding sequence in the gene have been removed after transcription. If even only a short segment of the amino acid sequence of the probe is known, it can be used to design a short nucleotide probe that will be complementary to the gene sequence, making it possible to identify the corresponding gene. There is redundancy in the genetic code, however, so the exact sequence of the probe cannot be specified. This can be circumvented by preparing a mixture of probes with all the possible nucleotides represented, so that at least a fraction of the probe molecules will have the correct sequence. Alternatively, less specific bases, such as inosine, can be used where the nucleotide sequence is uncertain.

RNA is much less stable and more difficult to handle than DNA, so the first step in isolating a messenger RNA is to convert it to DNA, using the enzyme reverse transcriptase (Section 6.2.C). Normally, all the messenger RNA of a cell is converted to cDNA and cloned, and the desired molecule is then identified. These steps use the tract of A nucleotides that most messenger RNAs have at their 3′-ends. The population of messenger RNAs can then be isolated by hybridizing them to an affinity column, a solid support containing complementary poly(T) oligonucleotides. All other molecules are removed by washing the solid support, before releasing the messenger RNAs. They are then used as templates for reverse transcriptase to synthesize DNA copies. Oligo-dT is used as the primer, so synthesis is initiated at the 3′-end of the messenger RNA. The resulting double-stranded DNA copy of the messenger RNA can be cloned into an appropriate vector.

Very often a cDNA copy is not complete, due to problems with the reverse-transcription step or because the messenger RNA molecules were not intact but degraded. Even short segments of a few hundred nucleotides can be very useful to identify an expressed gene, however, and they are known as **expressed sequence tags**.

Subtractive hybridization and construction of cDNA libraries. B. Blumberg & J. C. Belmonte (2008) *Methods Mol. Biol.* **461**, 569–587.

Identification of microRNAs and other small regulatory RNAs using cDNA library sequencing. M. Hafner *et al.* (2008) *Methods* **44**, 3–12.

6.3.C. Restriction Enzymes

Manipulating the very long molecules of DNA that occur naturally requires that they be cleaved into smaller fragments of manageable size, but not so small as to be indistinguishable from each other. **Restriction enzymes are endonucleases that recognize and cleave DNA molecules at specific sequences of 4–8 nucleotides**; they are the workhorses of molecular biology. They occur naturally in many bacteria, where they are used to recognize and cleave foreign DNA, such as that introduced by a bacterial virus. The initial fragmentation makes the DNA susceptible to further degradation by general exonucleases. **The host DNA is not cleaved because it has been modified covalently by methylation of specific cytosine or adenosine nucleotides** within the sequence recognized by the restriction enzyme, using a specific methyltransferase enzyme. One class of methyltransferases methylates the C5 carbon of cytosine, to form 5-methyl cytosine, whereas the other class methylates the 4-amino group of cytosine or the 6-amino group of adenine. The endonuclease and the methyltransferase act as a pair; in some cases, the two activities are catalyzed by a single protein molecule. Some bacteriophages have evolved systems to avoid destruction by restriction enzymes of their host by (1) producing proteins that inhibit the restriction enzymes, (2) stimulating the methylation system of the host, (3) modifying their DNA and (4) eliminating from their genomes sites recognized by the restriction enzymes.

Almost all restriction enzymes cleave at DNA sequences that are inverted repeats, in that they have the same sequence on both strands and a dyad symmetry axis (Section 2.2.C.6). Examples of such restriction sequences are the target sites for the restriction enzymes *Eco*R1 (GAATTC) and *Fin*II (CCGG).

Restriction enzymes generally cleave both strands of the DNA molecule, within or near the target sequence. The resulting fragments have a phosphate group at the new 5′-end and a hydroxyl group at the new 3′-end. Often, the bonds cleaved are not at the center of the target sequence, so cleaving between the same sequence on each strand generates fragments with protruding 5′- and 3′-single strands. For example, the *Eco*R1 restriction enzyme cuts between the G and A of the sequence GAATTC:

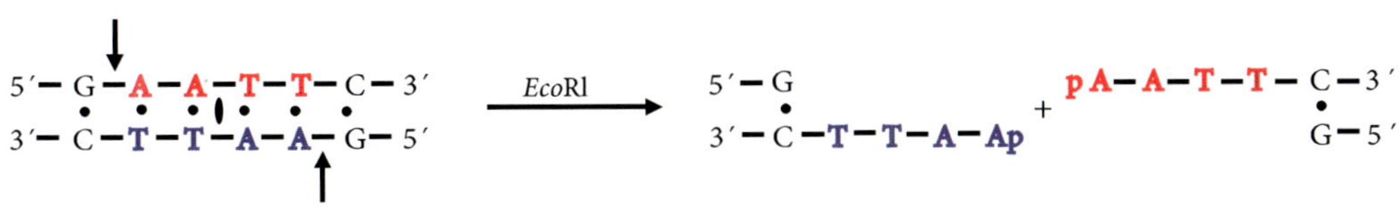

(6.3)

The protruding single strands are known as **staggered**, **sticky** or **cohesive** ends. They are extremely useful, as their base pairing with complementary sequences, perhaps generated by the same restriction enzyme, makes it easy to insert the DNA fragment into another DNA molecule (Figure 6-11). Such manipulations are central to the isolation and cloning of DNA.

Other restriction enzymes cleave at the center of the target sequence and thereby cleave at the phosphodiester bonds on opposite sides of the DNA helix. The resulting fragments do not have single-stranded ends but are known as **blunt-ended**. Lacking any sequence complementarity at the cloning site, they insert into other DNA molecules much less readily but there are no restrictions on what fragments can be cloned.

More than 2600 different restriction enzymes are known, with more than 230 cleavage specificities, and most are available commercially. Different restriction enzymes that recognize and cut at the same position in identical target sequences are referred to as **isoschizomers**. For example, the *Eco*RI and *Rsr*I restriction enzymes both recognize and cut between the G and A of the sequence GAATTC (Equation 6.3). Other restriction enzymes that recognize the same sequence but cut at different positions are known as **neoschizomers**. For example, restriction enzymes *Asp*718I and *Kpn*I both recognize the sequence GGTACC but one cleaves between the two Gs, the other between the two Cs.

Restriction enzymes are very specific for their target sequences under normal physiological conditions and in the presence of Mg^{2+}, but they become less specific under different conditions of pH, salt concentration or Ca^{2+} or Mn^{2+} in place of Mg^{2+}. Any activity towards alternative sequences is known as the **star activity**.

How restriction enzymes became the workhorses of molecular biology. R. J. Roberts (2005) *Proc. Natl. Acad. Sci. USA* **102**, 5905–5908.

The crystal structure of the rare-cutting restriction enzyme *Sda*I reveals unexpected domain architecture. G. Tamulaitiene *et al.* (2006) *Structure* **14**, 1389–1400.

DNA looping and translocation provide an optimal cleavage mechanism for the type III restriction enzymes. N. Crampton *et al.* (2007) *EMBO J.* **26**, 3815–3825.

Use of plasmon coupling to reveal the dynamics of DNA bending and cleavage by single EcoRV restriction enzymes. B. M. Reinhard *et al.* (2007) *Proc. Natl. Acad. Sci. USA* **104**, 2667–2672.

6.3.D. Restriction Maps

To elucidate the structure and sequence of large DNA molecules, it can be useful to **map the positions of restriction sites within the molecule**. This is accomplished by cleaving the large DNA molecule with two or more restriction enzymes that produce a reasonable number of fragments of convenient size. With very large DNA molecules, the restriction enzymes must have target sequences that occur rarely, whereas less specific restriction enzymes are useful with smaller molecules. The fragments resulting from cleavage by either or both restriction enzymes are separated according to their size by gel electrophoresis. The shorter fragments migrate more rapidly, and their mobilities can be compared with those of standard DNA molecules and used to estimate the length of each DNA molecule. This information is then used to construct a **restriction map, which gives the order of the fragments in the original DNA molecule**.

The procedure is illustrated using a short linear molecule of DNA in Figure 6-13. The DNA was digested with the restriction enzymes *Eco*RI and *Bam*H1, which cleave at G/AATTC and G/GATCC sequences, respectively (/indicates the bond at the site of cleavage). Each generated three fragments individually, and five in combination, indicating that the original DNA molecule had two restriction

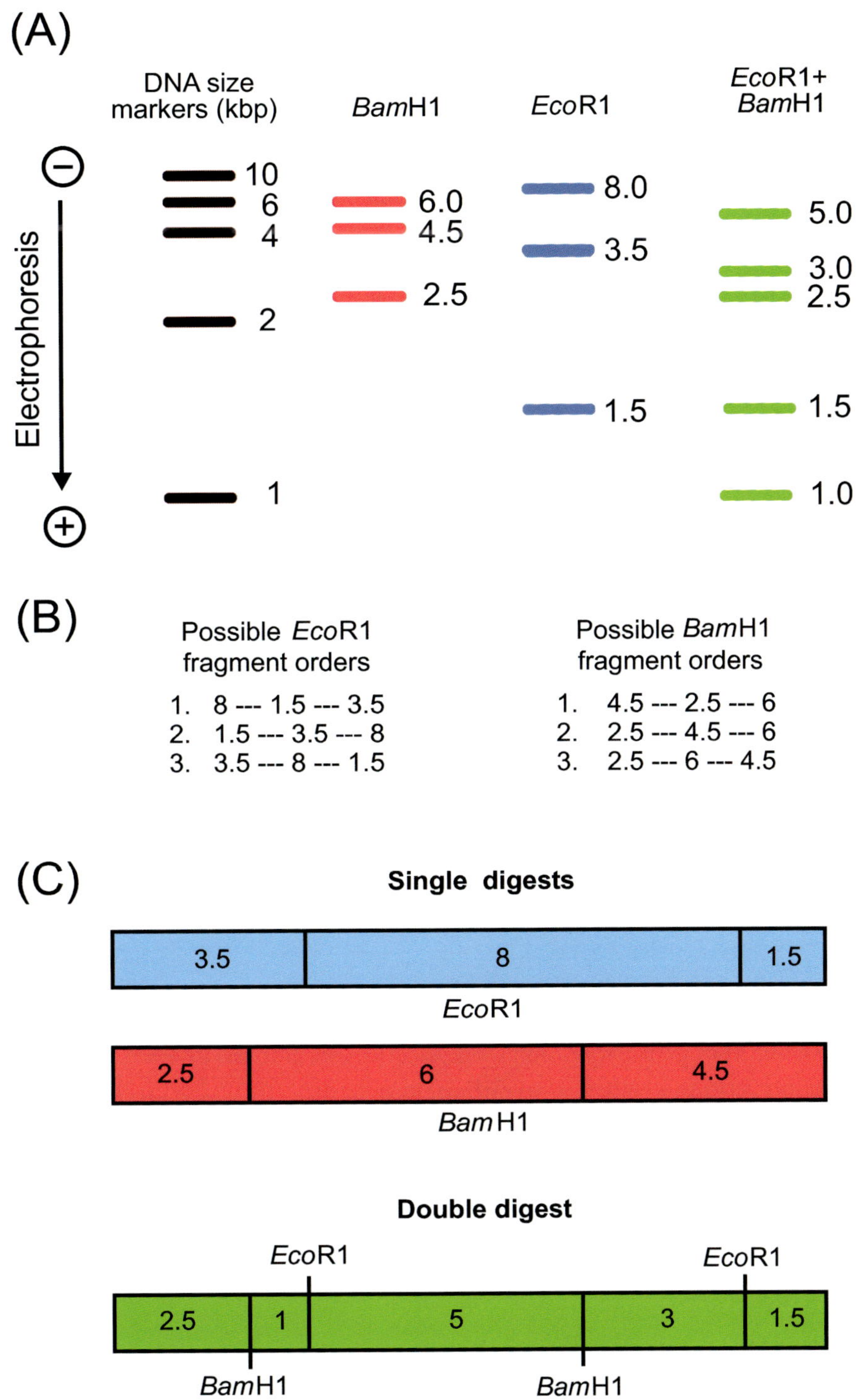

Figure 6-13. Constructing a restriction map. (A) Schematic illustration of the idealized pattern obtained after electrophoresis of the DNA fragments generated by cleavage of a 13,000-base pair DNA molecule with the restriction enzymes *Eco*R1 and *Bam*H1 individually or together. Electrophoresis was from the *top* to the *bottom*, with the smaller fragments migrating more rapidly. Their sizes are estimated by comparison to the standards in lane 1 on the *left*. The lengths of the fragments are given in thousands of base pairs (kbp). Each restriction enzyme produced three fragments, so there were two restriction sites for each enzyme. (B) All possible orders in the original DNA molecule of the three restriction fragments from each restriction enzyme. The orders (3) from both enzymes combine to reproduce the sizes of the fragments observed with the combined enzymes. (C) The positions of the restriction sites indicated by the data.

sites cleaved by each enzyme. The sizes of the five fragments generated by the two enzymes in combination indicate the positions of these sites within the original DNA molecule. This simple illustration is straightforward, but the procedure is much less so with large molecules containing many restriction sites. In cases of ambiguity, the original DNA molecule can be labeled at one end and the time-course of cleavage by the restriction enzymes monitored. A similar approach can be used with polypeptides. The labeled molecules all contain one end of the molecule, and fragments corresponding to all the restriction sites should occur at least transiently.

Restriction maps are also important in comparing the genomes of related individuals and species. Genetic mutations can alter the restriction sites or insert or delete genetic material between the sites.

An algorithm for assembly of ordered restriction maps from single DNA molecules. A. Valouev *et al.* (2006) *Proc. Natl. Acad. Sci. USA* **103**, 15770–15775.

6.4. SEQUENCING DNA

A most striking characteristic of natural DNA molecules is their lengths, which can be enormous. The genome of the relatively simple common bacterium *E. coli* is a single double-stranded DNA molecule (closed in a circle) in which each strand contains 4.6 million nucleotides. This corresponds to 9.2 million bits or 1.15 megabytes of information: each residue can be one of four bases and $4 = 2^2$, so each residue contains two bits of information, and 8 bits = 1 byte. DNA molecules from higher organisms can be much larger: each chromosome generally contains a single double-stranded DNA molecule. The human genome consists of approximately 3 billion nucleotides divided amongst 24 molecules (chromosomes). Individual chromosomes can consist of as many as 1 billion nucleotides that, if extended, would stretch 0.33 m in length. Yet **it is possible to determine exactly the covalent structures of these enormous molecules** in terms of their nucleotide sequences, and this is one of the most important accomplishments of molecular biology. The sequence information can then be used to deduce the function of the DNA segment, map its chromosomal location, determine how it might be involved in regulation of gene expression or replication, or elucidate how it interacts with proteins. Furthermore, the nucleotide sequences of the regions coding for protein can be translated, using the genetic code, into the amino acid sequences of the proteins, and these protein sequences can be compared with the sequences of all known proteins, giving clues to their functions and evolutionary origins.

Molecular engineering approaches for DNA sequencing and analysis. X. Bai *et al.* (2005) *Expert Rev. Mol. Diagn.* **5**, 797–808.

Faster, cheaper DNA sequencing. L. Sage (2005) *Anal. Chem.* **77**, 415A–416A.

Towards rapid DNA sequencing: detecting single-stranded DNA with a solid-state nanopore. H. Yan & B. Xu (2006) *Small* **2**, 310–312.

6.4.A. Isolating the DNA Fragments to be Sequenced

DNA sequencing requires that the DNA molecules be relatively short and homogeneous. Initially specific DNA segments were isolated by cloning them into vectors derived from the M13 bacteriophage; these vectors contain a single-stranded DNA chromosome that can accommodate inserts of up to 5000 bases of foreign DNA. The cloning produces homogeneous clones with a single DNA insert. Pure single-stranded DNA can be isolated from this phage easily, and this method is still in common use. Virtually any plasmid can be used in a similar way for DNA sequencing, and essentially any clone of up to about 200,000 bases can be isolated by cloning and then sequenced. DNA is frequently isolated for sequencing by using PCR (Section 6.1.C). Any segments of genomic DNA can be amplified for direct sequencing in a matter of hours.

6.4.B. Chain-termination, Sanger Method

The chain-termination method of DNA sequencing was first described by Sanger in 1977 and is now used almost universally. It involves the enzymatic synthesis of a DNA strand by a DNA polymerase *in vitro*, using the DNA strand to be sequenced as the template (Figure 6-1). Synthesis is initiated by DNA polymerases only at the 3′-end of a primer that is base paired to the DNA template, so synthesis is initiated at only that one site (Figure 6-2). The growing chain is terminated by the occasional random incorporation of a 2′,3′-**dideoxynucleoside triphosphate** (ddNTP), which has no free hydroxyl group and does not permit further DNA synthesis (hence the name *chain-termination*).

Most sequencing applications use as the primer a short synthetic oligonucleotide 18–35 nucleotides in length and complementary in sequence to the template at a unique position. This position is adjacent to, and determines, the region to be sequenced. The sequence of this part of the template must be known in advance; it can, fortunately, be part of the plasmid or virus in which the DNA segment was cloned. The primer is hybridized to the template at an appropriate temperature to ensure that it binds only in the correct position (Figure 6-2). After formation of this duplex, the primer is extended by the DNA polymerase in the presence of the four deoxynucleoside triphosphates: dGTP, dATP, dTTP and dCTP, which are often designated collectively as **dNTPs**.

Extension of the new strand requires a free 3′-hydroxyl group (Figure 6-4). One will be available for continued polymerization of the growing chain so long as normal dNTPs are incorporated (Figure 6-1). The ddNTPs, however, lack a 3′-hydroxyl group, and their incorporation into the DNA terminates chain elongation. Synthesis directed by a sample of template DNA with a unique primer in the presence of all four dNTPs, plus a small concentration of one ddNTP, will produce a population of molecules that have the same 5′-end but different 3′-ends; all will end with the same terminal dideoxynucleotide base, but at different positions. Consequently, the chains synthesized will have a distribution of sizes depending on the sites at which the ddNTP was incorporated (Figure 6-14) and the sizes of the various fragments are determined by the sequence of the template. Four reactions are performed, each with a different ddNTP. The products of the four reactions can be analyzed by electrophoresis on a denaturing polyacrylamide gel, which separates the products by size and can resolve DNA strands differing in length by a single nucleotide. The sequence of the template can be determined from the order of the bands on the gel (Figure 6-15). The shortest oligonucleotides migrate most rapidly, and the sequence can be read off starting there by the presence of bands in the mixtures with the four different ddNTPs. Of course, this sequence is of the strand being synthesized and is the complement of the template strand.

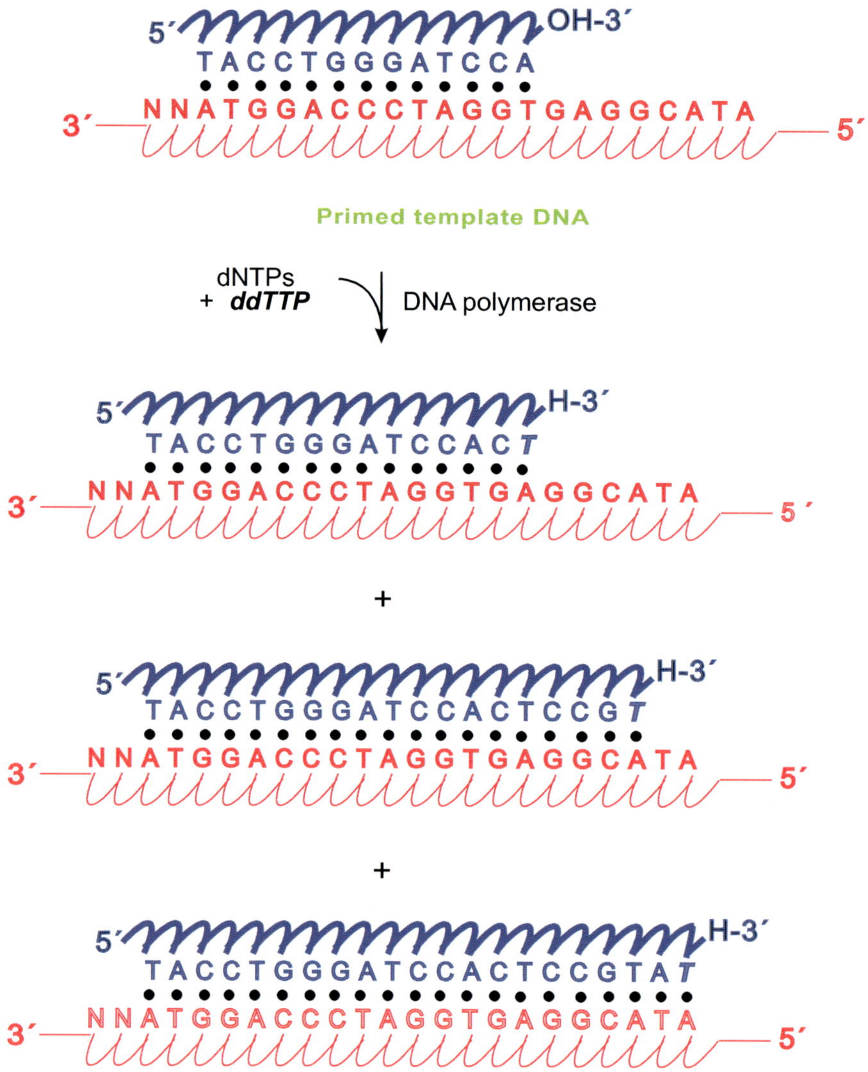

Figure 6-14. Chain-termination by incorporation of a ddNTP. Copying the primed template from Figure 6-2 in the presence of all four deoxynucleoside triphosphate and one dideoxynucleoside triphosphate (ddNTP, in this case ***T***, indicated by *bold italic*) yields a population of extended primer molecules with the same 5′-ends but different 3′-ends lacking a hydroxyl group, depending on the site at which a ddNTP was incorporated. Consequently, the lengths of the extended primer molecules indicate the positions in the sequence of T nucleotides (A on the other strand).

The **DNA polymerase** enzyme is the most critical component. Many DNA polymerases and reverse transcriptases have been used for sequencing. Most have been modified genetically or chemically to eliminate exonuclease activities or improve their utilization of the ddNTPs. The most popular polymerases designed for DNA sequencing are heat-stable because they originate from the thermophile *Thermus aquaticus* and are known as ***Taq* DNA polymerase**. The 5′–3′-exonuclease activity of natural *Taq* DNA polymerase can degrade the sequencing primers, so it is usually removed. This activity is due to the first 300 amino acid residues at the N-terminus of this enzyme, which can be deleted or a few crucial residues altered. The limited ability of *Taq* DNA polymerase to incorporate ddNTPs is improved more than 10^4-fold simply by changing residue Phe667 to Tyr. The sequencing

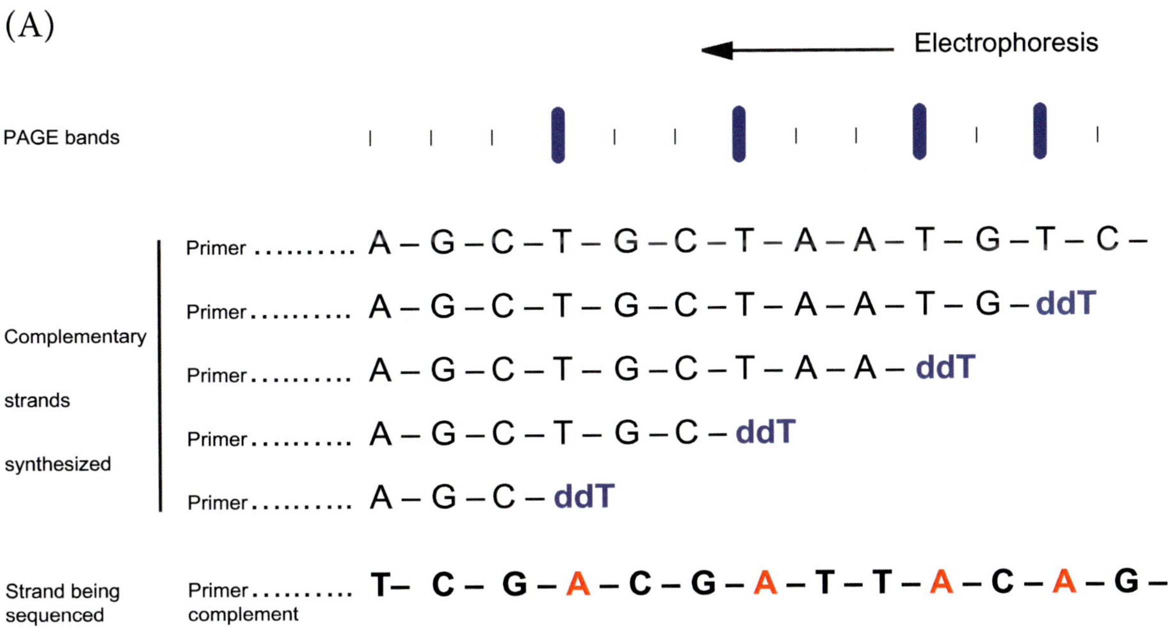

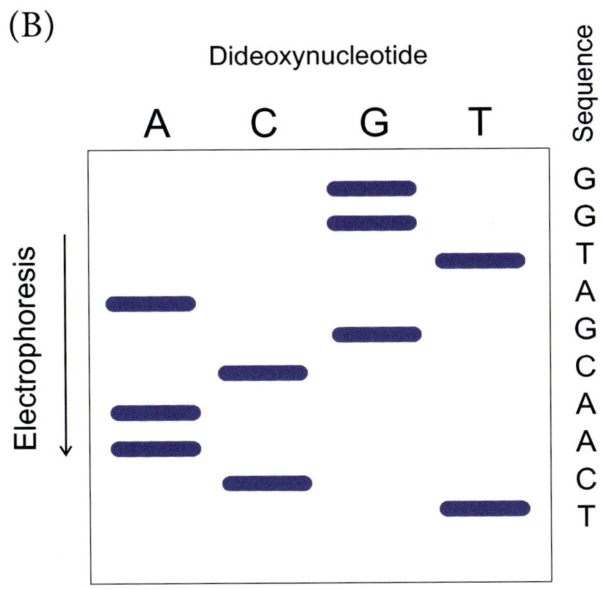

Figure 6-15. Sequencing DNA using the chain-termination method. (A) Complementary strands are synthesized enzymatically from the primer, using the DNA strand as template (Figure 6-1). The complementary strands are elongated until a dideoxynucleotide is incorporated in place of the normal one, which terminates strand elongation (Figure 6-14). A set of complementary strands with their lengths determined by the positions of these nucleotides will result. The same mixtures but using the other three dideoxynucleotides would be analyzed at the same time (B). The complementary strands synthesized are separated by electrophoresis that can resolve strands differing in length by a single nucleotide. The shorter strands migrate more rapidly. The positions of the bands indicate the positions of the four nucleotides in the synthesized DNA strand, and the complete sequence of both strands can be determined.

bands produced should have the same intensity, but those produced by natural *Taq* DNA polymerase vary in intensity more than 15-fold and depend on the nearby sequence. In contrast, those produced by the Tyr667 polymerase vary in intensity less than three-fold, which makes interpretation of the sequencing results much more accurate. The DNA polymerase from phage T7 naturally has a Tyr at the corresponding position and functions similarly, but it is not heat-stable.

Cycle sequencing uses repeated cycles of thermal denaturation and polymerization to increase the amount of product in a DNA sequencing reaction and the sensitivity of the sequencing experiment. During each cycle, the thermostable DNA polymerase extends the annealed primer molecules. The mixture is then heated to 95°C to dissociate the extended primer from the template. The mixture is cooled, allowing another molecule of the excess primer to anneal to the template, and the elongation reactions are repeated. The amount of product DNA increases linearly with the number of cycles, and it can become much greater than the amount of template used originally. The dissociation step enables double-stranded templates to be used directly for sequencing. Cycle sequencing is usually more reliable over a wider range of template concentrations than are other methods. It is used for most large-scale DNA-sequencing projects.

All DNA-sequencing methods require that the strands synthesized be labeled. The label used originally was α-^{32}P dATP. It was simply added to the chain-termination reaction so that the newly synthesized DNA molecules were labeled uniformly with ^{32}P and could be detected by autoradiography. Subsequently, the isotopes ^{33}P and ^{35}S have been used because their radioactive emissions have lower energy and they consequently produce higher resolution autoradiograms. ^{35}S is used in the form of α-thio-dATP, in which an O atom of the phosphate group is replaced by S. Limiting the radioactivity to the terminating ddNTPs offers the advantage that the only DNA chains labeled are those that were elongated and terminated specifically. This diminishes the background radioactivity and produces extremely clean DNA sequencing autoradiograms.

DNA-sequencing methods that use fluorescence rather than radioactivity have been automated and have become essential for large-scale sequencing projects. Either the primers (dye primers) or the dideoxynucleotides (dye terminators) can be made fluorescent and used to label the DNA chains produced by chain-termination. The four bases can be distinguished if they fluoresce with different colors, which makes it possible to use all four in the same sequencing mixture. Fluorescent dye primers can use fluorescence energy transfer to optimize the absorption and emission properties of the label. For example, they might have a fluorescein derivative at the 5′-end that serves as a common fluorescence donor, with the fluorescence acceptor being a rhodamine derivative attached to a modified thymidine within the primer sequence. The spacing between fluorescence donor and acceptor can be varied by the position of the modified thymidine in the primer sequence, altering the efficiency of fluorescence energy transfer. Four distinguishable primers can be devised, one for each of the four bases: all absorb strongly at the same excitation wavelength of 488 nm provided by the fluorescein acceptor, but differ in the wavelength of the fluorescent light they emit; for example, those for the four different nucleotides can emit light at 525, 555, 580 and 605 nm. All four primers can be combined in a single sequencing reaction, improving the sensitivity and accuracy as well as providing the complete sequence of all four nucleotides.

ddNTP terminators labeled with four different fluorescent dyes have also been used extensively for DNA sequencing and are commercially available. Like the radioactive terminators, they have the advantage of labeling only DNA chains that have been elongated and terminated specifically, so background bands are minimized.

Emerging technologies in DNA sequencing. M. L. Metzker (2005) *Genome Res.* **15**, 1767–1776.

Microfabricated bioprocessor for integrated nanoliter-scale Sanger DNA sequencing. R. G. Blazej *et al.* (2006) *Proc. Natl. Acad. Sci. USA* **103**, 7240–7245.

Four-color DNA sequencing by synthesis using cleavable fluorescent nucleotide reversible terminators. J. Ju *et al.* (2006) *Proc. Natl. Acad. Sci. USA* **103**, 19635–19640.

6.4.C. Separating the DNA Fragments by Size

All sequencing methods require the separation of DNA fragments on the basis of their size, and **it must be possible to separate molecules differing in length by a single nucleotide**. A sieving gel is usually necessary to separate the DNA fragments by size; otherwise, DNA molecules of different lengths have identical charge-to-mass ratios and the same electrophoretic mobility in aqueous solution. In the original DNA sequencing experiments using radioactive labeling, separation was by electrophoresis in 4–8% (w/v) cross-linked polyacrylamide gels, containing 7–8 M urea to keep the DNA molecules unfolded. The gels were 0.2–0.4 mm thick and cast between glass plates that were 40–80 cm long; each gel could accommodate up to 96 samples. After electrophoresis for 2–18 hours, the gels were separated from the glass plates and subjected to autoradiography. Usually 200–400 nucleotides of the sequence could be determined from a single sample. These procedures are labor-intensive, however, so there was much incentive to improve the separation methods.

Current separation methods monitor continuously the fluorescence due to the labeled DNA molecules as they migrate electrophoretically past a fixed position on the gel. Noncross-linked gels in capillaries that are 50–100 μm in diameter and 0.4 –0.7 m long provide faster and higher resolution electrophoresis separations, due to the efficient heat transfer of these narrow capillaries. A sequence of 500 or more nucleotides can be determined from a single experiment. A major advantage is that the results are collected and evaluated directly by computer, saving human labor.

End-labeled free-solution electrophoresis overcomes the need for a gel by using a label (or 'drag-tag') to cause the free solution mobility of the DNA to depend upon the size of an oligonucleotide. Attachment of an uncharged drag-tag molecule of a fixed size to various lengths of DNA in the sample selectively slows down smaller DNA chains, because they have less electrophoretic force to pull the drag-tag than do longer DNA molecules. Other variations involve separating the fragments by mass spectrometry.

To obtain the sequence of a very large DNA molecule, such as that of a complete genome, requires short fragments to be sequenced and then reassembled into the complete molecule using overlaps of the sequences of the various segments. A logical approach would be to sequence, say, 400 residues of one segment, then use a short sequence of it to design a primer to sequence the next 400 nucleotides, and so forth, but this is relatively slow. Other approaches to sequencing large genomes fragment the DNA into relatively short pieces of DNA randomly and insert all of them into an appropriate cloning vector; the individual clones are purified and all are sequenced. The original sequence is assembled by finding the same sequence in the various overlapping clones. This approach of random **shotgun sequencing** requires that each nucleotide be sequenced at least twice, and usually many more times, but it can be more efficient than trying to do the sequencing in a rational, logical order.

Recent developments in mass spectrometry for the characterization of nucleosides, nucleotides, oligonucleotides, and nucleic acids. J. H. Banoub *et al.* (2005) *Chem. Rev.* **105**, 1869–1915.

Mass-spectrometry DNA sequencing. J. R. Edwards *et al.* (2005) *Mutation Res.* **573**, 3–12.

DNA sequencing based on intrinsic molecular charges. T. Sakata & Y. Miyahara (2006) *Angew. Chem. Int. Ed. Engl.* **45**, 2225–2228.

Novel design of multicapillary arrays for high-throughput DNA sequencing. A. Tsupryk *et al.* (2006) *Electrophoresis* **27**, 2869–2879.

Scale-up development of high-performance polymer matrix for DNA sequencing analysis. F. Wan *et al.* (2006) *Electrophoresis* **27**, 3712–3723.

6.4.D. Alternative Approaches

Another general sequencing method, known as the **chain-cleavage** or **Maxam–Gilbert** method, was developed early and used extensively. This method also maps the DNA sequence to the sizes of DNA molecules, but it does so by degrading existing DNA chains rather than synthesizing new ones. A sample of purified DNA to be sequenced is first labeled at one end. Labeling the 5′-end is usually done with [γ-^{32}P]ATP and T4 **polynucleotide kinase**. Labeling the 3′-end of either DNA or RNA can be accomplished by adding a radioactive pNp to the 3′-hydroxyl group using the enzyme T4 **RNA ligase**. Then the end-labeled DNA is subjected to a chemical treatment that will break some of the DNA molecules at random, but only at places where one or two bases occurs. Dimethyl sulfate at acid pH methylates both A and G bases, whereas only G is modified at neutral pH (Figure 2-33). Hydrazine ($H_2N–NH_2$) normally reacts with both C and T bases, but only with C in 1.5 M NaCl. These modifications make the adjacent phosphodiester linkages very susceptible to chemical cleavage. The result of each treatment is a population of labeled molecules whose sizes are determined by the sites of cleavage in the sequence. Determination of the sizes in samples cleaved with these four sequence-specific cleavage treatments can yield complete sequence information, as in chain-termination sequencing (Figure 6-15). Unfortunately, it is difficult to find chemical cleavage conditions that consistently give unambiguous, clean cleavage products, and few labeling methods provide labels stable enough for these treatments. For these and other technical reasons, chain-cleavage methods of sequencing are rarely used today.

In the **reverse Sanger** sequencing approach, small amounts of α-thiophosphate forms of the four dNTPs are included while the DNA is copied normally. The α-thiophosphate dNTPs have an O atom of the α-phosphate group replaced by S. They are incorporated readily into the nascent DNA molecules, so a complete polynucleotide chain is synthesized. It is then treated with an exonuclease that removes nucleotides sequentially from the 3-terminus, until an α-S-nucleotide is reached. The lengths of the resistant molecules indicate the positions of the thiophosphate-containing nucleotides, so the use of all four α-thiophosphate dNTPs in separate reaction mixtures generates a sequence ladder.

A very novel method for sequencing DNA relies on the motion of single RNA polymerase molecules when they are transcribing the sequence of a single-stranded DNA molecule into a complementary RNA molecule. When low concentrations of one of the four types of nucleotides limit the rate of transcription, RNA polymerase molecules pause at positions corresponding to that base. This pausing is monitored with an optical trapping apparatus that is capable of resolving base pairs during transcription. Using limiting amounts of each of the four nucleotides, the DNA sequence can be determined from the aligned patterns of pauses recorded from as few as four molecules.

Sequencing end-labeled DNA with base-specific chemical cleavages. A. M. Maxam & W. Gilbert (1980) *Methods Enzymol.* **65**, 499–560.

Single-molecule, motion-based DNA sequencing using RNA polymerase. W. J. Greenleaf & S. M. Block (2006) *Science* **313**, 801.

6.5. SEQUENCING RNA

RNA is much more difficult to handle than DNA, due to its relative instability, single-stranded structure, tendency to adopt folded conformations, and the presence of modified bases. **Consequently, the majority of RNA sequences are obtained by sequencing DNA copies**. The DNA can be either (1) cloned DNA that codes for RNA genes or (2) the cDNA prepared by transcription of the RNA with the enzyme reverse transcriptase (Section 6.2.C). This enzyme synthesizes a cDNA strand using the order of nucleotides in the template RNA strand. The DNA sequences of RNA genes do not, however, provide direct information about all the post-transcriptional modifications of natural RNAs. The initial messenger RNA transcripts are subjected to splicing, cleavage, addition of a 5′-cap moiety, 3′-polyadenylation and RNA editing that alter the sequence. In addition, the primary transcripts of eukaryotic and prokaryotic RNAs are often processed by specific modifications of the nucleotide residues. Roughly 100 modifications of the bases and ribose moieties of the standard four ribonucleotides are known (Figure 4-3) and they tend to be located primarily in the regions of RNAs that are functionally important. Consequently, methods have been developed for sequencing RNA directly and rapidly, comparable to those for DNA sequencing, and special chemical, enzymatic and spectrophotometric techniques have been devised to identify modified nucleotides.

The use of reverse transcriptase for efficient first- and second-strand cDNA synthesis from single- and double-stranded RNA templates. I. E. Tzanetakis *et al.* (2005) *J. Virol. Methods* **124**, 73–77.

Rapid genome sequencing of RNA viruses. T. Mizutani *et al.* (2007) *Emerg. Infect. Dis.* **13**, 322–324.

6.5.A. Direct Sequencing of Oligoribonucleotides

The earliest methods for sequencing RNA were developed for transfer RNAs and relied largely on relatively nonspecific ribonucleases (RNases). The RNA was cleaved into sets of many short oligoribonucleotides using the relatively nonspecific RNases. The fragments produced were separated by column chromatography. Their short sequences were determined from their nucleoside compositions and by stepwise degradation. The overlaps of a large number of these partial sequences could be used to reconstruct the sequence of the original RNA.

An improved method resulted from labeling the RNA molecule at one end and subsequently detecting only molecules containing the label. The 5′-terminus of RNA is usually labeled by removing any existing 5′-phosphate groups of the RNA using the enzyme alkaline phosphatase, followed by complete phosphorylation with [γ-^{32}P]ATP catalyzed by phage T4 polynucleotide kinase. Alternatively, the 3′-end can be phosphorylated by adding the nucleotide cytidine-3′,5′[^{32}P] diphosphate using T4 RNA ligase in the presence of ATP. The end-labeled RNA is cleaved partially into random fragments using a nonspecific RNase, and the products separated by two-dimensional **fingerprinting** using electrophoresis and chromatography (Figure 6-16). Only molecules containing the labeled end are observed. A complete series of all the possible fragments can be viewed as a ladder produced by removing nucleotides sequentially from the end of the RNA molecule that is not labeled. The shift in positions of two neighboring fragments upon removing one nucleotide depends on the pK_a values of that nucleotide and identifies it. This approach makes it possible to determine the sequences of 8–15 nucleotides in one experiment.

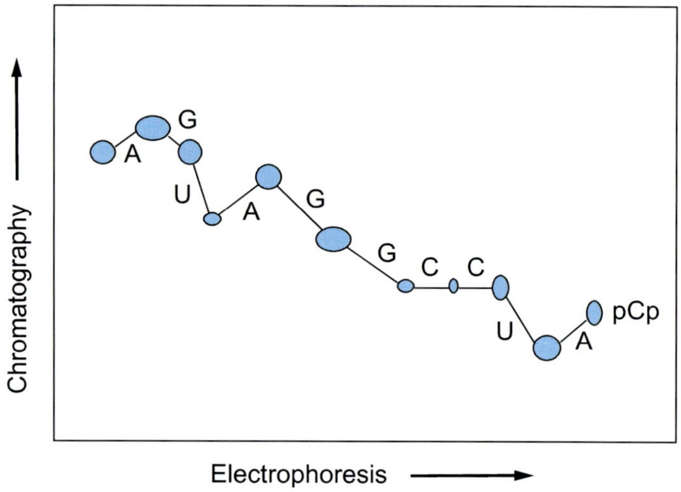

Figure 6-16. Two-dimensional separation of oligoribonucleotides generated by partial cleavage by an endonuclease of an RNA radioactively labeled at its 5′-end. Only those fragments with this radioactivity are detected using autoradiography. The smallest fragments migrate most rapidly during electrophoresis; the smallest in this example is the nucleotide pCp, which would have been at the 5′-end. The shift in mobility upon removal of each 3′-terminal nucleotide gives its identity. The sequence in this hypothetical example would be 5′–CAUCCGGAUGA–3′.

Sequencing methods comparable to the partial cleavage at specific nucleotides of end-labeled DNA (Section 6.4.D) have been adapted to RNA. Enzymatic and chemical procedures are available for cleaving the polynucleotide chain at only certain bases (Table 6-1). The oligoribonucleotides generated are separated by electrophoresis on the basis of their lengths, and only the bands with the end-label are detected by autoradiography. A sample that is nonspecifically cleaved is usually prepared for comparison as a control. The sites of cleavage by the various reagents identify the nucleotide at that position.

Table 6-1. Specificities of cleavage reactions used in RNA sequencing

	Specificity
Enzymatic cleavage by	
Ribonuclease U from *Ustilago sphaerogena*	… ApN …
Ribonuclease T_1 from *Aspergillus oryzae*	… GpN …
Ribonuclease A from bovine pancreas	… UpN … and …CpN…
Ribonuclease from *Bacillus cereus*	… UpN … and …CpN …
Ribonuclease Phy M from *Physarum polycephalum*	… ApN … and … UpN …
Chemical cleavage in aniline at pH 4.5 by	
Dimethyl sulfate + $NaBH_4$	… GpN …
Hydrazine	… UpN … > … pCpN …
Hydrazine in 3 M NaCl	… CpN … > … UpN …
Diethyl pyrocarbonate	… ApN … > … GpN …

N depicts any nucleotide. Data from A. Simoncsits *et al.* (1977) *Nature* **269**, 833–836; H. Donis-Keller *et al.* (1977) *Nucleic Acids Res.* **4**, 2527–2538; D. H. Peattie (1979) *Proc. Natl. Acad. Sci. USA* **76**, 1760–1764.

RNA can also be sequenced by methods comparable to the chain-termination method used with DNA (Section 6.4.B). A synthetic oligodeoxyribonucleotide complementary to a known sequence within the RNA molecule is used as primer for DNA synthesis, which is catalyzed by AMV reverse transcriptase in the presence of radioactively labeled dNTPs and one of the four nucleoside 5′-triphosphate dideoxynucleotide analogs (ddNTPs). The sequence is then read from the ladders produced by polyacrylamide gel electrophoresis (PAGE) of the four mixtures using each of the four ddNTPs. The procedure is, however, complicated by the presence of modified nucleotides, because they affect the reverse-transcriptase activity. DNA synthesis is frequently terminated at bulky bases, and the corresponding positions are detected as enhanced bands on the PAGE sequencing ladder. On the other hand, this is a useful indicator of the presence of a modified nucleotide.

Reverse Sanger sequencing is easier with RNA than with DNA (Section 6.4.D). The RNA is copied by reverse transcriptase in the presence of α-S-ribonucleoside triphosphates, in which a phosphate oxygen is replaced by sulfur, that are added to the reaction mixture at a ratio of 1:3 to the normal dNTPs. The phosphorothioate nucleotides are incorporated randomly into the DNA strand synthesized. Four separate mixtures are produced, using the four different α-S-ribonucleoside triphosphates. The full-length DNA transcript is isolated and purified. It is then cleaved with an exonuclease like **snake venom phosphodiesterase**, which removes nucleotides sequentially from the 3′-terminus, until an α-S-nucleotide appears there. The resulting fragments are separated on the basis of their size, and the lengths of the resistant molecules indicate the positions of the thiophosphate-containing nucleotides.

Reverse Sanger sequencing of RNA by MALDI-TOF mass spectrometry after solid phase purification. B. Spottke *et al.* (2004) *Nucleic Acids Res.* **32**, e97.

Shotgun sequencing of the negative-sense RNA genome of the rhabdovirus Maize mosaic virus. S. E. Reed *et al.* (2005) *J. Virol. Methods* **129**, 91–96.

6.5.B. Identifying Modified Nucleotides

Modified nucleotides affect the chemical properties of the RNA molecules, especially their modification by enzymes, and complicate the sequencing reactions. The different reactivities of modified nucleotides can cause the fragments terminating there to be irregular, missing or enhanced on the sequencing gels. This provides hints about the presence of modified nucleotides in RNA, but they still need to be identified. Once isolated, modified bases can be identified by thin-layer or column chromatography techniques, comparing them with synthetic standards (Figure 6-17), or by electrospray ionization mass spectrometry.

The nucleotide present at each site of cleavage during sequencing can be labeled specifically and then identified (Figure 6-18). A mixture of oligoribonucleotides is prepared by partial cleavage of the denatured RNA chain. The 5′-hydroxyl groups of the oligoribonucleotides generated are then radiolabeled by T4 RNA kinase and [γ-^{32}P]ATP. After separation by gel electrophoresis according to their size, bands of the different oligoribonucleotides are digested to mononucleotides by ribonuclease T2. Only the terminal nucleotide of each band is ^{32}P-labeled. It can be identified separately by one- or two-dimensional chromatography (Figure 6-17).

Detection and quantification of modified nucleotides in RNA using thin-layer chromatography. H. Grosjean *et al.* (2004) *Methods Mol. Biol.* **265**, 357–391.

A novel method for sequence placement of modified nucleotides in mixtures of transfer RNA. T. M. Wagner *et al.* (2004) *Nucleic Acids Symp. Ser.* **48**, 263–264.

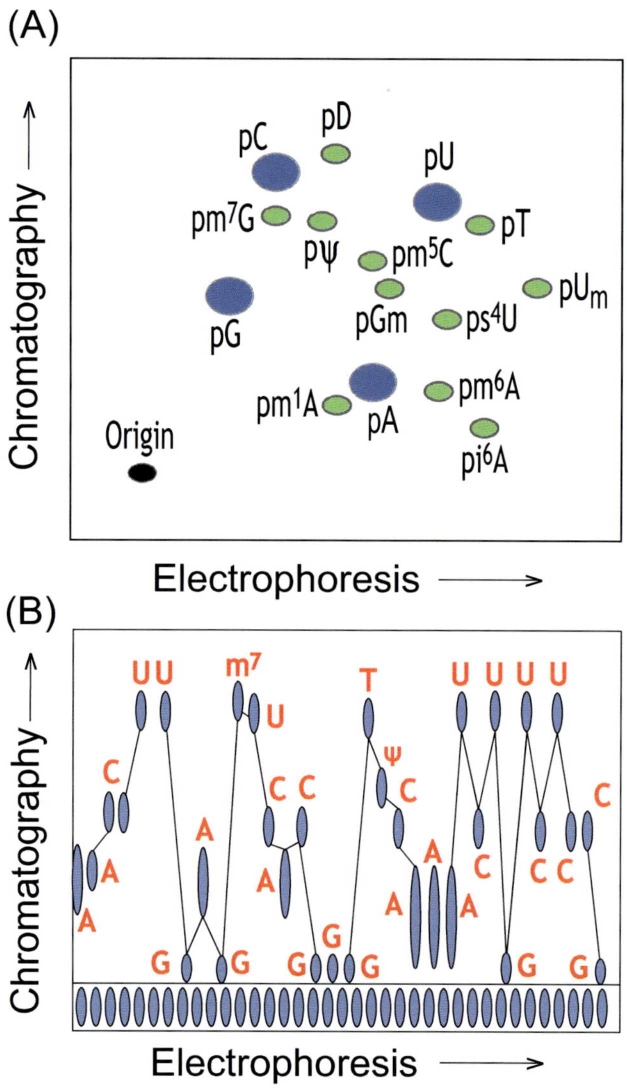

Figure 6-17. Identification of ribonucleotides by chromatography and electrophoresis. (A) Two-dimensional fingerprint of ribonucleoside 5′-phosphates. The major ribonucleotides pA, pU, pG and pC are indicated by the *blue spots*. The others are the minor nucleotides (Figure 4-3): i^6A, N^6-isopentenyladenosine; U_m, 2′-O-methyluridine; m^6A, N^6-methyladenosine; s^4U, 4-thiouridine; m^1A, 1-methyladenosine; G_m, 2′-O-methylguanosine; m^5C, 5-methylcytosine; Ψ, pseudouridine; D, 5,6-dihydrouridine; m^7G, 7-methylguanosine; T, 5-methyluridine (thymidine). Data from Y. Kuchino *et al.* (1987) *Methods Enzymol.* **155**, 379–396. (B) Identification of nucleotides at the 5′-ends of random fragments that had been labeled with ^{32}P at the terminal phosphate group. The fragments had been separated by size by electrophoresis in the horizontal dimension. They were then degraded to individual nucleotides and separated by chromatography in the vertical dimension. The labeled nucleotide was detected by autoradiography. If the electrophoretic bands differ only in their lengths at the 5′-end, the labeled nucleotides will give the sequence of the original RNA molecule, as indicated. Data from J. Stanley & S. Vasilenko (1978) *Nature* **274**, 87–89.

6.6. CHEMICAL SYNTHESIS OF DNA

Many methods in molecular biology require the chemical synthesis of oligonucleotides of defined sequences. One example is the primer that is required for sequencing DNA next to a site with a known sequence (Figure 6-1). Others involve the detection and identification of specific stretches of DNA with a known sequence (Section 5.3.C) and the introduction of changes in nucleotide sequence (Section 6.6.E). Not surprisingly, very powerful methods for synthesizing such oligonucleotides have been developed and it is now a routine procedure, largely performed by automated instruments.

Single-stranded oligonucleotides can be assembled by chemically linking two nucleotides with a

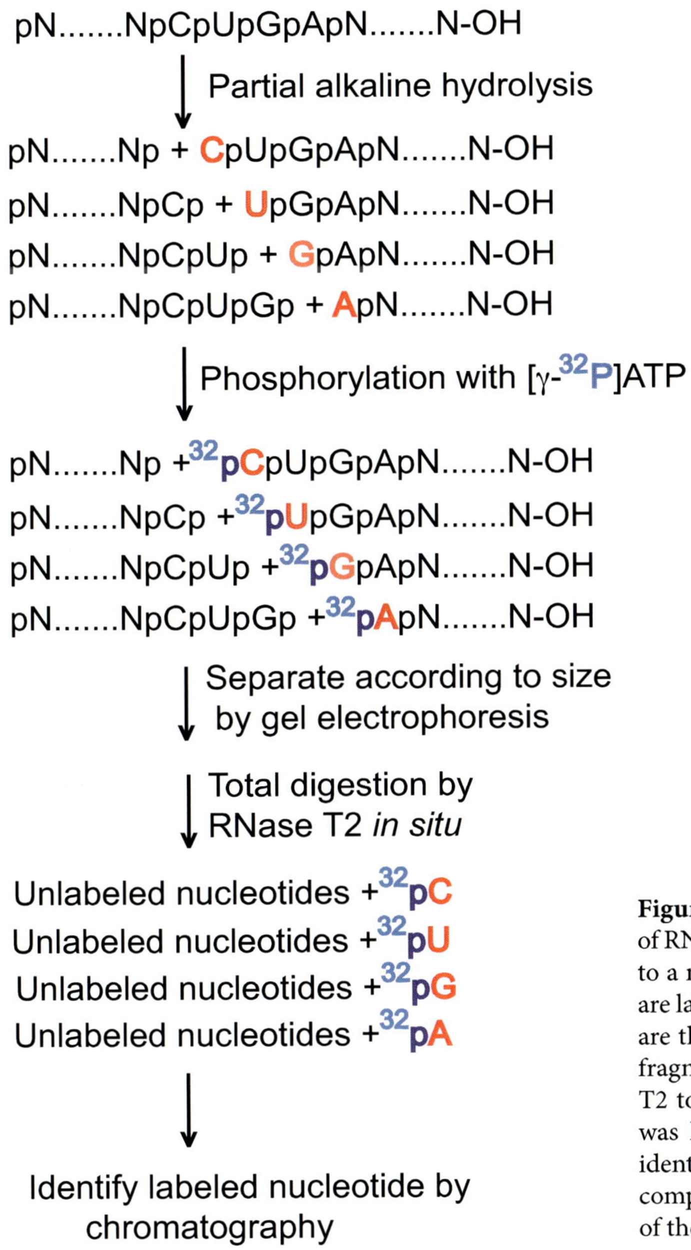

Figure 6-18. The post-labeling method of sequencing of RNA. Limited alkaline hydrolysis of the RNA leads to a mixture of random oligoribonucleotides, which are labeled radioactively at the 5′-end. The fragments are then separated by PAGE according to size. Each fragment is digested completely by ribonuclease T2 to the individual nucleotides. Which nucleotide was labeled, and therefore was at the terminus, is identified by chromatography, as in Figure 6-17-B. A complete set of fragments can provide the sequence of the RNA.

phosphodiester bond and then adding individual nucleotides to one of the ends. The process is similar to that used in replicating DNA, except that no enzymes or template are used, so any sequence can be produced. The process is repetitive, so the efficiency of adding each nucleotide determines what length of oligonucleotide can be produced with a reasonable yield.

DNA nanotechnology: novel DNA constructions. N. C. Seeman (1998) *Ann. Rev. Biophys. Biomol. Structure* **27**, 225–248.

3′-modified oligonucleotides by reverse DNA synthesis. C. D. Claboe *et al.* (2003) *Nucleic Acids Res.* **31**, 5685–5691.

A simple, rapid, high-fidelity and cost-effective PCR-based two-step DNA synthesis method for long gene sequences. A. S. Xiong *et al.* (2004) *Nucleic Acids Res.* **32**, 398.

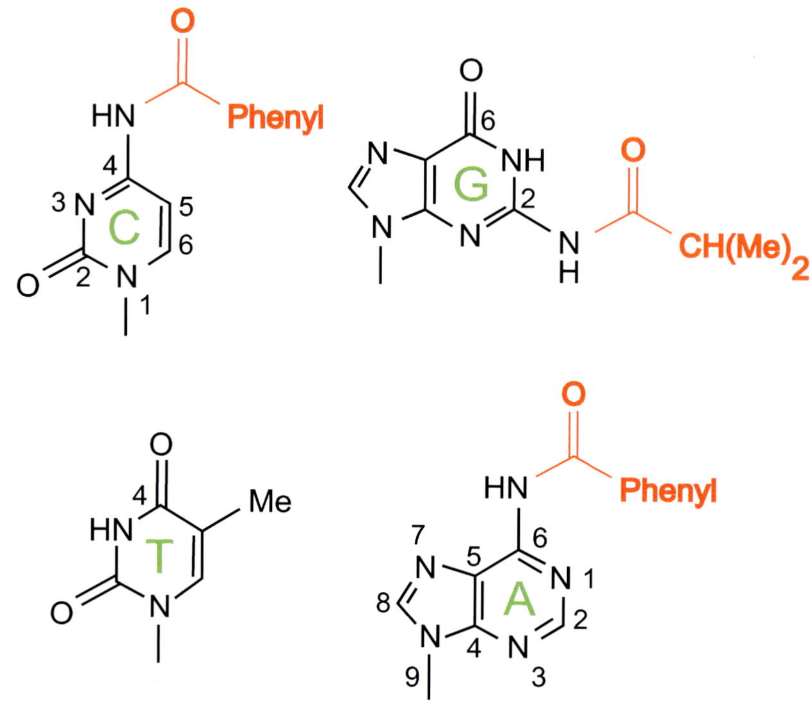

Figure 6-19. The forms of the cytosine (C), guanine (G), thymine (T) and adenine (A) bases used in deoxynucleosides for DNA synthesis. The protecting groups are *colored*. Thymine requires no protecting group.

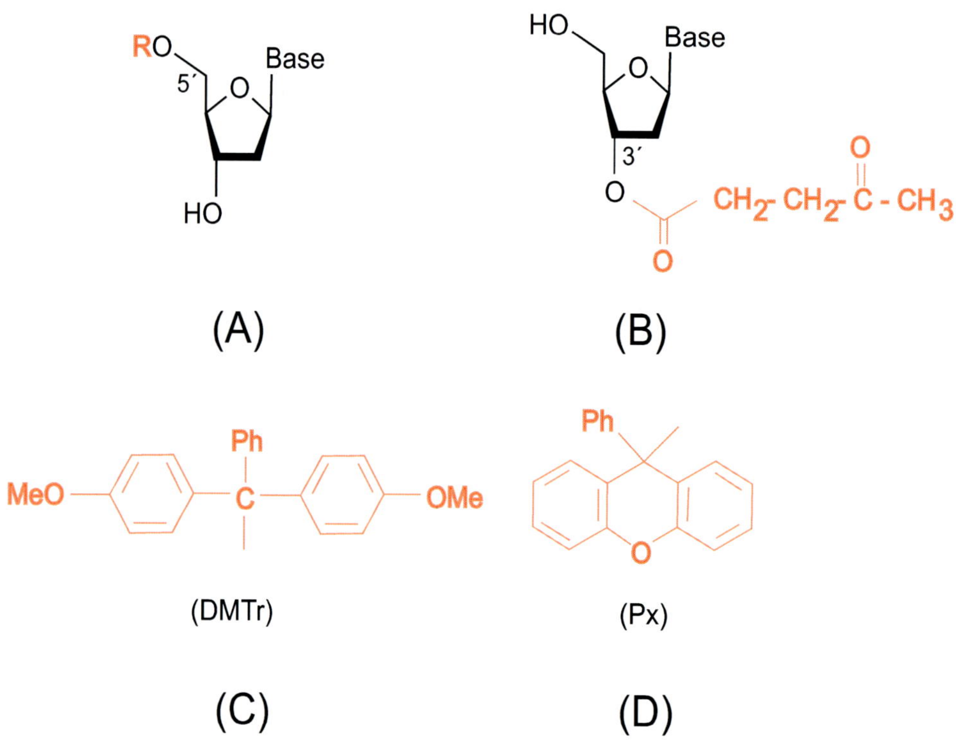

Figure 6-20. Protection of the 5′- (A) and 3′- (B) hydroxyl groups of deoxyribonucleosides. The 5′-hydroxyl is protected with R groups such as (C) (di-*p*-anisyl)phenylmethyl (DMTr) or (D) 9-phenylxanthen-9-yl (Px); Ph is a phenyl group. The protecting groups are *colored*.

6.6.A. Protecting Groups for 2′-Deoxynucleosides

Only one pair of functional groups on the 2′-deoxyribonucleoside building blocks is to be linked covalently, which requires that all other potentially reactive sites be protected reversibly. The protecting groups should be easy to introduce, remain intact throughout the assembly of the DNA sequences and be removed completely under conditions in which the assembled DNA molecule is stable. The nature of the protecting groups is one of the most important factors to be considered in designing a protocol for DNA synthesis. Two other factors to consider are the chemical procedures for generating the phosphodiester linkages between the nucleotides and for purifying the final synthetic DNA molecules.

The glycosidic linkages between the base and the sugar, especially those of purine deoxyribosides, are relatively sensitive to acid. Consequently, the protecting groups used should be removable without the need for very acidic conditions. The functional groups of the adenine, cytosine and guanine bases are generally protected with acyl groups (Figure 6-19). Thymine residues are often left unprotected, although certain procedures may require that their O4 atoms be protected, along with the O6 atoms of *N*-2-acylguanine residues.

The 5′-hydroxyl group is best protected with (di-*p*-anisyl)phenylmethyl (DMTr) or 9-phenylxanthen-9-yl (Px) groups (Figure 6-20). Acyl groups are generally used to protect the 3′-hydroxyl groups (Figure 6-20-B). They have also been used to protect the 5′-hydroxyl group, especially in the solution-phase synthesis of DNA. Both *N*-acyl- and *O*-acyl-protecting groups can be removed at the end of the synthesis by treatment with ammonia, conditions where DNA is very stable.

6.6.B. Coupling Methods

Three different types of coupling methods have been used successfully. Each will be illustrated with a single cycle, in which one nucleotide is attached to another. To synthesize an oligonucleotide of *N* nucleotides requires that these reaction sequences be repeated $N - 1$ times.

1. Phosphotriester Procedure

All of the internucleotide linkages are protected as phosphotriesters throughout the assembly of the target DNA sequences using this approach, which is the basis for its name. A contrasting method, known as the **phosphodiester approach**, leaves the internucleotide linkages unprotected throughout the assembly of the DNA sequences, but it is now obsolete. The phosphotriester approach couples the free 5′-hydroxyl group of a protected nucleoside or oligonucleotide component to a slight excess of the 3′-(2-chlorophenyl) phosphate moiety of the other nucleoside or oligonucleotide (Figure 6-21). Coupling usually involves a condensing agent such as 1-(mesitylene-2-sulfonyl)-3-nitro-1,2,4-(1*H*)-triazole (MSNT) in anhydrous pyridine solution:

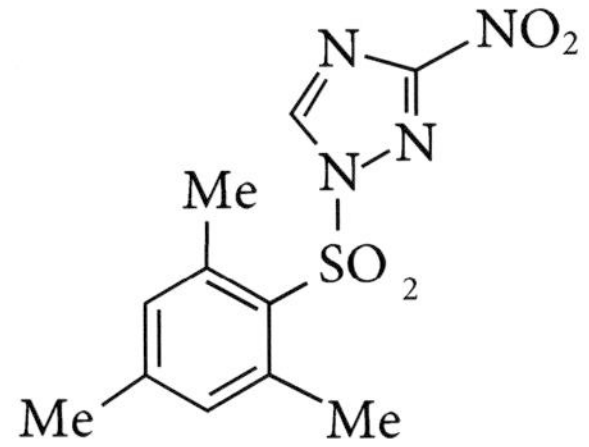

(6.4)

The MSNT also acts as a dehydrating agent and is used in excess.

The fully assembled DNA sequence is unblocked by treatment with an excess of an appropriate oxime (such as *E*-2-nitrobenzaldoxime) and N^1,N^1,N^3,N^3-tetramethylguanidine:

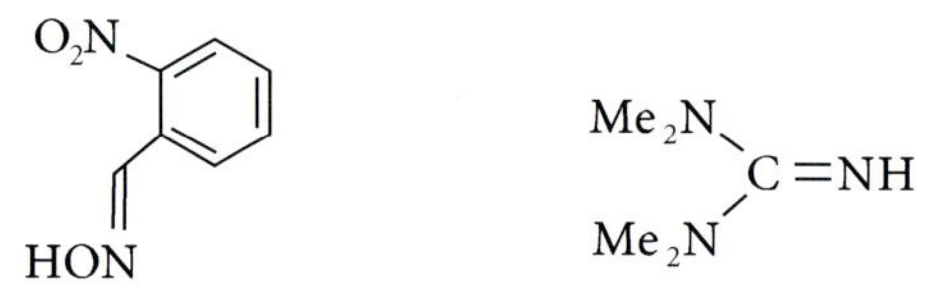

E-2- nitrobenzaldoxime Tetramethyl guanidine (6.5)

All of the 2-chlorophenyl-protecting groups are removed without cleaving the internucleotide linkages. The *N*- and *O*-acyl-protecting groups (Figure 6-19 and 6-20-B) are subsequently removed by treatment with concentrated aqueous ammonia. The 5′-terminal-protecting group (Figure 6-20-A) is acid-labile and is removed under mild conditions to generate the final DNA sequence. The phosphotriester approach is extremely versatile and can be used in solution or solid-phase and by stepwise addition of mononucleotides or fusing of oligonucleotides.

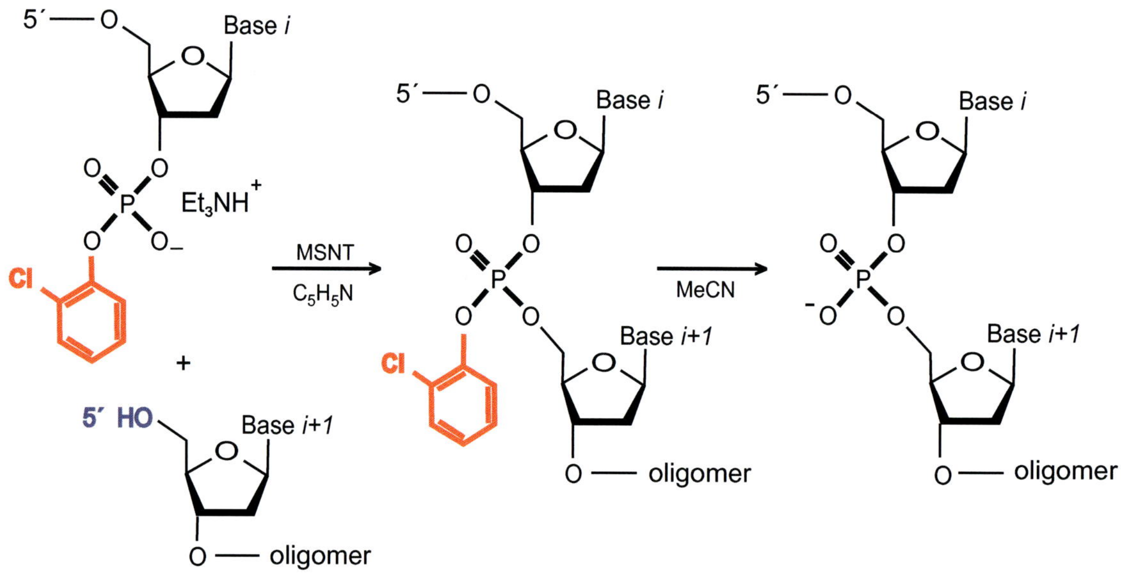

Figure 6-21. The phosphotriester approach to coupling deoxynucleotides. The 5′-reactant is present as the 3′-(2-chlorophenyl) phosphate derivative. Coupling uses a reagent such as MSNT (Equation 6.4) in pyridine, while removal of the protecting groups in the second step uses an oxime and tetramethyl guanidine (Equation 6.5) in acetonitrile (MeCN).

2. Phosphoramidite Procedure

Protected nucleoside 3′-(2-cyanoethyl) *N*, *N*-di-isopropylphosphoramidites are used in the phoroamidite approach. In the presence of relatively weak acids, such as 1*H*-tetrazole, they react readily with the free 5′-hydroxyl group of protected nucleoside or oligonucleotide components (first step in Figure 6-22) to produce protected phosphite triesters. These phosphite triesters are unstable intermediates, so usually they are immediately oxidized by treatment with iodine and a weak base to the corresponding phosphotriesters, as in the second step of Figure 6-22. The 2-cyanoethyl and *N*- and *O*-acyl-protecting groups (Figure 6-19) are removed from the phosphates of the fully assembled DNA sequences by treatment with concentrated aqueous ammonia.

The coupling yields tend to be higher with the phosphoramidite approach than the phosphotriester approach (Figure 6-21) because activated derivatives of phosphorous acid are much more reactive than the corresponding activated derivatives of phosphoric acid. Consequently, **the phosphoramidite approach is widely used for automated solid-phase synthesis of DNA sequences** and in the chemical synthesis of relatively small quantities of the DNA, which are sufficient for most procedures in molecular biology.

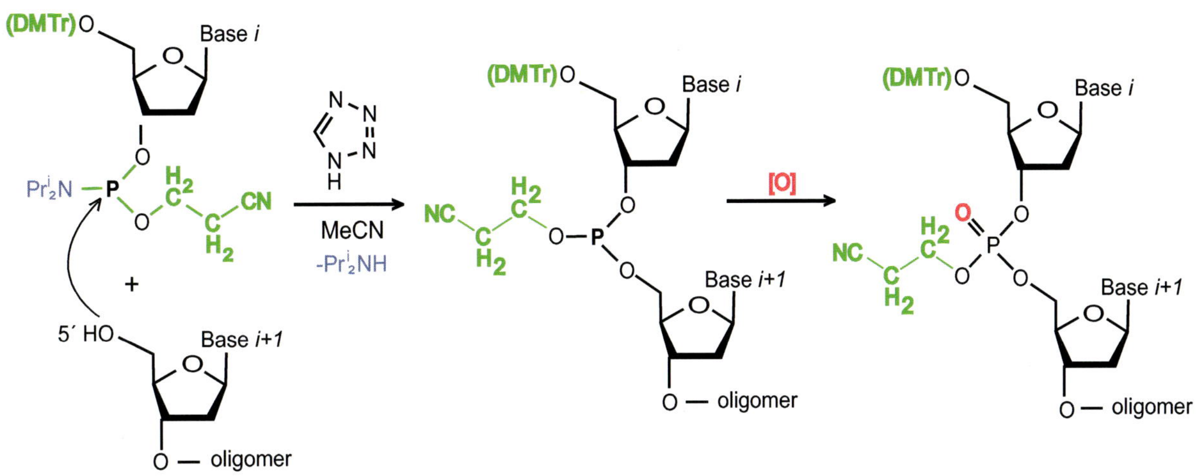

Figure 6-22. The phosphoramidite approach to coupling a deoxynucleotide to the 5′-end of a DNA oligomer (*bottom, left*). The nucleotide to be added is presented as the 3′-(2′-cyanoethyl) *N,N*-di-isopropylphosphoramidite, protected with a DMTr group, and 1*H*-tetrazole in MeCN is a catalyst of the reaction. Pr^i is the isopropyl group and [O] is an oxidant. The protecting groups must be removed from the product of the reaction by treatment with ammonia before another nucleotide can be added to its 5′-end.

Linker phosphoramidite reagents for the attachment of the first nucleoside to underivatized solid-phase supports. R. T. Pon & S. Yu (2004) *Nucleic Acids. Res.* **32**, 623–631.

Internucleotide-linkage formation via the phosphoramidite method using a carboxylic acid as a promoter. M. Tsukamoto *et al.* (2004) *Nucleic Acids Symp. Ser.* **48**, 25–26.

Development of new N-unprotected phosphoramidite building blocks having a silyl-type linker. A. Ohkubo *et al.* (2005) *Nucleic Acids. Symp. Ser.* **49**, 127–128.

---

3. *H-Phosphonate Procedure*

This approach uses protected nucleoside 3′-(*H*-phosphonate) monoesters (Figure 6-23) as building blocks. It has the advantage that they are easy to prepare and more stable and easier to handle than the building blocks of the phosphoramidite approach. A nucleoside 3′-(*H*-phosphonate) building block reacts readily with the 5′-hydroxyl group of a protected nucleoside or oligonucleotide (the first reaction in Figure 6-23) in the presence of a condensing agent such as pivaloyl chloride in pyridine solution. After completion of the synthesis, the *H*-phosphonate diester groups are oxidized, usually with iodine in the presence of a base in aqueous tetrahydrofuran, to generate unprotected phosphodiester internucleotide linkages, as in the second reaction of Figure 6-23. The *N*- and *O*-acyl-protecting groups can then be removed by treatment with concentrated aqueous ammonia.

The *H*-phosphonate approach is best suited for use in solid-phase synthesis (Section 6.6.D). While it has some merits and the potential to be an important synthetic method, it is more susceptible to side-reactions and is not as efficient in solid-phase synthesis as the phosphoramidite approach.

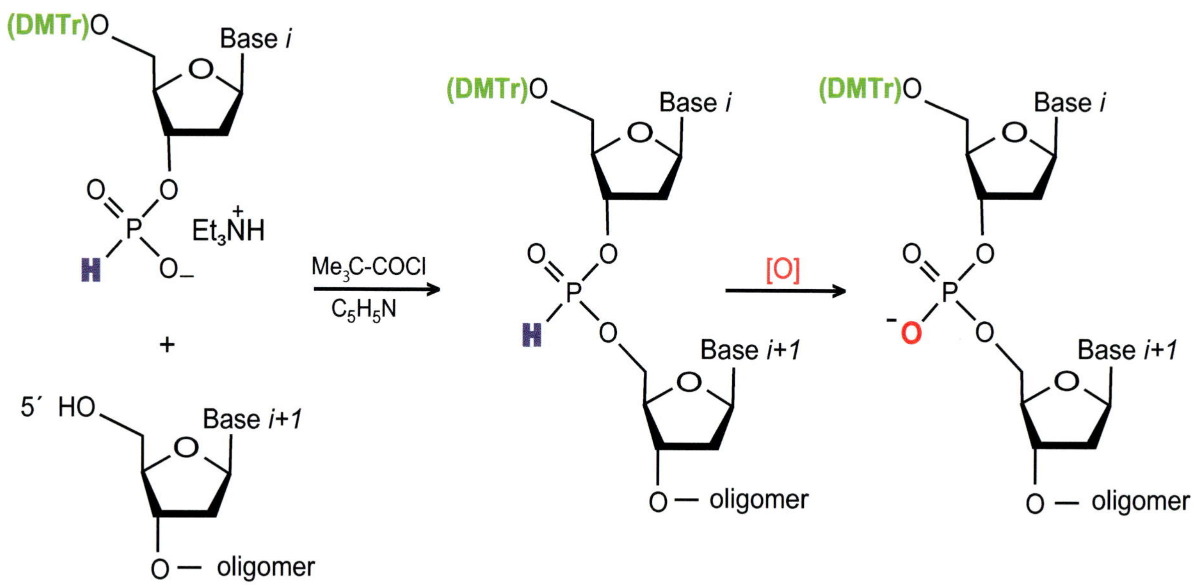

Figure 6-23. The *H*-phosphonate approach to coupling a deoxynucleotide to the 5′-terminus of a DNA oligomer (*bottom, left*). The nucleotide to be added is the 3′-(*H*-phosphonate) monoester (*top, left*) protected with the DMTr group. Coupling occurs in the first step in the presence of a coupling agent, such as pivaloyl chloride (Me_3C–CO–Cl) in pyridine (C_5H_5N). In the second step, the *H*-phosphonate diester is oxidized to give the phosphodiester linkage.

Di- and oligonucleotide synthesis using H-phosphonate chemistry. J. Stawinski & R. Stromberg (2005) *Methods Mol. Biol.* **288**, 81–100.

Developing synthetic methods for bioactive phosphorus compounds using H-phosphonate chemistry: a progress report. T. Johansson *et al.* (2005) *Nucleosides Nucleotides Nucleic Acids* **24**, 353–357.

Stereocontrolled synthesis of H-phosphonate DNA. N. Iwamoto *et al.* (2006) *Nucleic Acids Symp. Ser.* **50**, 159–160.

Efficient preparation of amine-modified oligodeoxynucleotide using modified H-phosphonate chemistry for DNA microarray fabrication. N. K. Kamisetty *et al.* (2007) *Anal. Bioanal. Chem.* **387**, 2027–2035.

6.6.C. Solution-phase DNA Synthesis

Solution-phase synthesis is used primarily when a large quantity of a specific sequence is required. It is laborious, in that it is often necessary to purify the products by chromatography after each coupling step, although this might be no more than filtration through a bed of silica gel. Of the three phosphorylation methods described in Figures 6-21, 6-22 and 6-23, only the phosphotriester approach (Figure 6-21) is very suitable for the synthesis of DNA sequences in solution. This method is very versatile, being useful for both the stepwise addition of individual nucleotides and the coupling of oligonucleotides, in which two or more nucleotide residues are added at a time. It has the advantage that the fully protected intermediates generated are soluble in organic solvents and may therefore be purified by conventional chromatographic techniques. Once all of the protecting groups have been removed, the unprotected DNA sequences generated may, if necessary, be further purified.

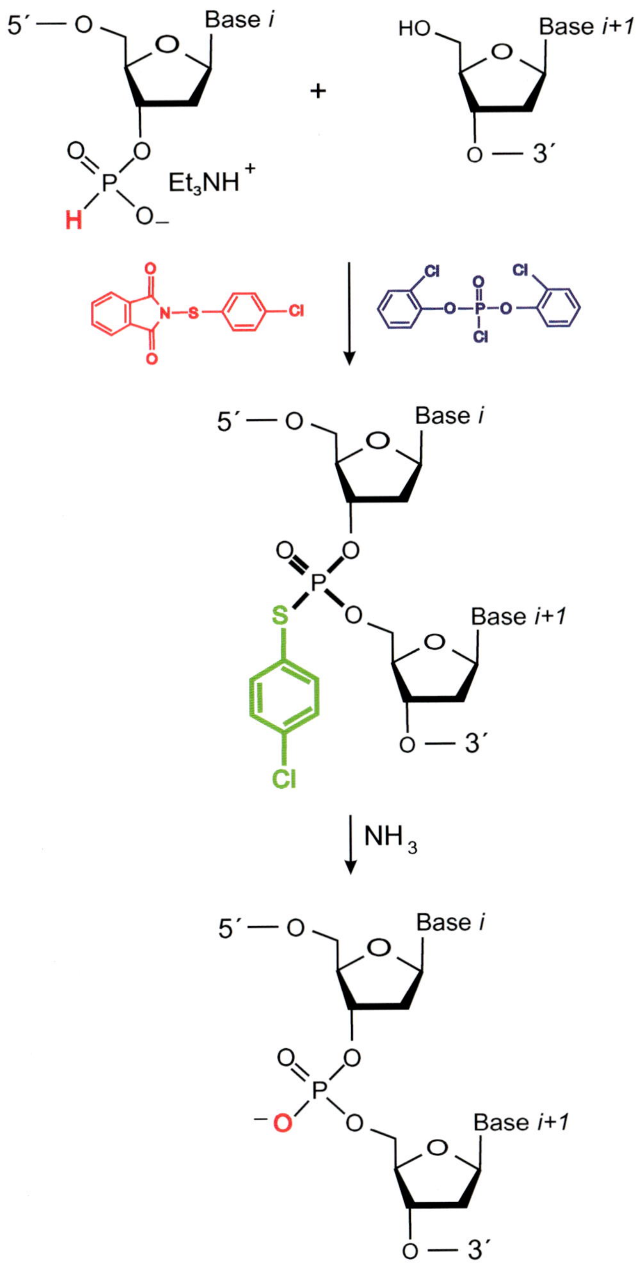

Figure 6-24. One cycle of coupling nucleotides in solution. The 5′-nucleotide is used as the *H*-phosphonate (Figure 6-23). The second step requires *E*-2-nitrobenzaldoxime and tetramethylguanidine in acetonitrile, followed by treatment with concentrated aqueous ammonia.

The reaction sequence described in Figure 6-24 is a modified phosphotriester approach involving *H*-phosphonate coupling at low temperature. It produces virtually quantitative coupling, with no side-reactions. The unstable *H*-phosphonate diester intermediates of Figure 6-23 are not isolated but immediately converted at low temperature into the much more stable *S*-(4-chlorophenyl) phosphorothioate intermediates by treatment with the phthalimide derivative described in Figure 6-24. Upon complete assembly of the desired DNA sequence, the *S*-(4-chlorophenyl) phosphorothioate groups are quantitatively converted into phosphodiesters by the oximate treatment described in Figure 6-21.

6.6.D. Solid-phase DNA Synthesis

Solid-phase synthesis is much less laborious than the solution methods and highly preferable, especially if only small (i.e. milligram to gram) quantities of DNA are required. The growing DNA sequence is attached to a solid support, such as controlled-pore glass, and all the reagents that remain after each reaction are washed away. The main advantages of solid-phase synthesis are that (1) it is very rapid, so a DNA sequence of 50 nucleotide residues can easily be produced within 1 day; (2) very high yields of all the chemical reactions are obtained by using large excesses of reagents; (3) the whole process may be fully automated in a DNA synthesizer; and (4) only one purification step is required, of the final product at the end of the synthesis.

All three of the coupling methods described in Figures 6-21, 6-22 and 6-23 have been tried with solid-phase synthesis, and the phosphoramidite approach of Figure 6-22 has emerged as the method of choice because the coupling efficiencies are very high and there are no significant side-reactions. Most commercial automatic synthesizers have been designed specifically for phosphoramidite chemistry.

The main steps in the addition of one nucleotide residue in a synthetic cycle are described in Figure 6-25. The synthesizer uses solutions in acetonitrile of phosphoramidites of each of the four protected deoxynucleosides (Figure 6-19). The solid support is usually porous glass or polystyrene, designated P. To it is attached the desired 3′-terminal nucleoside residue through a succinoyl linker (Figure 6-25); this material is placed in a small column that is inserted into the synthesizer. The first step in the synthetic cycle involves treatment with acid to remove the 5′-*O*-DMTr-protecting group from the 3′-terminal nucleoside residue. The next step is the reaction between the released 5′-hydroxyl group and the appropriate phosphoramidite, catalyzed by 1*H*-tetrazole to give a phosphite triester. Large excesses of both the phosphoramidite and 1*H*-tetrazole are delivered automatically, which can produce a coupling efficiency of up to 98.5%. Any remaining 5′-hydroxyl functions are 'capped' by reaction with a large excess of acetic anhydride to ensure that any molecules that did not react are not involved in subsequent reactions. Consequently, most of the incorrect sequences at the end of the synthesis are of significantly lower molecular weight than the target DNA sequence, which facilitates their removal.

After capping, the phosphite triester group is oxidized to a phosphotriester group, which completes the first synthetic cycle. Removal of the 5′-*O*-DMTr-protecting group produces the final product of Figure 6-25; it is also the first step of the second synthetic cycle that will add a second nucleotide. This cycle is repeated for each successive nucleotide added to the chain. A number of washing steps are involved, but an entire cycle can be completed within 8 minutes.

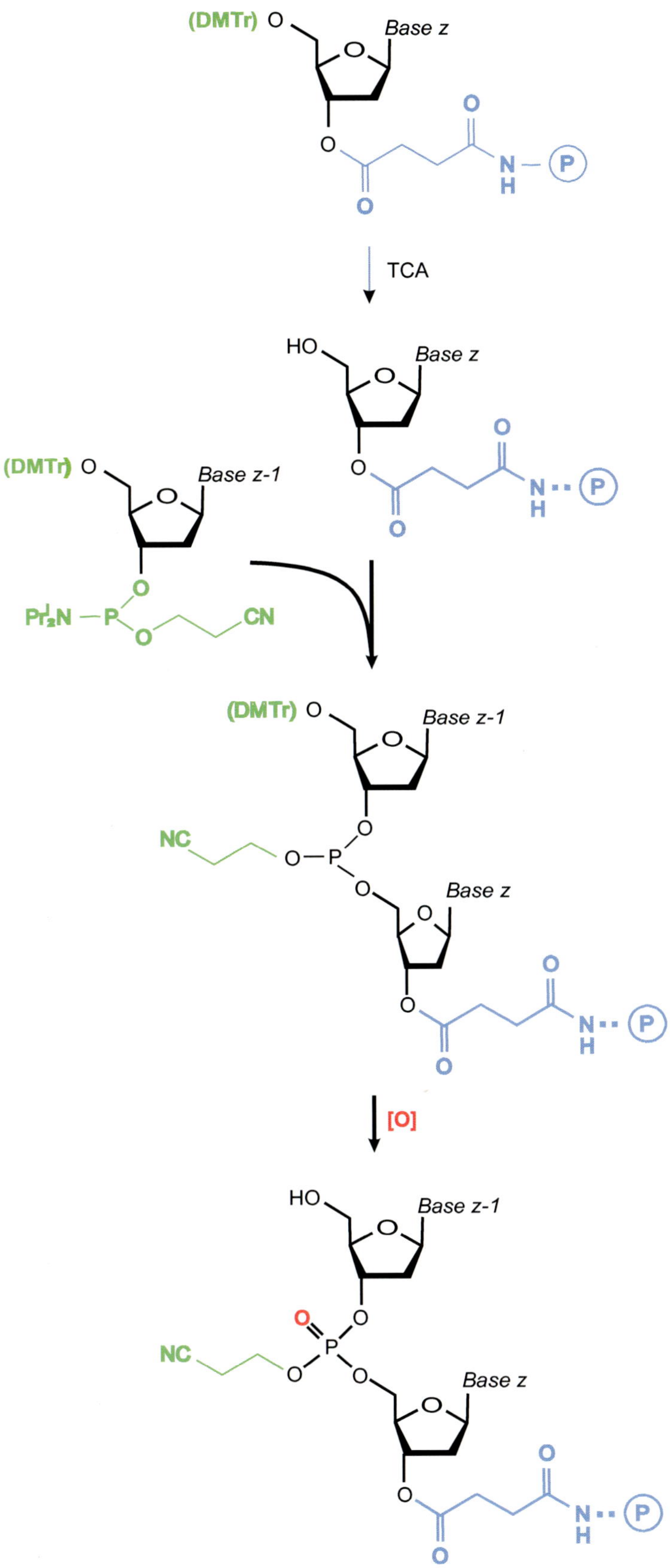

Figure 6-25. One cycle of coupling a nucleotide to the 5′-terminus of a growing DNA chain during synthesis on a solid support (designated by the P in a *circle*). The nucleotide is added as the 3′-(2′-cyanoethyl) *N,N*-diisopropylphosphoramidite protected by the DMTr group (Figure 6-20). TCA is trichloroacetic acid and [O] is an oxidant, such as I_2.

The efficiency of each coupling step can be estimated by measuring spectrophotometrically the amount of colored (di-*p*-anisyl)phenylmethyl ($DMTr^+$) cations that are released at the beginning of each synthetic cycle. If the average coupling efficiency is 98.5%, complete DNA sequences containing 50 and 100 nucleotide residues should be produced with roughly 47% and 22% overall yields, respectively, so DNA sequences containing up to approximately 100 nucleotide residues can be prepared in this way. If the average coupling efficiency is only 97.5%, however, the overall yields would decrease to about 29% and 8%, respectively. Therefore, the coupling procedure needs to be as efficient as possible.

The desired DNA sequence of N nucleotides should be complete after $N - 1$ synthetic cycles but the products must be released from the solid support and all the protecting groups removed. In one unblocking procedure, the 5′-terminal DMTr-protecting group is removed in what would be the first step of synthetic cycle N. The synthetic DNA is then detached from the solid support by treatment with concentrated aqueous ammonia at about 55°C; this also removes all of the 2-cyanoethyl-protecting groups from the internucleotide linkages and all of the *N*-acyl-protecting groups from the bases. The 'capped' truncated sequences are usually removed by column chromatography or PAGE. In another procedure, an ammonolysis step is carried out prior to removal of the 5′-terminal DMTr-protecting group. This procedure is preferable if purification of the crude products is to be carried out by reverse-phase chromatography, because the target DNA sequence is still protected with a 5′-*O*-DMTr group whereas the capped truncated sequences are not; this difference makes it easier to isolate the correct molecules. The DMTr-protecting groups are finally removed by acid hydrolysis under very mild conditions.

6.6.E. Site-directed Mutagenesis

The nucleotide sequence of a fragment of cloned DNA very often needs to be altered. It is possible to change only a single base pair or a few adjacent base pairs, or to create more extensive changes, such as deletions and insertions. There are many procedures for doing this efficiently. In each case, the desired change in sequence must be produced by synthesizing an oligonucleotide with that sequence.

Cassette mutagenesis replaces a restriction fragment from a cloning vector with another fragment with the desired nucleotide sequence (Figure 6-26-A), using restriction enzymes as in Figure 6-11. The new DNA fragment has usually been synthesized chemically. This procedure requires two restriction sites to be available so that the restriction fragment is not too large for DNA synthesis. If not, the alternative method of **enzymatic extension of a mutagenic oligonucleotide** is generally used (Figure 6-26-B). A single-stranded oligonucleotide is synthesized that contains the new sequence flanked on each side by 10–20 nucleotides that are exactly complementary to the DNA strand to be altered. This oligonucleotide base pairs with the original DNA fragment, in spite of the different sequence in the middle. It serves as a primer to initiate copying of the remainder of the original sequence by an appropriate DNA polymerase enzyme. Use of the enzyme thereby avoids the need to synthesize chemically the entire DNA molecule.

Many variations on these methods have been devised, especially using PCR (Section 6.1.C).

Restriction enzyme-mediated integration (REMI) mutagenesis. A. Kuspa (2006) *Methods Mol. Biol.* **346**, 201–209.

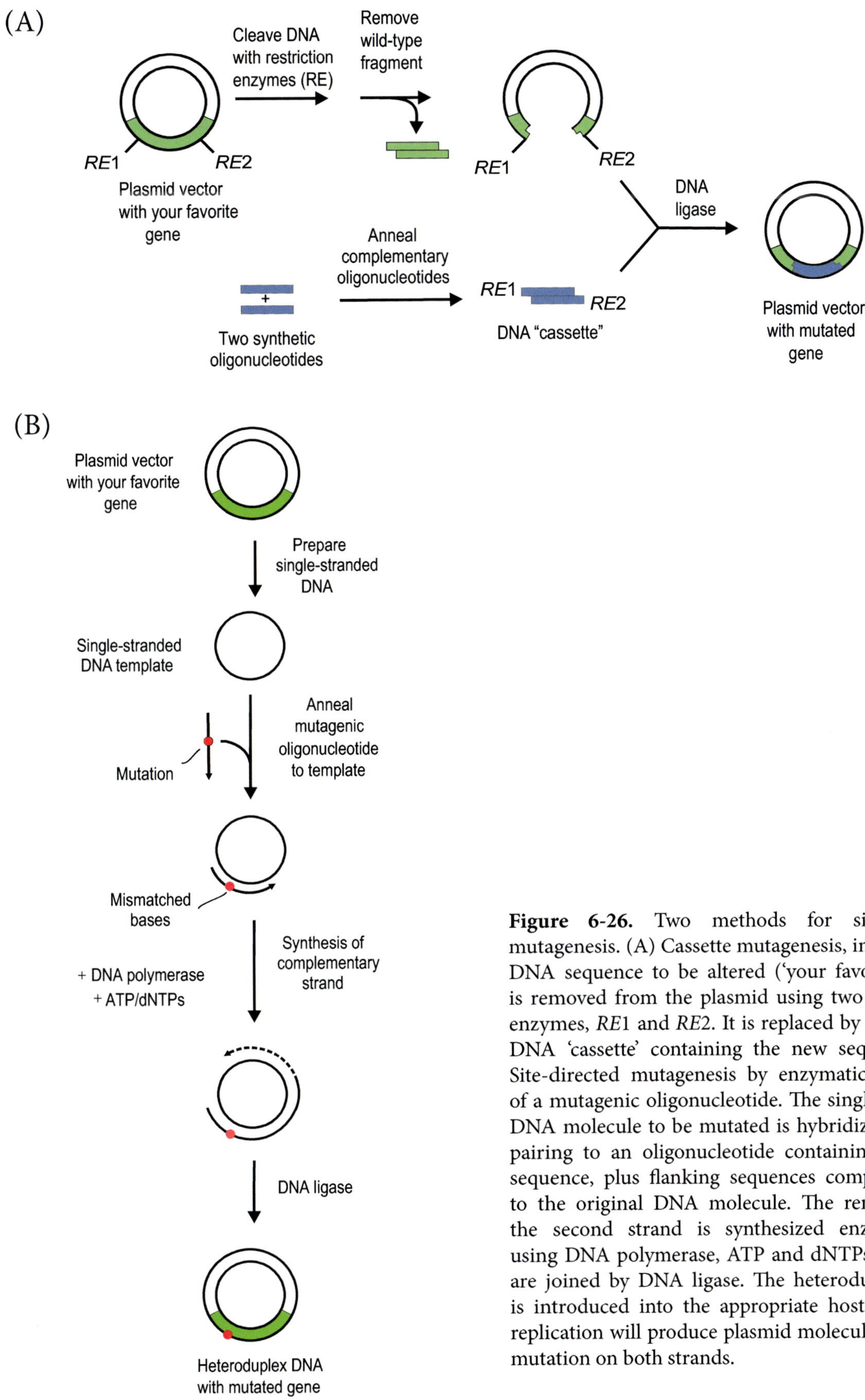

Figure 6-26. Two methods for site-directed mutagenesis. (A) Cassette mutagenesis, in which the DNA sequence to be altered ('your favorite gene') is removed from the plasmid using two restriction enzymes, *RE*1 and *RE*2. It is replaced by a synthetic DNA 'cassette' containing the new sequence. (B) Site-directed mutagenesis by enzymatic extension of a mutagenic oligonucleotide. The single-stranded DNA molecule to be mutated is hybridized by base pairing to an oligonucleotide containing the new sequence, plus flanking sequences complementary to the original DNA molecule. The remainder of the second strand is synthesized enzymatically, using DNA polymerase, ATP and dNTPs. The ends are joined by DNA ligase. The heteroduplex DNA is introduced into the appropriate host, where its replication will produce plasmid molecules with the mutation on both strands.

6.7. CHEMICAL SYNTHESIS OF RNA

The chemical synthesis of RNA is inherently more complicated than that of DNA (Section 6.6) because it requires the use of an additional protecting group, for the 2′-hydroxyl group. With this exception, **the principles of the chemical synthesis of RNA and DNA are virtually the same**. Similar protecting groups are used for the bases and internucleotide linkages, and the same three principal coupling methods (i.e. the phosphotriester, phosphoramidite and *H*-phosphonate approaches; Section 6.6.B) are used in the synthesis of RNA sequences. Likewise, only the phosphotriester approach is suitable for solution-phase synthesis of RNA sequences, and the phosphoramidite approach is again the method of choice for automated solid-phase synthesis.

General deoxyribozyme-catalyzed synthesis of native 3′-5′ RNA linkages. W. E. Purtha *et al.* (2005) *J. Am. Chem. Soc.* **127**, 13124–13125.

Nonenzymatic template-directed RNA synthesis. M. Hey & M. Gobel (2005) *Methods Mol. Biol.* **288**, 305–318.

Synthesis and properties of ENA oligonucleotides targeted to human telomerase RNA subunit. M. Horie *et al.* (2005) *Nucleic Acids Symp. Ser.* **49**, 171–172.

Synthesis, characterization, and biological properties of small branched RNA fragments containing chiral (Rp and Sp) 2′,5′-phosphorothioate linkages. R. Mourani & M. J. Damha (2006) *Nucleosides Nucleotides Nucleic Acids* **25**, 203–229.

Chemical synthesis of RNA including 5-taurinomethyluridine. T. Ogata & T. Wada (2006) *Nucleic Acids Symp. Ser.* **50**, 9–10.

6.7.A. Protecting the 2′-Hydroxyl Group

The protecting group for the 2′-hydroxl group must remain intact until the final unblocking step at the end of the synthesis, and it then must be possible to remove it under very mild conditions without the released 2′-hydroxyl group attacking the adjacent (3′-5′)-internucleotide phosphodiester linkages (Figure 4-2), which will cause their cleavage or migration.

Protecting groups for the 2′-hydroxyl group can be removed by hydrolysis under either basic or acidic conditions. Basic conditions can permit cleavage, but not migration, of the internucleotide linkages (Figure 4-2). The 2′,3′-cyclic phosphate cleavage product formed initially undergoes further hydrolysis to produce a mixture of the corresponding 2′- and 3′-phosphates. Removing the 2′-protecting groups under acidic conditions can produce both cleavage and migration of the internucleotide linkages (Figure 6-27). While cleavage of the internucleotide linkages is clearly undesirable, phosphoryl migration is much more serious in that it produces molecules with one or more 2′-5′-internucleotide linkages. Removing such isomeric contaminants is virtually impossible, even in the case of relatively short oligoribonucleotides. Consequently, acid-labile protecting groups must be removed under conditions sufficiently mild to avoid phosphoryl migration.

Of the 2′-hydroxyl-protecting groups used most widely and successfully, the Thp, Mthp, Ctmp and Fpmp* groups are acetal systems that can be removed under mild acidic conditions:

* *Thp, tetrahydropyran-2-yl; Mthp, 4-methoxytetrahydropyran-4-yl; TBDMS, tert-butyldimethylsilyl; Ctmp, 1-[(2-chloro-4-methyl)phenyl]-4-methoxypiperidin-4-yl; Fpmp, 1-(2-fluorophenyl)-4-methoxypiperidin-4-yl.*

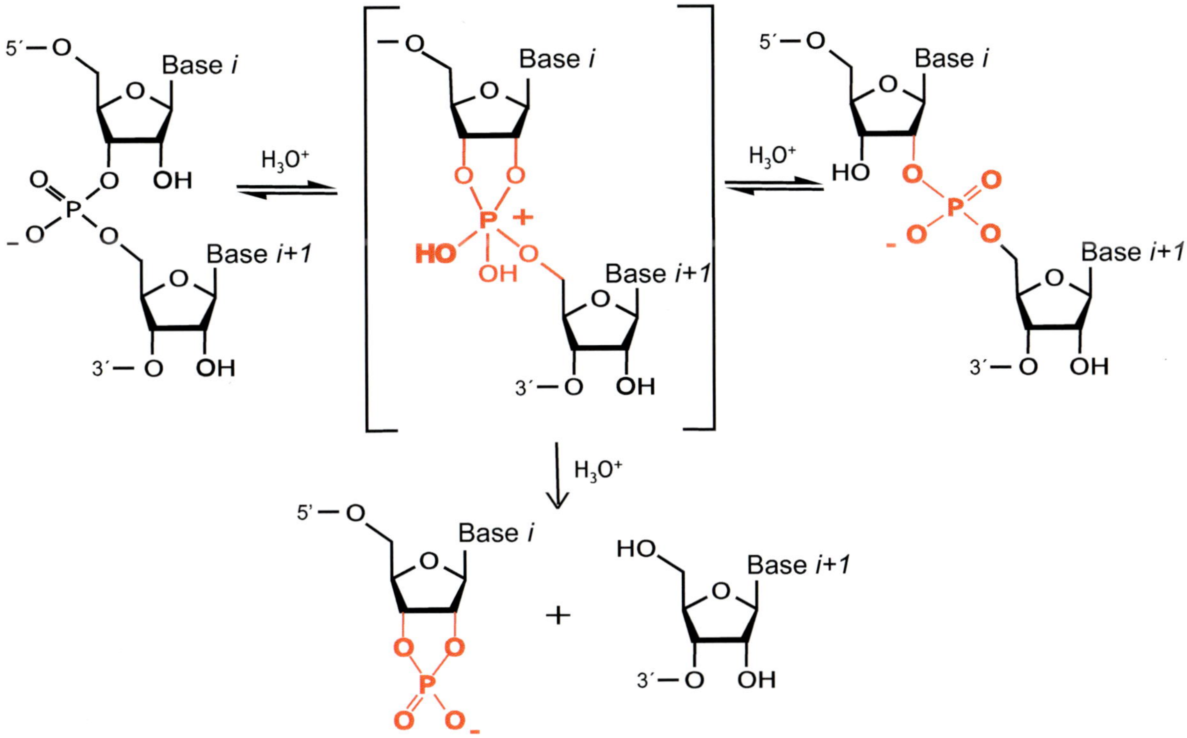

Figure 6-27. Cleavage and migration of the RNA internucleotide linkage under acidic conditions. The migration interconversions are indicated by the horizontal reactions, whereas cleavage occurs in the vertical reaction.

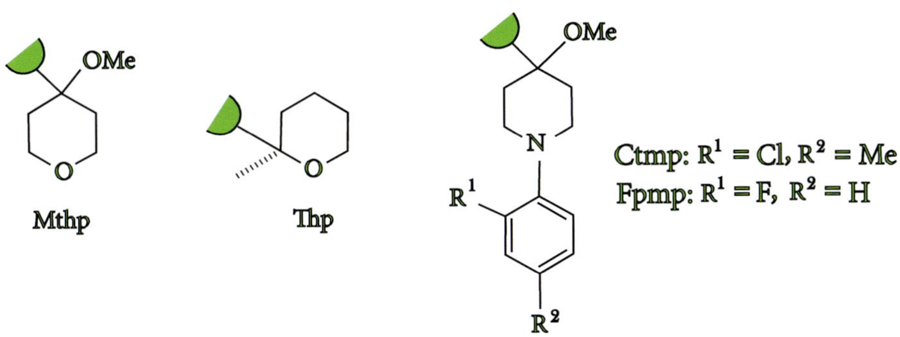

(6.6)

They are completely stable under the basic conditions of concentrated aqueous ammonia at 55°C that are needed to remove virtually all of the other protecting groups from fully assembled RNA sequences. The 2-nitrobenzyl-protecting group:

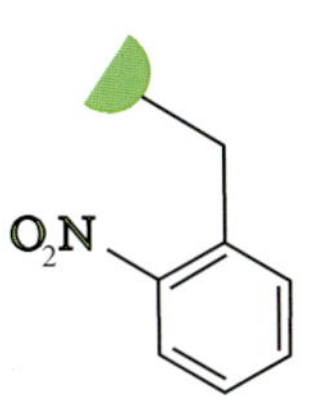

2-Nitrobenzyl

(6.7)

is removed by photolysis at wavelengths >280 nm. It is also stable to the alkaline treatment. TBDMS:

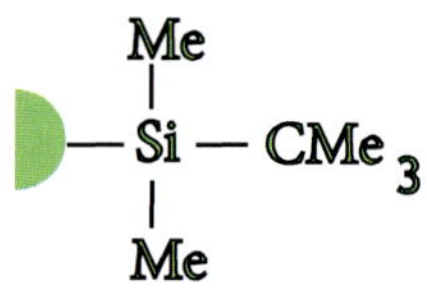

TBDMS

(6.8)

is used widely in the automated solid-phase synthesis of RNA sequences. It has the advantage that it can be removed in the final unblocking step without using acidic conditions. Under the standard basic unblocking conditions, however, it is partially removed and tends to undergo migration between the 2′- and 3′-hydroxyl groups. Nevertheless, it is used widely in automated solid-phase synthesis.

RNA synthesis using 2′-O-(tert-butyldimethylsilyl) protection. B. S. Sproat (2005) *Methods Mol. Biol.* **288**, 17–32.

Synthesis of RNA using 2′-O-DTM protection. A. Semenyuk *et al.* (2006) *J. Am. Chem. Soc.* **128**, 12356–12357.

A new method for RNA synthesis by use of the cyanoethyl group as the 2′-hydroxyl protecting group. H. Saneyoshi *et al.* (2006) *Nucleic Acids Symp. Ser.* **29**, 125–126.

6.7.B. RNA Synthesis in Solution

As in the case of DNA (Section 6.6.C), solution-phase synthesis is labor-intensive and used only if relatively large quantities of RNA sequences are required.

Relatively long RNA sequences have been synthesized in solution using the phosphotriester approach. In one strategy (Figure 6-28), the 2′-hydroxyl groups are protected with the achiral Mthp group (Equation 6.6) and the methoxymethylene group is used to protect both the 2′- and 3′-hydroxyl groups at the 3′-end that is not to be joined. The 5′-terminal hydroxyl groups are protected with acyl groups, such as the 2-(dibromomethyl)benzoyl (Dbmb) group:

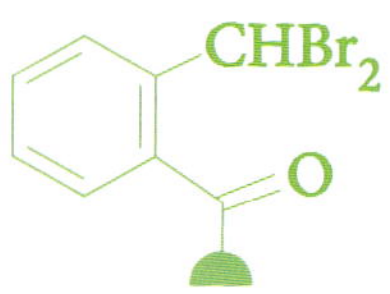

Dbmb

(6.9)

The Dbmb groups can be removed under very mild basic conditions. Internucleotide linkages are protected with 2-chlorophenyl groups:

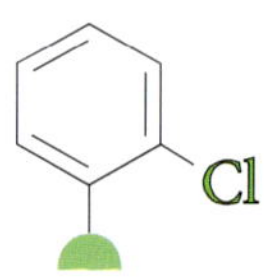

(6.10)

and the bases are protected as:

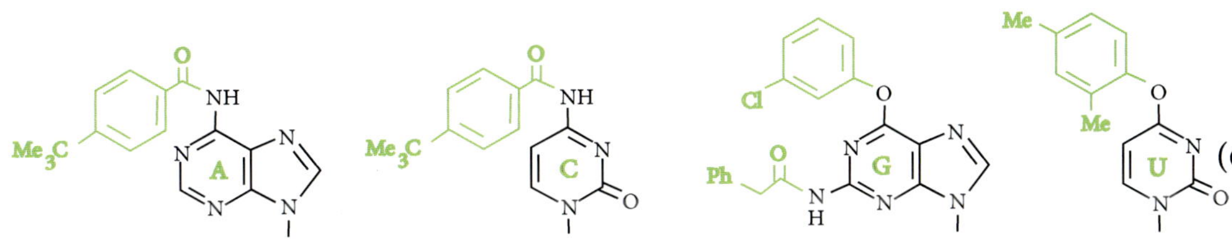
(6.11)

The guanine and uracil residues are protected with aryl groups on O6 and O4, respectively, to prevent the occurrence of side-reactions during the coupling steps. The coupling reactions use MSNT (Equation 6.4) in dry pyridine solution.

Once the fully protected RNA sequences have been assembled, they are unblocked in three steps. First, the 2-chlorophenyl-protecting groups are removed from the internucleotide linkages, and the O6 and the O4 aryl groups are removed from the guanine and uracil residues by treatment with *E*-2-nitrobenzaldoxime and tetramethyl-guanidine (Equation 6.5). Secondly, all the acyl-protecting groups are removed using concentrated aqueous ammonia. Finally, the Mthp and 2′,3′-terminal methoxymethylene-protecting groups are removed by hydrolysis under mild acidic conditions.

RNA oligonucleotide synthesis via 5′-silyl-2′-orthoester chemistry. S. A. Hatsel *et al.* (2005) *Methods Mol. Biol.* **288**, 33–50.

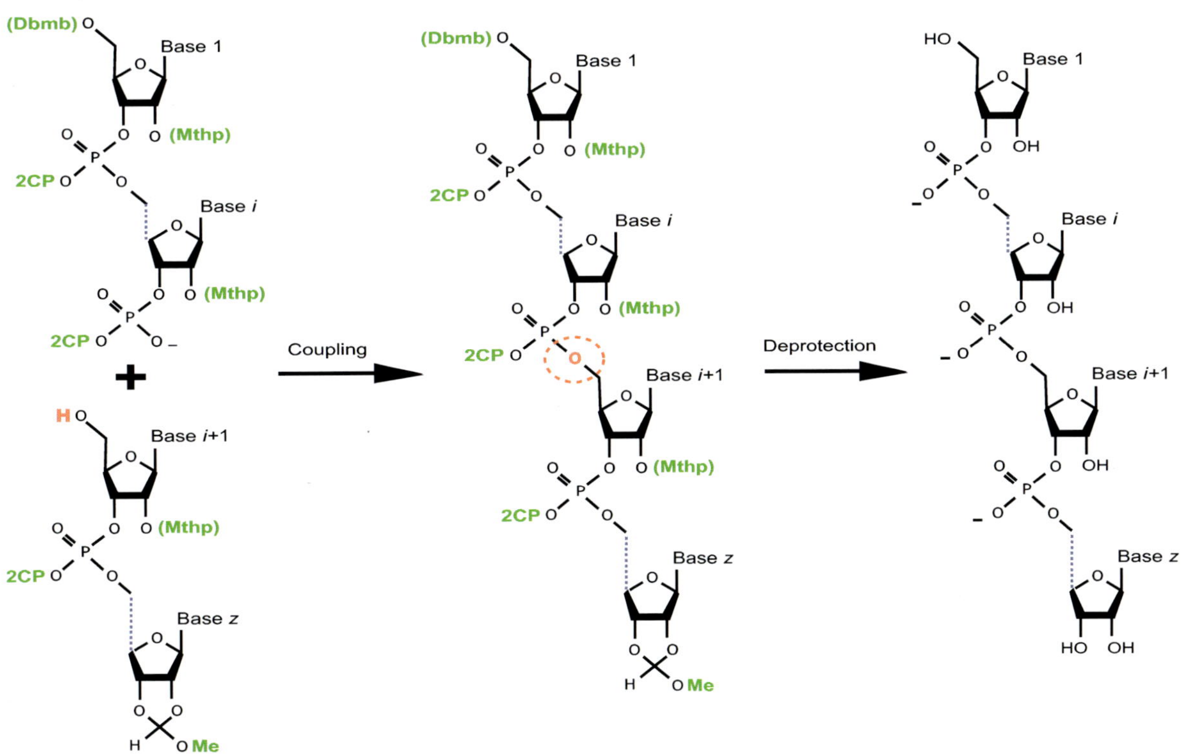

Figure 6-28. Coupling two protected oligoribonucleotides in solution, followed by removal of the protecting groups. The two *dashed* bonds indicate that there are other nucleotides in the chain at these points. The Mthp-protecting group is described in Equation 6.6 and Dbmb in Equation 6.9. 2CP is the 2-chlorophenyl group (Equation 6.10). The methyl group (Me) can be on either the 2′- or 3′-hydroxyl group at the 3′-terminus.

6.7.C. Solid-phase RNA Synthesis

The advantages of the preparation of relatively small (i.e. milligram to gram) quantities of RNA using solid-phase techniques are basically the same as those described with DNA (Section 6.6.D). The protocols involved are also closely similar, apart from the final unblocking procedures. The first step in each cycle is removal of the 5′-terminal protecting group (Figure 6-29, step 1), which must be both rapid and quantitative. The DMTr and Px groups (Figure 6-20) are frequently used as the 5′-protecting

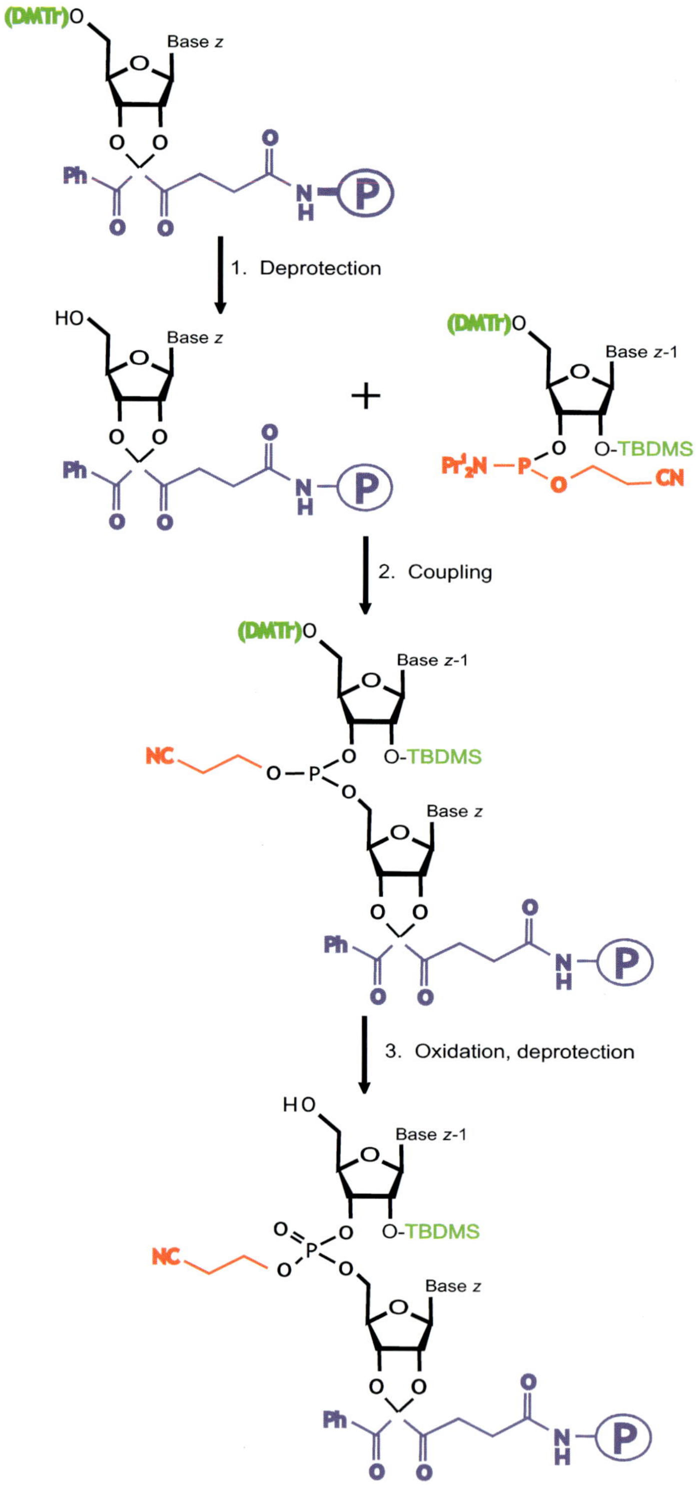

Figure 6-29. One cycle of solid-phase synthesis of RNA. P in a *circle* is the polymer support. The protecting groups DMTr and TBDMS are described in Figure 6-20 and Equation 6.8. The two protecting groups linked to the 2′- and 3′-hydroxyl groups of nucleotide *z* can be attached to either. The deprotection of the 5′-hydroxyl group of ribonucleotide *z* (which will be at the 3′-end of the final polymer) in step 1 takes place in 3% trichloroacetic acid in dichloromethane. The coupling in step 2 requires 1*H*-tetrazole in acetonitrile, followed by acetic anhydride, a weak base and 1-methylimidazole. Step 3 requires 0.1 M iodine in water-pyridine, followed by 3% trichloroacetic acid in dichloromethane. The oligonucleotide is now ready to have another nucleotide added, in another step 2.

groups because they are especially acid-labile. They also have the advantage that it is easy to measure the released groups spectrophotometrically and thereby monitor the efficiency of each coupling step. As in DNA synthesis, the adenine, cytosine and guanine base residues are protected with *N*-acyl groups (Figure 6-19) and the uracil moieties are left unprotected. The bases of any modified ribonucleosides need to be protected in an analogous manner. 2-Cyanoethyl phosphoramidite building blocks are generally used, as in DNA synthesis (Figure 6-22). The solid support P is usually controlled-pore glass or polystyrene, and the 3′-terminal ribonucleoside residue is usually attached through a 3′- or 2′-succinoyl linker (Figure 6-30). It does not matter which of the two hydroxyl groups is actually linked to the solid support, and the other 2′- or 3′-hydroxyl group is conveniently protected with an acyl group, such as benzoyl.

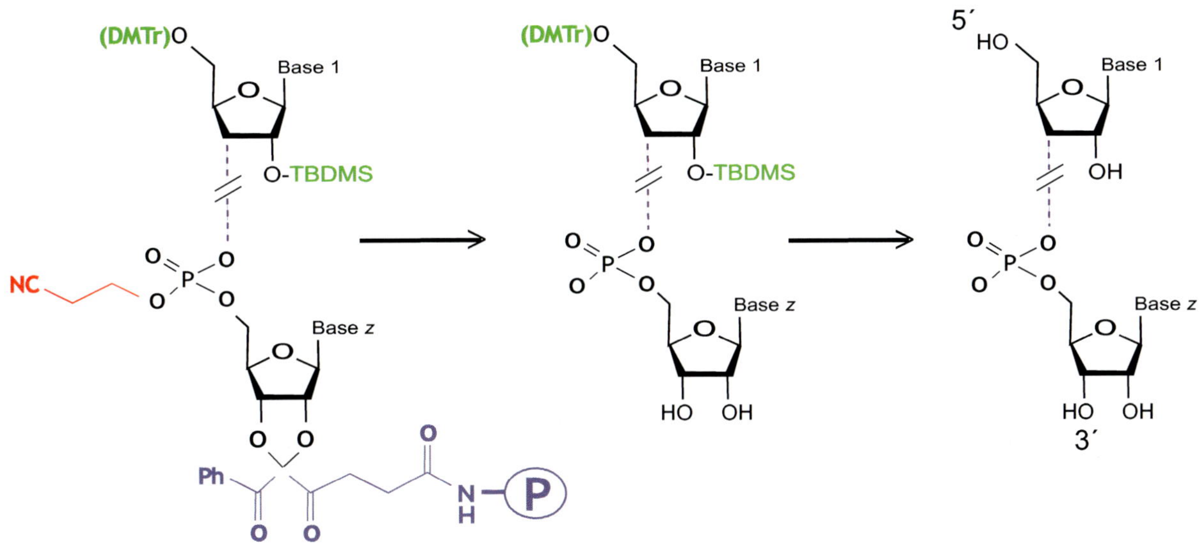

Figure 6-30. Release from the polymer and deprotection of the assembled RNA molecule of *z* ribonucleotides. The first step takes place in 3% trichloroacetic acid in dichloromethane followed by ammonia in aqueous ethanol; step 2 takes place in triethylamine trihydrofluoride. P is the polymer support.

The 2′-protecting groups must be completely stable to repeated treatment with acid in the steps for removing the DMTr or Px groups at the beginning of each synthetic cycle (Figure 6-29, step 1), as well as to the alkaline conditions used for detachment of the assembled RNA sequences from the solid support and for removal of the *N*-acyl-protecting groups from the base moieties and of the 2-cyanoethyl-protecting groups from the internucleotide linkages in the penultimate unblocking step (Figure 6-30). TBDMS (Equation 6.8) and Fpmp (Equation 6.6) are the most suitable groups for protection of the 2′-hydroxyl groups, although they require somewhat different methods to release and deprotect the RNA molecule. The RNA sequences protected with Fpmp are partially deprotected by treatment with concentrated aqueous ammonia, without any loss of the Fpmp-protecting groups. This has the considerable advantage that the 2′-protected RNA is completely stable to base-catalyzed hydrolysis and any contaminating ribonucleases and can readily be purified and freed from truncated sequences before it is fully unblocked. Cleavage and migration of the internucleotide linkages during the final unblocking step can be avoided by using mild acidic conditions.

Both methods have produced relatively high-molecular weight RNA sequences. The rates of coupling and coupling efficiencies, however, are somewhat lower than those for the corresponding DNA phosphoramidites.

Recent advances in the high-speed solid phase synthesis of RNA. W. S. Marshall & R. J. Kaiser (2004) *Cur. Opinion Chem. Biol.* **8**, 222–229.

~ CHAPTER 7 ~

NUCLEIC ACID–PROTEIN INTERACTIONS

The interactions between proteins and nucleic acids are some of the most important in biology. RNA binds to proteins mainly for structural reasons, although in some cases the RNA is also involved in catalysis (Chapter 8). Binding to DNA serves two principal functions: to (1) organize and compact the DNA of chromosomes and (2) effect and regulate the processes of gene transcription, DNA replication and DNA recombination. These two functions are very different and have different structural and binding requirements. **Chromosomal DNA is organized by abundant proteins that can bind to many sites irrespective of the sequence** (Sections 7.3.F and 7.3.G). At the other extreme, **regulation of the enzymatic processes that manipulate DNA require precise targeting to particular DNA sequences**, which involves specific recognition of the nucleotide sequence by one or more specific proteins. Many such proteins act as repressors or activators of gene expression, either by themselves or in combination with corepressors or coactivators. A repressor usually acts simply by binding to a specific site on the DNA adjacent to a gene and blocking access of the proteins that would transcribe the gene into messenger RNA, which then gets translated into protein. Others do not prevent access but block further progression along the DNA of the RNA polymerase enzyme (Section 6.2). An activator usually acts by being part of the site next to the gene that is recognized and bound by the transcription apparatus; it assists in the binding of the other components of the transcription machinery. Most prominent are the **transcription factors**, which are part of the complex machinery for controlling the transcription of DNA into messenger RNA. In these cases, the specificity of binding is paramount.

It is not straightforward for a protein to distinguish between different nucleotide sequences in double-stranded DNA, for the nucleotides of the two antiparallel strands are base-paired within the interior of the double helix. The exterior surface of the double helix is almost independent of its nucleotide sequence, being comprised primarily of the constant phosphate–sugar backbone. Only the edges of the nucleotides are accessible to the solvent and a protein, primarily within the major groove of the DNA double helix (Figure 7-1). Nevertheless, this is sufficient for **the nucleotides to be discriminated by the different polar groups that are accessible**.

The interactions between proteins and nucleic acids can be very strong. They must often be sufficient for a protein to find a single specific DNA binding site in a cell. A typical eukaryotic cell has a diameter of 10 μm, which corresponds to a volume of 5×10^{-10} liters, so **one or two specific binding sites on a genome would be present at a concentration of only about 10^{-12}** M; the dissociation constant must be no greater than this for binding to be stable. **One consequence of such tight binding is that**

the complex usually dissociates very slowly. A further consequence is that the association reaction can also be very slow at the very low reactant concentrations approaching the dissociation constant, because the rates of association and dissociation are equal at equilibrium.

Nucleic acids, especially DNA, can be very large molecules, with many sites at which proteins can bind. One must then distinguish between (1) relatively nonspecific binding, in that it occurs with many different nucleotide sequences and at many different sites on the DNA molecule, and (2) specific binding to a certain sequence of nucleotides. **Proteins often bind to long DNA molecules by binding**

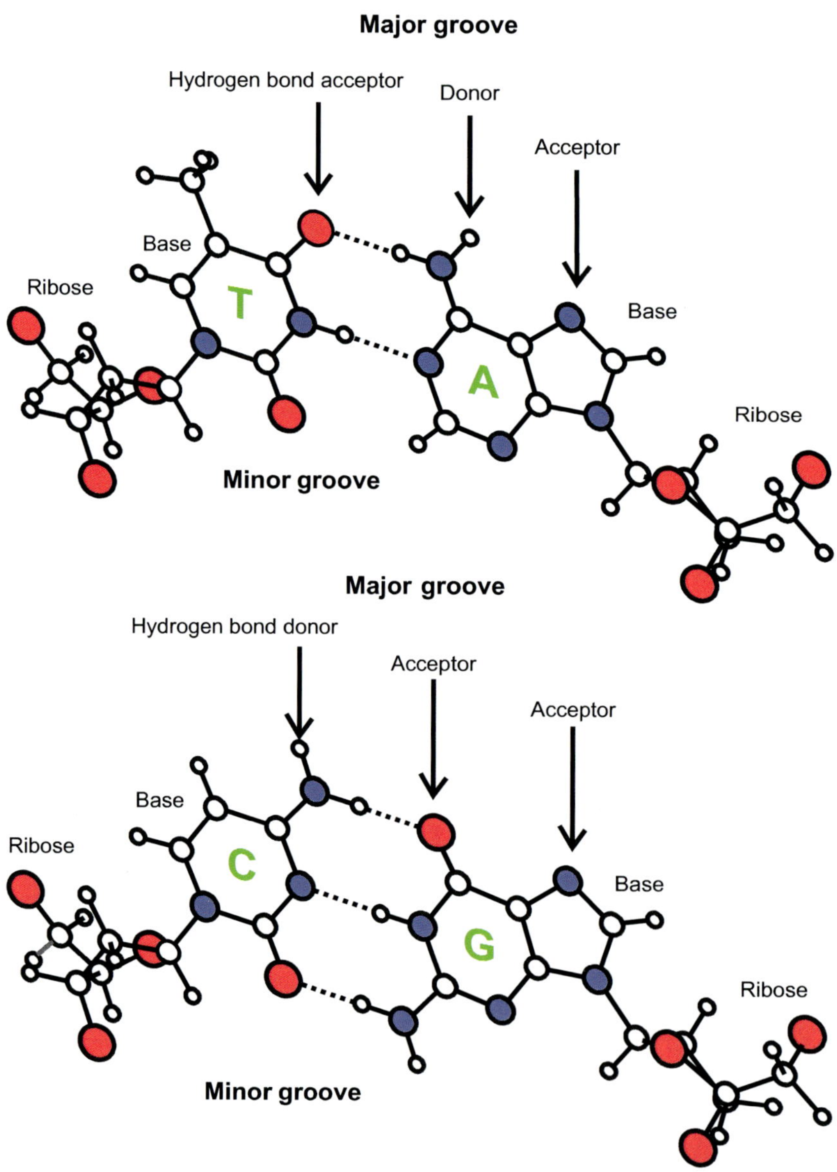

Figure 7-1. Hydrogen-bond interactions that can be used to distinguish between the various base pairs in double-helical B-form DNA. The upper edges of the bases are exposed in the major groove, and each has a unique pattern of hydrogen bond donors and acceptors. Each base pair can be reversed, leading to two more different patterns. The *red circles* represent O atoms, the *blue circles* are N, and the *smallest circles* are H atoms; the others are carbons.

loosely and nonspecifically at the first site they encounter, then diffusing one-dimensionally along the double helix until they encounter their specific binding site; they bind much more tightly to it, and consequently remain bound to this site. This is the basis for the seemingly paradoxical observation that a protein can find a specific binding site on a DNA molecule more rapidly if the DNA molecule is elongated with nonspecific sequences.

Finding a single site on genomic DNA is not a trivial matter; the human genome consists of 3×10^9 base pairs. The sequence of nucleotides that make up genomic DNA is not very regular, and to a first approximation the sequence can be taken to be random. It is then possible to estimate how many nucleotides must make up a binding site in order to have a reasonable likelihood of being unique and occurring only once in the genome. Each residue in DNA can be occupied by four different nucleotides (or base pairs), so a segment of N nucleotides can exist as 4^N different sequences. This number must be at least as large as the number of base pairs in the genome for it to have a reasonable chance of being unique. A 16-nucleotide segment can exist in 4.29×10^9 ($= 4^{16}$) different sequences, so **a binding site comprising 16 nucleotides has a reasonable chance of being unique in the human genome**. Human proteins that recognize a single site on the genome would be expected to recognize 16 base pairs specifically. The genomes of prokaryotes are smaller, that of *Escherichia coli* comprising 4×10^6 base pairs, but 11 nucleotides ($4^{11} = 4.19 \times 10^6$) are still probably needed to define a unique site.

Understanding the interactions between proteins and nucleic acids requires that the sequence specificity be determined, that the affinities of the protein for this and other sequences be measured, and that the structure of the complex be determined.

Protein–DNA recognition patterns and predictions. A. Sarai & H. Kono (2005) *Ann. Rev. Biophys. Biomolec. Structure* **34**, 379–398.

Recognizing DNA. R. Lavery (2005) *Quart. Revs. Biophys.* **38**, 339–344.

An extra dimension in nucleic acid sequence recognition. K. R. Fox & T. Brown (2005) *Quart. Revs. Biophys.* **38**, 311–320.

Identification and characterization of DNA-binding proteins by mass spectrometry. E. Nordhoff & E. Lehrach (2007) *Adv. Biochem. Eng. Biotechnol.* **104**, 111–195.

7.1. TECHNIQUES FOR MEASURING PROTEIN–DNA INTERACTIONS

Some of the techniques that are used to characterize the binding of ligands to macromolecules are also applicable to nucleic acids and proteins. Spectroscopic measurements, such as ultraviolet (UV) absorbance, fluorescence and circular dichroism, can usually detect the difference between free and associated nucleic acids or proteins, but the signals are often small and they do not reveal which binding sites are being used. Affinity chromatography can be very useful, as either the nucleic acid or the protein can be bound covalently to a column or blotting matrix.

The interactions between nucleic acids and proteins differ from those of other ligands in that both reactants are macromolecules and the complex frequently dissociates only slowly. Consequently, techniques such as ultracentrifugation can distinguish between free and bound protein and nucleic acid, and the stoichiometry of the interaction can often be inferred from the change in the sedimentation coefficient. Both free reactants and their complexes can often be observed by electron microscopy, or those in a complex can be cross-linked covalently. Some of the techniques for small ligands, such as equilibrium dialysis, are not applicable with two macromolecules, but other techniques take advantage of the special properties of nucleic acids and can be extremely informative.

Most important is that specific complexes with proteins dissociate only very slowly, because such complexes are very stable, usually having dissociation constants of 10^{-7}–10^{-14} M. In this case, if the rate constant for binding is 10^{6} s^{-1} M^{-1}, the rate constant for dissociation will be between 10^{-1} and 10^{-8} s corresponding to half-times for dissociation of between 7 s and 7×10^{7} s (2 years). Consequently, tight complexes can be separated from the free components by many techniques, such as chromatography, filtration and electrophoresis.

Protein–DNA interactions. V. Dotsch (2001) *Methods Enzymol.* **339**, 343–357.

Circular dichroism for the analysis of protein × DNA interactions. M. L. Carpenter *et al.* (2001) *Methods Mol. Biol.* **148**, 503–510.

Ultraviolet crosslinking of DNA–protein complexes via 8-azidoadenine. R. Meffert *et al.* (2001) *Methods Mol. Biol.* **148**, 323–335.

Analysis of DNA–protein interactions by intrinsic fluorescence. M. L. Carpenter *et al.* (2001) *Methods Mol. Biol.* **148**, 491–502.

Surface plasmon resonance applied to DNA–protein complexes. M. Buckle (2001) *Methods Mol. Biol.* **148**, 535–546.

Scanning transmission electron microscopy of DNA–protein complexes. J. S. Wall & M. N. Simon (2001) *Method Mol. Biol.* **148**, 589–601.

Covalent trapping of protein–DNA complexes. G. L. Verdine & D. P. G. Norman (2003) *Ann. Rev. Biochem.* **72**, 337–366.

Crystallization of protein–DNA complexes. T. Hollis (2007) *Methods Mol. Biol.* **363**, 225–2237.

7.1.A. Filter-binding Assays

Complexes that dissociate slowly can be measured using filter-binding assays, which are simple, rapid, sensitive and versatile. They are based on the empirical observation that **most proteins, but not double-stranded DNA molecules, bind to the surfaces of nitrocellulose membrane filters**. Therefore, when a DNA-binding protein is incubated with a specific DNA sequence and then passed through a filter, the DNA•protein complex that was formed and any excess protein will bind to the membrane filter, while uncomplexed DNA will pass through. If the DNA used is radiolabeled, the amount of the DNA•protein complex bound to the membrane filter can be determined easily from the amount of radioactivity retained. Denatured single-stranded DNA, however, also binds to nitrocellulose and cannot be studied in this way.

The use of highly radioactive DNA makes it possible to use very low concentrations of DNA and protein, and the protein need not be pure. The lower limit for the concentration of radioactive DNA is usually about 10^{-11} M. A typical experiment involves titration of a constant amount of radiolabeled DNA with the DNA-binding protein of interest. The amount of radioactivity retained on the membrane increases with increasing protein concentration until all the radioactive DNA that can bind to that DNA-binding protein is depleted from the incubation mixture. The binding affinity can be determined by measuring the level of radioactive complexes formed in the presence of a nonradioactive competitor DNA of known concentration. The kinetics of association of a specific DNA•protein complex can also be studied in this way, by mixing the components at time zero and then measuring the time–course of binding by filtering aliquots at various times; the binding reaction can be quenched by the addition of an excess of unlabeled DNA to the aliquots to bind to the free protein molecules. The kinetics of dissociation are measured by mixing radioactive DNA with an excess of the protein, to saturate all the DNA binding sites; all the radioactivity of an aliquot will bind to the filter. At time zero, an excess is added of the same DNA, but not radioactive. As protein dissociates from the radioactive complexes, it is effectively sequestered by binding to the nonradioactive DNA and the radioactivity of aliquots retained by the filter will decrease with time and reflect the rate of dissociation of the complex.

Filter-binding assays cannot differentiate between dissimilar DNA•protein complexes, so this assay is most useful when only a single DNA•protein complex is formed. Most proteins bind to nitrocellulose but their individual binding affinities vary. The optimum conditions for binding a DNA•protein complex to the nitrocellulose membrane will differ for each protein and must be determined for each case.

Filter-binding assays. P. G. Stockley (1994) in *Methods in Molecular Biology*, Vol. 31 (G. G. Kneate, ed.), Humana Press, NJ, pp. 251–262.

Is nitrocellulose filter binding really a universal assay for protein–DNA interactions? S. Oehler *et al.* (1999) *Anal. Biochem.* **268**, 330–336.

7.1.B. Gel Retardation Assay

The gel retardation assay is also known as the **bandshift assay** or **electrophoretic mobility shift assay**. It can be used to detect proteins that bind to DNA or RNA in crude cell extracts or to study the binding activities of purified proteins. The assay is based on the observations that protein•DNA complexes remain intact when separated gently by nondenaturing gel or capillary electrophoresis and that they migrate as distinct bands more slowly than does the free DNA fragment (Figure 7-2). The gel matrix can be polyacrylamide for DNA fragments shorter than 500 base pairs, agarose gels for fragments longer than 1000 base pairs, and mixtures of polyacrylamide and agarose for the intermediate lengths.

The assay is simple, quick and very sensitive when highly radioactive DNA is used. It is useful for both sequence-specific proteins and nonspecific proteins, such as histones, so long as the complexes have dissociation constants $<10^{-7}$ M and do not dissociate during the electrophoretic separation. Including unlabeled nonspecific DNA in the binding mixture to sequester any proteins that would bind nonspecifically ensures that only specific complexes are observed. Any complex observed should be shown to be specific by demonstrating that radiolabeled complex is not formed in the presence of an excess of the specific DNA that is not radioactive.

Gel retardation assays can be used quantitatively to estimate dissociation constants of protein•DNA complexes, so long as dissociation is negligible during the electrophoretic separation. Additionally, such assays can detect interactions between one protein that binds DNA and another protein that does not. The binding of such a second protein to a protein•DNA complex to form a triple complex

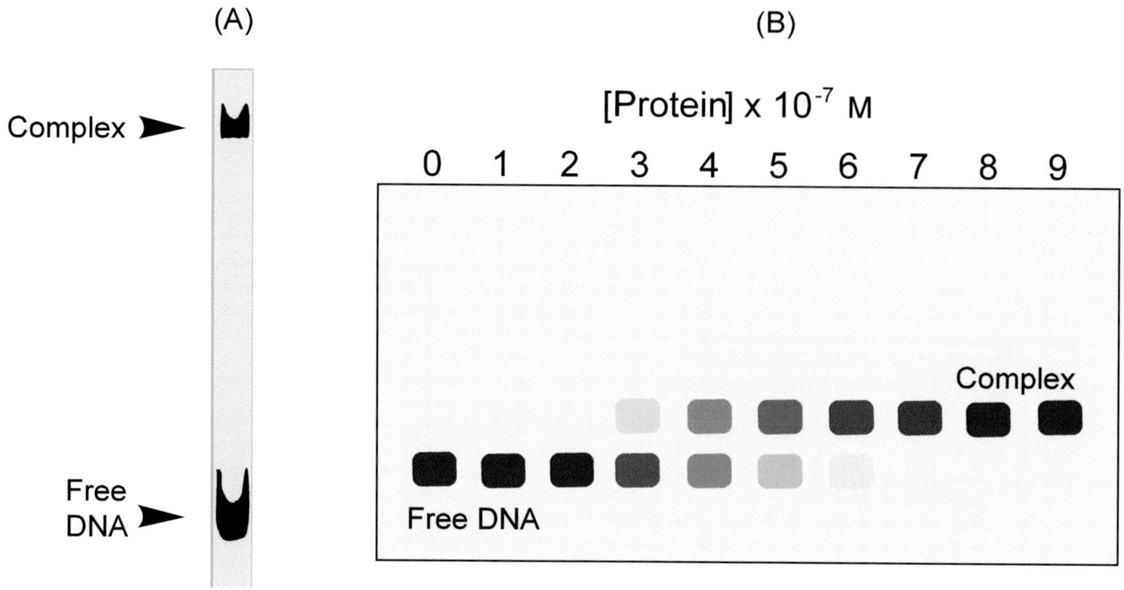

Figure 7-2. Idealized bandshift assays. Electrophoresis in 0.7% (w/v) agarose electrophoresis gels is from *top* to *bottom*. The DNA molecules are radioactive, and only they are detected using autoradiography. (A) Binding of a 150-kDa protein causes the radioactive 62-base pair DNA fragment containing the binding site to migrate as a distinct band, at the top of the gel, much more slowly than the unbound DNA at the bottom of the gel. (B) Titration of a 21-base pair oligonucleotide with increasing amounts of a binding protein. The DNA concentration was 1×10^{-8} M and the protein concentration was varied as shown. When the DNA concentration is less than the dissociation constant, the value of the dissociation constant can be estimated by the concentration of protein that produces 50% binding, in this case approximately 4×10^{-7} M.

will be apparent by a further retardation of mobility that is called a **supershift**. Supershift experiments can also be used to assay the binding of a second DNA-binding protein to the same DNA molecule or the binding of a second molecule of DNA to the protein. What proteins are present in a complex can be determined in purified systems by excising the band of the complex and then separating the proteins by SDS–polyacrylamide gel electrophoresis or mass spectrometry.

The gel retardation assay can also detect bending of DNA induced by a DNA-binding protein. When binding of a protein to a DNA fragment of length 100–250 base pairs causes it to bend, the complex migrates more slowly through the pores of the gel. The degree of retardation depends on the site where the protein binds. The retardation is greatest when the binding site is in the middle of the DNA molecule. This method can be used to obtain an estimate of the degree of DNA bending induced by a protein by comparing it with proteins that induce known bends.

Disadvantages of this assay are that the electrophoretic separation is relatively slow, at least compared with the filter-binding assay, so some complexes might dissociate during the separation. Also, electrophoresis requires that the buffer have a low ionic strength, although this will probably stabilize most complexes by enhancing electrostatic interactions. Often protein–nucleic acid complexes are observed to persist within gels for much longer than would be expected on the basis of their free solution lifetimes. Their lifetimes in free solution may be limited by their frequency of encounter with other molecules of DNA or with protein•DNA complexes, which will be decreased in gels.

Electrophoretic mobility shift assays for the analysis of DNA–protein interactions. M. A. Laniel *et al.* (2001) *Methods Mol. Biol.* **148**, 13–30.

Capillary DNA–protein mobility shift assay. J. Xian (2001) *Methods Mol. Biol.* **163**, 355–367.

A modified quantitative EMSA and its application in the study of RNA–protein interactions. Y. Li *et al.* (2004) *J. Biochem. Biophys. Methods* **60**, 85–96.

7.1.C. Footprinting

Footprinting is a method for identifying the protein-binding sites on DNA *in vivo* and *in vitro*. A bound protein molecule can protect DNA from cleavage or modification by a nuclease or chemical reagent (Section 2.6), so the binding site is apparent as a region of the DNA with reduced reactivity in the presence of the protein. This is determined by analyzing the products of a partial cleavage reaction by denaturing polyacrylamide gel electrophoresis (Figure 7-3). The cleavage pattern is usually simplified by labeling the DNA molecule with radioactivity at only one end; only the DNA fragments containing this end are detected using autoradiography. Only partial cleavage will produce a collection of DNA fragments of different lengths that reflect the sites at which cleavage occurred. **The 'footprint' of the protein appears as a gap in the normal pattern of cleavage.** In some instances, the bound protein can cause intensification of a band, implying that the bound protein made that DNA site hyperreactive, usually by changing its local conformation.

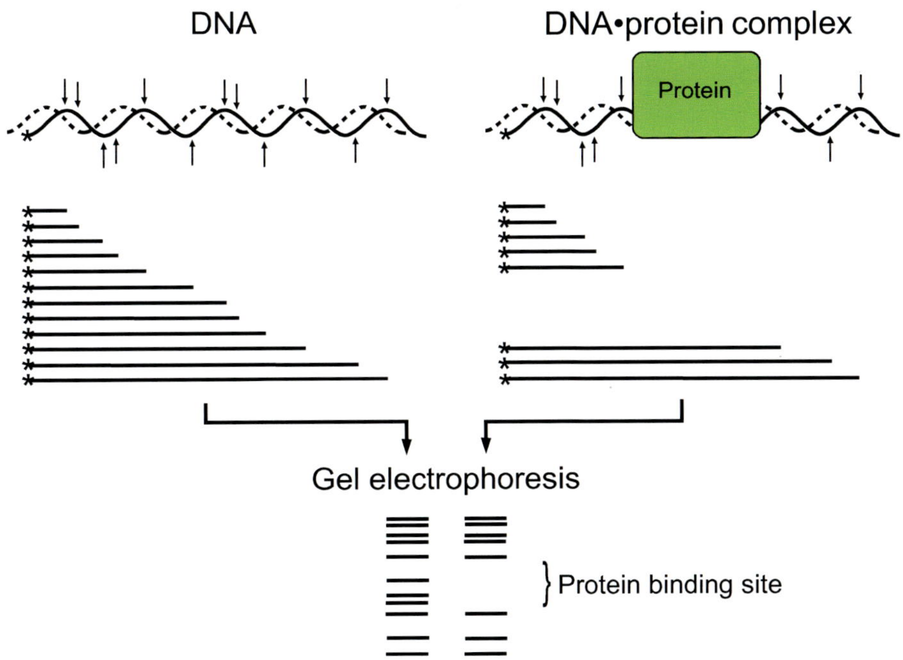

Figure 7-3. Schematic description of the footprinting assay. A naked DNA fragment and its fragmentation pattern produced by cleavage at the *arrows* is shown at the *left*. The DNA is labeled radioactively at one end, so only the fragments containing this end of the DNA are visualized by autoradiography after gel electrophoresis under denaturing conditions (*bottom*). With protein bound to a specific segment of DNA, potential cleavage sites within this segment are protected and cleavage does not occur. This results in a gap in the gel electrophoresis pattern corresponding to the binding site for the protein. Adapted from D. Rhodes & L. Fairall (1997) in *Protein Function: a practical approach*, 2nd edn (T. E. Creighton, ed.), IRL Press, Oxford, p. 228.

The extent of the cleavage reaction is usually chosen so that, on average, a given DNA molecule will have at most one cleavage. The nuclease DNase I is commonly used *in vitro* for footprinting, while dimethylsulfate and hydroxyl radicals are used for footprinting using chemical cleavage methods. **DNase I** cleaves preferentially in the minor groove of DNA, also depending on the local conformation of the bound DNA; cleavage is enhanced when the minor groove is on the outside of DNA wrapped around a protein. It is usually used to cleave only one of the two DNA strands, rather than both. **Dimethylsulfate** methylates the N7 position of guanine in the major groove and the N2 position of adenine in the minor groove, which sensitizes the sugar–phosphate backbone to mild alkaline hydrolysis and makes it possible to determine the sites of reaction. For **hydroxyl-radical** footprinting, the $OH^{\bullet}$ radicals are generated by the **Fenton reaction**, in which hydrogen peroxide reacts with a complex of Fe^{2+} and EDTA:

$$[\mathrm{Fe\ (EDTA)}]^{-2} + \mathrm{H_2O_2} \rightarrow [\mathrm{Fe\ (EDTA)}]^{-} + \mathrm{OH}^{-} + \mathrm{OH}^{\bullet} \tag{7.1}$$

The $OH^{\bullet}$ radicals react with deoxyribose moieties of the DNA backbone, which can cause cleavage of the backbone. This reagent attacks primarily the minor groove irrespective of the nucleotide sequence. It is, however, sensitive to the width of the groove and cleaves preferentially where the minor groove is wide. The opposite selectivity is exhibited by a similar type of reagent, **Mn-porphyrin**, which cleaves DNA preferentially where the minor groove is narrow.

The modification reaction need not cleave the DNA double helix but merely modify it so that it can no longer be copied by an enzyme like DNA polymerase. In this case, a radioactive oligonucleotide primer complementary to an appropriate sequence on the DNA molecule is used to initiate synthesis of the complementary strand, as in the chain-termination method of sequencing DNA (Section 6.4.B). The newly synthesized complementary strand will be extended until a site of modification is reached. The length of the strand synthesized indicates the site of the modification.

The various cleavage methods can provide complementary information. For example, the enzyme DNase I is a large molecule, whereas a hydroxyl radical is very small and thus can provide a more detailed picture. On the other hand, DNase I cleavage can provide information about the local DNA conformation, whereas cleavage by hydroxyl radicals tends to be too uniform over all the possible sites in the DNA molecule.

Footprinting techniques can also be used to probe the three-dimensional (3-D) structures of nucleic acids, especially RNA (Section 4.2).

Mapping nucleic acid structure by hydroxyl radical cleavage. T. D. Tullius & J. A. Greenbaum (2005) *Curr. Opinion Chem. Biol.* **9**, 127–134.

DNase I footprinting of small molecule binding sites on DNA. C. Bailly *et al.* (2005) *Methods Mol. Biol.* **288**, 319–342.

Visualising DNA: footprinting and 1-2D gels. A. R. Urbach & M. J. Waring (2005) *Mol. Biosyst.* **1**, 287–293.

Radiolytic footprinting with mass spectrometry to probe the structure of macromolecular complexes. K. Takamoto & M. R. Chance (2006) *Ann. Rev. Biophys. Biomol. Structure* **35**, 251–276.

7.2. PRINCIPLES OF PROTEIN–DNA RECOGNITION

Individual sequence-specific proteins usually recognize DNA sequences that are 3–15 nucleotides long. They use two types of interactions.

(1) The principal basis for sequence selectivity is direct contact between the polypeptide chain of the DNA-binding protein and the exposed edges of the DNA base pairs, primarily in the major groove of B-DNA. There the four different Watson–Crick base pairs can be distinguished by their distinctive pattern of hydrogen-bond donor and acceptor groups (Figure 7-1). **There is, however, no clear one-to-one correspondence between amino acid residues and the bases that they recognize**. The contacts may involve either hydrogen bonds (Figure 7-4) or van der Waals interactions, which usually detect the methyl group of thymine nucleotides. Small molecules, such as water molecules that are tightly bound to both proteins and DNA and integral components of the macromolecular structure, may also participate in these interactions and so provide binding specificity by proxy. The interfaces between DNA and protein are as tightly packed as are the interiors of proteins and DNA, so the interactions are probably as energetically favorable in all these cases. Cavities in the DNA–protein interfaces tend to be larger but are more likely to be occupied by water molecules.

(2) The specificity of the direct interactions can be supplemented by the bendability or deformability of the DNA molecule, which limits the energetically favorable conformations of a particular binding site. The conformational flexibility of the DNA double helix depends upon the nucleotide sequence, so it provides additional information about the sequence. Bending of DNA is achieved by a variety of mechanisms, notably the induction of bends by spatial constraints on a rigid protein surface, the insertion of hydrophobic residues between adjacent bases in the DNA duplex, and charge neutralization on one face of the double helix. **Compared with direct contacts to the base pairs, bending of the DNA is a second-order effect that modulates the affinity for the whole binding site**.

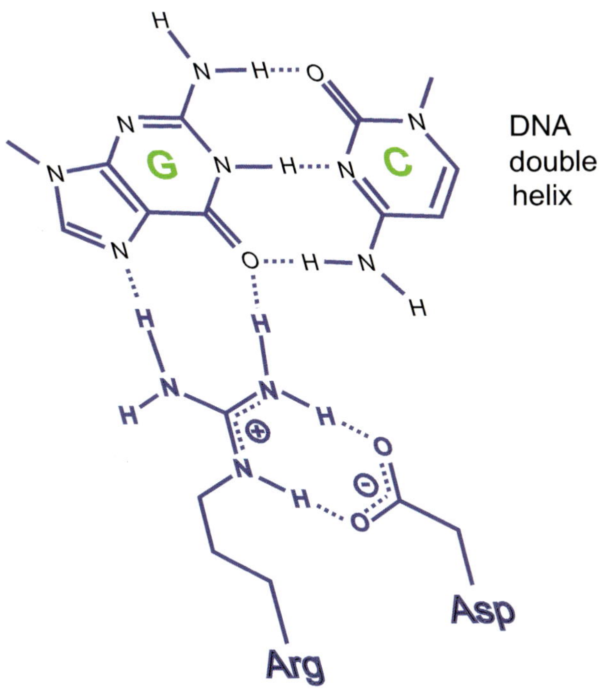

Figure 7-4. Hydrogen bonds involved in the recognition of guanine bases by zinc fingers. The Arg and Asp residues are involved in an ion pair, which may help to align the guanido group of the Arg residue to form two hydrogen bonds to the guanine base. Adapted from N. P. Pavletich & C. O. Pabo (1991) *Science* **252**, 809–817.

The binding energy available from direct specific interactions with the base pairs of binding sites of average length, although significant, is not by itself in general sufficient to produce a stable bound complex. The additional binding energy required results from other interactions between the protein and DNA that are independent of the nucleotide sequence, especially direct electrostatic interactions between basic amino acid residues and the negatively charged sugar–phosphate backbone. The spatial constraints imposed by these types of interaction may also constrain the conformation of the DNA when bound to a protein. Both components of the interaction contribute to the overall binding constant, but the first, sequence-specific, interactions determine the specificity of a protein for its binding of DNA.

Recognition of specific nucleotide sequences by proteins is a complex and subtle phenomenon that is not well understood. Consequently, **it is not yet possible to predict the specific nucleotide sequences that a given protein will recognize**.

Multiprotein–DNA complexes in transcriptional regulation. C. Wolberger (1999) *Ann. Rev. Biophys. Biomol. Structure* **28**, 29–56.

DNA mimicry by proteins and the control of enzymatic activity on DNA. D. T. Dryden (2006) *Trends Biotechnol.* **24**, 378–382.

7.2.A. Specificity of DNA–Protein Binding

The binding of proteins to specific DNA sequences is fundamentally different from the problem of how to discriminate between various small-molecule ligands, which can be based at least in part on their size and shape. Binding specific sequences is unique in that the ligand is a short segment of DNA that is immersed in a sea of other DNA sequences that are part of the same molecule and very similar chemically and structurally to the specific ligand. **All proteins that bind to specific DNA sequences also have significant affinity for nonspecific sites**, binding that is equally probable at any particular point along the DNA and independent of the sequence. The two phenomena of specific and nonspecific binding are usually measured by the apparent dissociation constants of the complex of a protein (P) with DNA that contains a DNA site that is specific (DNA^S) or completely unrelated (DNA^N):

$$P + DNA^S \longrightarrow P{\bullet}DNA^S \quad K_d^S = [P][DNA^S]/[P{\bullet}DNA^S] \tag{7.2}$$

$$P + DNA^N \longrightarrow P{\bullet}DNA^N \quad K_d^N = [P][DNA^N]/[P{\bullet}DNA^N] \tag{7.3}$$

The dissociation constants for specific and nonspecific binding are compared in Table 7-1 for a number of different types of DNA-binding proteins. Each pair of specific and nonspecific binding constants was measured under the same conditions. Most pertinent is their ratio, because the absolute values depend upon a number of variables. The dissociation constants measured depend on the conditions, such as the temperature, pH and concentration and type of cations and anions. In addition, the stability of a specific complex can depend on the DNA sequences flanking the specific binding site, varying as much as 17-fold even if the flanking sequences do not contact the protein directly. Furthermore,

Table 7-1. Specificities of DNA-binding proteins

Protein	Specific site	K_d^N (M)	K_d^S (M)	K_d^N/K_d^S	$-\Delta\Delta G°$ (kcal/mol)
Transcription factors					
CRP	*lac* promoter	1.4×10^{-6}	1.2×10^{-11}	1.2×10^{5}	6.8
E12	CAGGTG	1.9×10^{-7}	1.4×10^{-8}	1.4×10^{1}	1.5
GCN4	ATGACTCAT	1.1×10^{-6}	3.5×10^{-8}	31	2.0
	ATGACGTCAT	1.1×10^{-6}	2.7×10^{-8}	41	2.2
IHF	λ attP H′	2.9×10^{-6}	1.6×10^{-9}	1.8×10^{3}	4.4
MASH-1	CAGGTG	1.9×10^{-8}	5.9×10^{-9}	3.2	0.7
MEF-2C	TATAAATA	$>1.2 \times 10^{-5}$	1.1×10^{-7}	$>1.1 \times 10^{2}$	>2.7
MyoD/E12	CAGGTG	7.9×10^{-8}	7.5×10^{-9}	11	1.4
TBP	TATAAAAG	$>1.0 \times 10^{-5}$	3.7×10^{-9}	$>2.7 \times 10^{3}$	>4.6
Repressors					
lac	O^{sym}	2.4×10^{-4}	8.3×10^{-12}	2.9×10^{7}	10.1
λ CI	O_R1	4.0×10^{-4}	1.2×10^{-10}	3.3×10^{6}	8.7
	O_R2	4.0×10^{-4}	6.7×10^{-10}	6.0×10^{5}	7.7
	O_R3	4.0×10^{-4}	6.3×10^{-9}	6.3×10^{4}	6.4
λ Cro	O_R1	1.5×10^{-6}	8.3×10^{-12}	1.8×10^{5}	6.6
	O_R2	1.5×10^{-6}	1.2×10^{-10}	1.3×10^{4}	5.1
	O_R3	1.5×10^{-6}	2.0×10^{-12}	7.5×10^{5}	7.3
MetJ	Met Box	1.6×10^{-6}	5.0×10^{-4}	3.1×10^{2}	3.4
P22 Arc	Left half-site	4.7×10^{-8}	2.7×10^{-11}	1.7×10^{3}	4.3
trp	Wild-type op.	4.2×10^{-5}	2.2×10^{-9}	1.9×10^{4}	5.8
Restriction endonucleases					
*Eco*RI	GAATTC	2.8×10^{-4}	5.9×10^{-12}	4.7×10^{7}	10.3
*Eco*RV	GATATC	1.0×10^{-5}	3.8×10^{-10}	2.6×10^{4}	6.0

Data from R. Alleman (1999) in *Encyclopedia of Molecular Biology* (T.E. Creighton, ed.), Wiley-Interscience, NY, p. 744.

the apparent dissociation constants of the protein complexes with nonspecific DNA depend on the length of the DNA probe, as the protein can bind to any of the nucleotides. Consequently, all the values for K_d^N in Table 7-1 were corrected for the length of the DNA probe by dividing the measured K_d^N by double the difference between the lengths of the specific and nonspecific DNA molecules. **The specificity of DNA binding is best expressed as the ratio K_d^N/K_d^S.**

The dissociation constants of the specific complexes vary over approximately six orders of magnitude and are as small as 10^{-12} M (Table 7-1). The complexes of bacterial proteins tend to be more stable than those of eukaryotic proteins, probably because the specific DNA binding sequences in the prokaryotes are longer. In contrast, the dissociation constants of the nonspecific complexes span only approximately four orders of magnitude, and the eukaryotic proteins bind more tightly. Therefore, prokaryotic proteins tend to bind DNA with greater specificity than do most eukaryotic transcription factors.

The longer specific sequences and the greater specificities in prokaryotes seem counterintuitive. The *E. coli* genome consists of only 4×10^6 base pairs, whereas those of eukaryotes tend to be much larger. Of the proteins listed in Table 7-1, the restriction enzyme *Eco*R displays the highest DNA-binding specificity, 4.7×10^7. This specificity is greater than the number of base pairs in the genome, so binding of the *Eco*R1 protein could occur predominantly to a single specific sequence on the chromosome, yet the specific DNA-binding site of *Eco*R1 has the sequence GAATTC, and such a hexamer sequence should occur statistically approximately 1000 times in the *E. coli* genome. Binding to a unique site in a human genome would require that the specificity be $>3 \times 10^9$, the number of base pairs in the human genome. This would require statistically a minimal length of 16 base pairs, but eukaryotic transcription factors that are involved in gene expression recognize much shorter sequences. For example, proteins that contain the basic helix–loop–helix motif (BHLH) (Section 7.3.D) bind to the sequence CAGGTG, which occurs approximately 7×10^5 times on a mammalian chromosome. **Most eukaryotic transcription factors bind to DNA sequences that are too short to be unique on the mammalian chromosome, and with insufficient specificity**. The reason is that transcription factors are designed to bind to a family of similar sequences within the genome, and they regulate multiple genes. Consequently, they need some flexibility in their specificity, so they usually bind significantly to several related DNA sequences. MyoD recognizes DNA through a BHLH domain and can activate myogenesis in a wide variety of cell types, including myoblasts and fibroblasts, while the BHLH protein MASH-1 promotes the differentiation of committed neuronal precursor cells. BHLH proteins need to bind to DNA with a specificity of approximately 4×10^3 in order to bind with equal probability to a nonspecific site and to one of the approximately 700,000 specific sites on the mammalian chromosome. But even then, MASH-1 would still activate transcription from MyoD target promoters and vice versa. Clearly, unique binding sites on chromosomes need to consist of many more than just a few sequential nucleotides. **A single DNA sequence recognition motif typically recognizes only three to four base pairs**, which is insufficient to allow highly selective discrimination between all the sequences in a genome.

In practice, the effective size of a recognition site can be increased by using a larger protein assembly or the conjunction of two sequence-specific DNA-binding motifs in the same polypeptide chain that recognize longer DNA sequences. Accordingly, **regulation of gene expression in higher organisms relies on multiprotein complexes with the potential for combinatorial interactions that involve a substantial segment of DNA**. The complexes often include a stable homo- or heterodimer; some dimerization surfaces, especially the coiled coils in bZIP proteins (Section 7.3.D), are sufficiently similar or pliable that two different polypeptide chains can interact to form a nonsymmetrical heterodimer. In other cases, however, cooperative interactions between proteins bound at contiguous or distant DNA sites are required. One example of such interactions is the cooperative binding of λ C_I repressor dimers to the leftward and rightward operators in lambda phage DNA, each of which contains three C_I binding sites. Stable binding to any two of these sites depends on interactions between separate dimers, so occupation is sensitive to small changes in the concentration of repressor molecules over a certain range (Section 7.3.A.2).

Interactions can also occur between distant binding sites on the DNA in that binding to both simultaneously generates a loop of intervening DNA. Loop formation of this type can involve only a single stable protein assembly, as in the case of the tetramer of the *lac* repressor containing four helix–loop–helix motifs (Figure 7-5). Alternatively, cooperative interactions between proteins bound at the separate sites may be required. **When bound at distant sites, the two proteins should be bound on the same face of the double helix**, because the torsional rigidity of DNA prevents bringing the two proteins into appropriate spatial register when they are bound initially to opposite faces of the double helix. An example is the binding of the *E. coli* AraC protein to two sites separated by about 231 base pairs on the *araBAD* promoter. Altering the separation of these sites by integral numbers of double-helical turns has little effect on loop formation. In contrast, altering the separation by 0.5, 1.5 or 2.5 turns severely impairs loop formation and regulatory function.

Thermodynamic and kinetic methods of analyses of protein–nucleic acid interactions. From simpler to more complex systems. W. Bujalowski (2006) *Chem. Rev.* **106**, 556–606.

---

1. Specific Interactions

As with other protein–ligand interactions, the precise recognition of a defined DNA sequence by a DNA-binding protein requires optimal shape and chemical complementarity between the interacting species. Consequently, a large number of noncovalent interactions occur at the interface between them. **Individual interactions often contribute only a small amount to the overall stability of the complex, but all of these interactions are important for the preferred binding of the protein to the specific DNA site**.

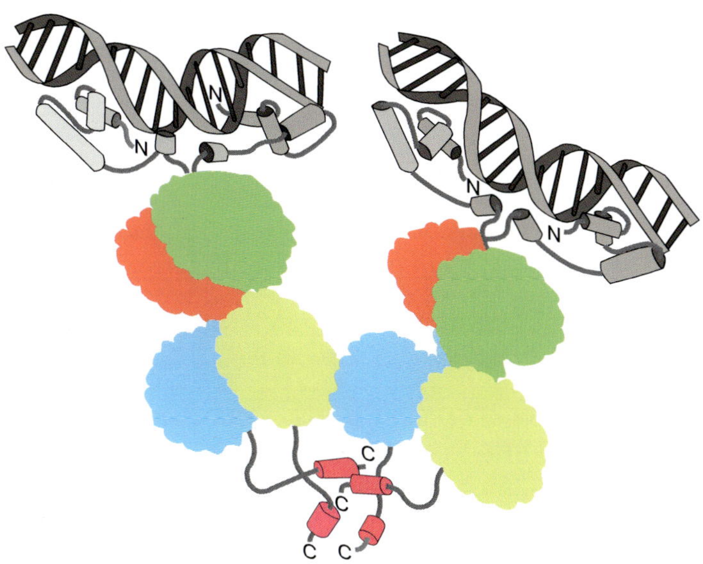

Figure 7-5. The tetrameric *lac* repressor bound to two DNA fragments, each with a palindromic sequence. Each arm of the V-shaped tetramer consists of a dimer containing a single DNA-binding site that has two-fold symmetry (i.e. is palindromic). The two helix–turn–helix motifs of each dimer bind in two successive major grooves, while shorter hinge helices bind adjacent to each other in the minor groove between the two major groove binding sites. The four subunits of the tetramer are held together by the C-terminal helices, which form a four-helix bundle. Adapted from M. Lewis *et al.* (1996) *Science* **271**, 12347–1254.

Initially the specificity of DNA•protein complexes was thought to be due to 'direct readout' of the interactions between the functional groups of the proteins and the DNA bases. The initial structures of DNA•protein complexes demonstrated such specific interactions to be due to hydrogen bonds, electrostatic interactions between charged groups and van der Waals interactions. Hydrogen bonds can be made between the polypeptide backbone or amino acid side-chains of the protein and the nucleobases of the DNA. Longer side-chains are more flexible, and the need to fix them should decrease their entropy and the stability of the interaction. However, they are often orientated in space by other side-chains or by making multiple contacts to the same base, adjacent bases or a base and phosphodiester group (Figure 7-4). In Figure 7-6, the carboxyl group of a Glu345 makes hydrogen bonds to both an adenine and adjacent cytosine, thereby specifying the first two bases of the binding site. A neighboring Arg residue orients the Glu side-chain in space through the formation of a 'clamp' between the phosphate backbone and peptide. If these groups are fixed in position before the DNA binds, they should contribute energetically to binding just as any other group. Further specificity arises from van der Waals interactions of the β- and γ-CH_2 groups of Glu345 with the methyl group of a nearby thymine base. Such van der Waals interactions appear to stabilize a complex by 0.5–2 kcal/mol.

The thermodynamics of the binding interactions between amino acid side-chains and DNA bases have been studied using site-directed mutation of both the protein and the DNA. The results from such experiments are difficult to interpret, as they appear to be complicated by subtle energetic changes that result from perturbation of the local structure of the DNA and/or the protein and changes in solvation. Nevertheless, **each hydrogen bond between the protein and the DNA appears to stabilize a complex by roughly 1.4 kcal/mol**, about the same value measured for hydrogen bonds in folded proteins.

DNA-binding proteins make extensive contacts to the phosphodiester groups of the DNA backbone through charged and uncharged side-chains, but it is difficult to measure the contributions of these interactions to the overall stability of the complex. One approach is to determine the contribution of the polyelectrolyte effect from the dependence of the reaction free energy on the concentration of univalent salts (Section 7.2.A.5) and add a contribution for the interactions between the DNA

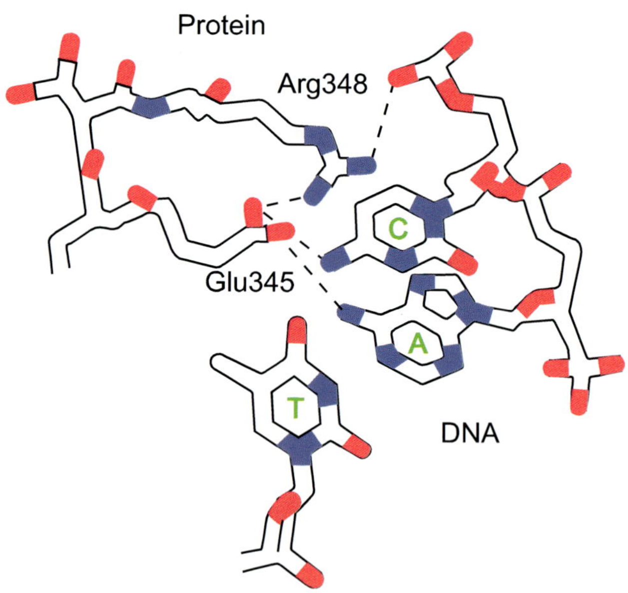

Figure 7-6. The network of hydrogen bonds between residues Glu345 and Arg348 of protein E47 and the CpA dinucleotide of the E-box DNA sequence. N atoms are *blue*, O atoms are *red*. The *dashed lines* are presumed hydrogen bonds. The carboxyl group of Glu345 is hydrogen bonded to N4 of cytosine and to N6 of adenine. Note the 'clamp' function of the side-chain of Arg348 connects the phosphate backbone to Glu345 and clamps the conformation of its side-chain. The β- and γ-CH_2 groups of Glu345 are in van der Waals contact with the methyl group of the thymine base. Adapted from a figure by R. Alleman.

phosphates and uncharged amino acid residues. The polyelectrolyte effect is, however, a consequence of the fact that DNA is a polyanion of high axial charge density (Section 2.3) and the association reaction is driven by the release of monovalent cations, at least at low salt concentrations. **The salt-dependence of binding to various specific and nonspecific DNA sequences is often different, indicating that protein–phosphate contacts are involved in determining the specificity of DNA binding.**

Anions also affect the dissociation constants of DNA•protein complexes, especially at high concentrations and most likely through the Hofmeister effect in which anions affect many types of interactions through their effects on the structure of water. The nature of the anion is important; for example, replacing chloride with glutamate can increase the affinity for a specific DNA sequence as much as 80-fold. At 25°C in a low salt buffer, the leucine zipper protein GCN4 (Section 7.3.D) binds with the same affinity to the sequences ATGAC**G**TCAT and ATGACTCAT. In a buffer containing 250 mM potassium glutamate, however, the affinity for the latter sequence is approximately one order of magnitude greater than for the other.

Is there a code for protein–DNA recognition? Probab(ilistical)ly. P. V. Benos *et al.* (2002) *BioEssays* **24**, 466–475.

Cation–π/H-bond stair motifs at protein–DNA interfaces. M. Rooman *et al.* (2002) *J. Mol. Biol.* **319**, 67–76.

Protein–DNA interactions: amino acid conservation and the effects of mutations on binding specificity. N. M. Luscombe & J. M. Thornton (2002) *J. Mol. Biol.* **320**, 991–1009.

A simple physical model for the prediction and design of protein–DNA interactions. J. J. Havranek *et al.* (2004) *J. Mol. Biol.* **344**, 59–70.

---

2. Nonspecific Complexes

Proteins that bind nonspecifically to DNA, or with little sequence specificity, recognize primarily the DNA backbone of phosphate and sugar groups. Electrostatic interactions with the ionized phosphate group of each nucleotide appear to provide the major driving force for binding and maintain a loose association between the protein and DNA. Both specific and nonspecific binding have been characterized with the ***lac* repressor**. It binds nonspecifically to any DNA with a K_d of 10^{-6} M at 0.1 M salt, but 10^8-fold more tightly to the correct sequence. Increased salt concentrations decrease the strength of the nonspecific binding, but specific binding is not so sensitive to the ionic strength. The interactions in the specific complex cause the *lac* repressor and DNA to bend by 25° and 36°, respectively, but the DNA maintains a canonical B-DNA conformation in the nonspecific complex, because the protein interacts mainly with its phosphate backbone. **The same set of residues of the *lac* repressor switch roles from a purely electrostatic interaction with the DNA backbone in the nonspecific complex to a highly specific binding mode with the base pairs of the specific sequence.**

Glucocorticoid receptor (Section 7.3.C.2) binds to DNA sequences containing inverted repeats separated by three base pairs. When the spacing is increased to four base pairs, one protein subunit binds specifically to one inverted repeat, while the other can not reach its specific binding site. Both subunits insert an α-helix into the major groove of the DNA double helix, but it is opened by approximately 2 Å in the specific complex, probably due to the greater number of interactions. The restriction enzyme ***Eco*RV** inserts a loop into the major groove of the DNA; in the nonspecific complex, it is found to be partially disordered and not so buried in the grooves; the surface area buried is 1570 Å^2 greater in the specific complex.

Nonspecific binding is very important for proteins that play a structural role for DNA irrespective of its nucleotide sequence, but it is also important for proteins that bind specifically to find those sites quickly. They bind nonspecifically and flexibly to the DNA molecule and diffuse along it one-dimensionally until they find their specific sequence.

Crystal packing interaction that blocks crystallization of a site-specific DNA binding protein–DNA complex. O. Littlefield & H. C. Nelson (2001) *Proteins* **45**, 219–228.

Structure and flexibility adaptation in nonspecific and specific protein–DNA complexes. C. G. Kalodimos *et al.* (2004) *Science* **305**, 386–389.

---

3. Water-mediated Contacts

Water molecules are observed at the interfaces of most protein•DNA complexes, where they appear to be involved in specific interactions between protein and DNA. They are believed to fill voids in the structure, screen electrostatic repulsions between like charges, act as linkers between complementary charges, and be intermediaries in hydrogen bonding.

Water-mediated contacts are the principal means for site-specific recognition between the *trp* repressor and its operator (Section 7.3.A.5). Of the 90 water molecules that are integral parts of the complex, 26 are located within the protein•DNA interface, with 13 in each half-site. Three of these 13 are involved in contacts between the nucleobases and the protein. These three water molecules are already in place in the free DNA, and two others are in place in the free protein. Such water molecules should therefore be considered as intrinsic parts of the DNA and protein structures.

In contrast, some DNA•protein interfaces are completely devoid of water, such as the perfectly complementary interface between TATA-binding protein and TATA box-containing DNA sequences (Section 7.3.B). The complex of the transcription factor GATA-1 with DNA (Figure 7-19-D) is devoid of water molecules that mediate contacts between the protein and the nucleobases, but several water molecules bridge the protein and the phosphate backbone.

Wet and dry interfaces: the role of solvent in protein–protein and protein–DNA recognition. J. Janin (1999) *Structure* **7**, R277–279.

The role of water in protein–DNA recognition. B. Jayaram & T. Jain (2004) *Ann. Rev. Biophys. Biomolec. Structure* **33**, 343–361.

---

4. Dehydration Effects

Sequence-specific DNA-binding reactions are characterized by relatively large negative changes in the heat capacity, ΔC_p. Consequently, the enthalpic ($\Delta H°$) and entropic ($T\Delta S°$) contributions to the free energy change (ΔG°) of the binding reactions vary with temperature in an almost parallel manner, making ΔG° nearly independent of temperature. Large negative changes of ΔC_p generally indicate that large amounts of nonpolar surface area are buried when the complex is formed, releasing water molecules. The areas buried in specific protein•DNA complexes vary between 1000 and 5500 Å^2, but in many cases these are too small to account for the observed values of ΔC_p. The discrepancies are thought to result from a series of temperature-dependent equilibria that are coupled to the binding equilibrium, such as ion binding, protonation and protein conformational changes.

Binding of proteins to DNA in a nonspecific fashion involves almost no changes in heat capacity, indicating that **the formation of nonspecific complexes does not involve major dehydration**. An interesting exception is provided by the DNA-binding domain of the transcription factor MASH-1, which appears to bind in only one manner. The association reaction between MASH-1 and an oligonucleotide containing the natural target of MASH-1 is characterized by a ΔC_p of −733 (±99) cal $mol^{-1}\ K^{-1}$, while the formation of a nonspecific complex with heterologous DNA results in a ΔC_p of −575 cal $mol^{-1}\ K^{-1}$. Unlike other DNA-binding proteins, all DNA complexes of MASH-1 demonstrate the thermodynamic characteristics of specific complexes, irrespective of the particular DNA sequence.

The number of water molecules released on formation of a complex between a protein and DNA has been estimated by varying the chemical activity of water, a_w. This can be accomplished through the addition of neutral salts and can be measured by the decrease in the vapor pressure of the solution. Varying a_w yields a value for the net number of water molecules released, i.e. the difference between the number of water molecules released and those taken up, which will include those resulting from changes in the exposure of surface area due to unfolding or refolding of the DNA or protein. For example, the interaction between a protein and DNA in the presence of a neutral solute (S) can be represented as:

$$\text{protein} + \text{DNA} \leftrightarrow \text{protein•DNA} + \nu_w\, H_2O + \nu_s\, S \tag{7.4}$$

where ν_w and ν_s are the net numbers of water and solvent molecules released from the protein and DNA. If molecules are taken up by the complex, the values of ν_w and ν_s will be negative. In general, for such a situation:

$$\nu_i = \frac{\partial \ln K_{obs}}{\partial \ln a_i} \tag{7.5}$$

where ν_i is the difference in the number of molecule i associated with reactants and products of the reaction with equilibrium constant K_{obs} and a_i is the activity (the concentration corrected for nonideality effects) of i. Consequently, the situation for Equation 7.4 will be:

$$\frac{d \ln K_{obs}}{d \ln a_w} = \nu_w + \nu_s \frac{d \ln a_s}{d \ln a_w} \tag{7.6}$$

Many osmolytes are preferentially excluded from the immediate surfaces of macromolecules, decreasing ν_s, so ν_w will be significantly greater than ν_s, and the term ν_w will predominate on the right side of Equation 7.6. Therefore, measurement of the effect of varying the concentration of such excluded osmolytes on the binding free energy (the left side of Equation 7.6) will yield the number of water molecules involved in the interaction, ν_w.

This method indicates that formation of the complex between the cyclic AMP receptor protein (CRP) and C1 site in the *lac* promoter (Section 7.3.A.6) is accompanied by the release of 79 ±11 water molecules. When CRP is transferred from the C1 site to a nonspecific site, 56 ±10 of these water molecules are taken up again. When gal repressor binds to the O_I operator site, between 100 and 180 water molecules are released, depending upon the neutral salt used. A further 6 ± 3 water molecules are released when the repressor moves from O_I to the O_E operator, to which it binds with greater affinity. Changing the water activity affects the DNA binding by ultrabithorax and deformed homoeodomains differently, even though they have closely similar amino acid sequences (Section 7.3.A.3); DNA binding by ultrabithorax releases 22–27 water molecules but only five are released in the case of deformed.

Complexes of lambda Cro repressor (Section 7.3.A.2) with varying DNA sequences and dissociation constants demonstrate that the number of water molecules sequestered by nonspecific complexes varies linearly with the binding free energy. One extra bound water molecule corresponds to the loss of approximately 150 cal/mol of binding free energy. Conversely, every 10-fold decrease in binding constant at constant salt and temperature is associated with eight to nine additional water molecules sequestered in the noncognate complex. The water seems to be sterically sequestered, because the measured number of water molecules does not depend upon the nature of the osmolyte used.

Clearly, the thermodynamics of the interactions between DNA and proteins are complex and not well understood.

Calorimetry of protein–DNA complexes and their components. C. M. Read & I. Jelesarov (2001) *Methods Mol. Biol.* **148**, 511–533.

Heat capacity effects of water molecules and ions at a protein–DNA interface. S. Bergqvist *et al.* (2004) *J. Mol. Biol.* **336**, 829–842.

Sequestered water and binding energy are coupled in complexes of lambda Cro repressor with non-consensus binding sequences. D. C. Rau (2006) *J. Mol. Biol.* **361**, 352–361.

5. *Release of Condensed Counterions*

An entropic driving force for proteins to bind to DNA is believed to come from the release of the condensed counterions that are bound electrostatically to a polyelectrolyte-like DNA (Section 2.3). This factor can be estimated by measuring the affinity at various salt concentrations and plotting the logarithm of the dissociation constant against the logarithm of the salt concentration. A straight line should be obtained, with slope Z reflecting the number of counterions released upon forming the complex; this is commonly interpreted as the number of ionic interactions generated in the complex. A large slope is taken to imply that the electrostatic interactions predominate, whereas a small value for Z or a negative slope indicates that they do not.

The affinities of most proteins for DNA decrease with increasing salt concentration, indicating that Z is positive. On the other hand, a TATA-binding protein from a halophilic organism adapted to living in conditions of high salt and temperature exhibits the opposite behavior, which has been attributed to only three amino acid residues of the protein. The halophilic complex is believed to sequester cations that become part of it, so the stability of this complex is increased by increasing salt concentrations.

Association of protein–DNA recognition complexes: electrostatic and nonelectrostatic effects. J. Norberg (2003) *Arch. Biochem. Biophys.* **410**, 48–68.

Halophilic adaptation of protein–DNA interactions. S. Bergqvist *et al.* (2003) *Biochem. Soc. Trans.* **31**, 677–680.

7.2.B. Changes in the Protein Conformation

The binding sites of many proteins that bind DNA specifically are remarkably flexible in the absence of the DNA, and binding to DNA induces specific local conformations. In other cases, the quaternary structure can alter. In general, complex formation does not result from a simple alignment of rigid, complementary surfaces, commonly known as the 'lock-and-key' model, but rather follows what is generally known as an **induced fit** mechanism. For example, the DNA-binding pocket of the endonuclease *Eco*RV is not accessible in the free protein and the cleft between the two protein subunits is too narrow. The pocket and cleft open to accommodate the DNA through a combination of tertiary and quaternary structure changes. The relative orientations of the amino and carboxy terminal domains of CRP change significantly on DNA binding. In extreme cases, the portion of the protein that will bind the DNA is unfolded until the DNA binds. The observed changes in conformation of the protein upon binding are believed to account for the discrepancy between the measured values of ΔC_p and those expected on the basis of the nonpolar surface buried in the complex.

Most DNA-binding proteins have their specific affinities for DNA modulated by another ligand, a corepressor or coactivator. This ligand is frequently bound to a protein domain other than that which binds to the DNA. In the one case where the mechanism is clearly known, the corepressor ligand alters the affinity of the *trp* repressor for DNA by altering the dimeric protein structure so that the spacing of the two DNA binding sites does or does not coincide with the spacing of the palindromic sequences of the DNA (Figure 7-16).

Use of urea and glycine betaine to quantify coupled folding and probe the burial of DNA phosphates in *lac* repressor–lac operator binding. J. Hong *et al.* (2005) *Biochemistry* **44**, 16896–16911.

Molecular flexibility in protein–DNA interactions. S. Gunther *et al.* (2006) *Biosystems* **85**, 126–136.

7.2.C. Changes in the DNA Conformation

Many proteins recognize regular B-DNA through the formation of a number of hydrogen bonds and van der Waals interactions between amino acid side-chains and functional groups of the bases, but others recognize and generate bent conformations of the DNA. A dramatic example is provided by the structure of **integration host factor (IHF)** bound to the H′ site of phage λ DNA, in which the 34-base pair piece of DNA is literally wrapped around the protein, creating a buried protein•DNA interface of 4600 Å^2 (Figure 7-7) and bending the DNA by nearly 180°. This bending completely reverses the direction of the DNA within a small number of base pairs.

The DNA-binding affinities of the **434** and **Cro repressors** (Section 7.3.A.2) depend upon the flexibility of bases at the center of the binding site that are not in direct contact with the protein.

Varying the nucleotide sequence of a binding site reveals a linear relationship between the free energy of the binding reaction with the proteins and the flexibilities of their central sequences. **The intrinsic sequence-dependent properties of the DNA are important for the formation of DNA•protein complexes.**

Double-stranded DNA often bends at the junctions between regions of G•C and A•T base pairs (Section 2.4). The transition from G•C to A•T base pairs causes this region of the DNA to be flexible and capable of potentially undergoing a bend. The crystal structure of the dodecamer d(CGCGAATTCGCG) has a bend of 18° at the GC/AT junction at one end of the helix, but not at the other. Although such sequences need not be bent, they usually are when proteins bind there. Changing the nucleotide sequence to disfavor bending destabilizes the complex, and there is a correlation between the affinity for a mutant site and its intrinsic bendability.

MEF-2C binds with maximal affinity to a DNA site with an alternating run of eight thymines and adenines, while the affinity for an intrinsically rigid sequence of only A nucleotides is reduced by more than one order of magnitude. Binding of DNA by MEF-2C is accompanied by DNA bending of approximately 70°, irrespective of the particular DNA sequence. When such a protein binds to a more rigid DNA sequence (with lower affinity), it often does not bend the DNA, or a substantial amount of the binding free energy must be used to bend the DNA target. Consequently, **proteins often target the most deformable base step within their target sequence**. The **purine repressor, PurR**, binds to runs of four consecutive adenines and thymines, interrupted by the dinucleotide CpG. PurR bends the DNA into the major groove by 45° through the insertion of a hinge into the minor groove and intercalation of the side-chains of two Leu residues into the CpG step. Purine–pyrimidine steps stack more stably than do purine–purine or pyrimidine–pyrimidine steps by approximately 0.52 kcal/mol, while the stacking energy of pyrimidine–purine steps is reduced by another 0.39 kcal/mol. Therefore, CpG is the step that can be unstacked most easily within the *pur* operator. Similarly, the bending of 22° of the PRDI site within the interferon-β promoter in the complex with interferon regulatory factor occurs at the CpA step, which is the most easily unstacked dinucleotide in the PRDI sequence ACTTT**CA**CTTCTC. The DNA complex of the chromosomal protein Sac7d from the hyperthermophile *Sulfolobus acidocaldarius* has the DNA kinked by 61° through the intercalation of two amino acid side-chains between the bases of the dinucleotide CpG in the sequence G**CG**ATCGC. This is the base step that is most easily unstacked.

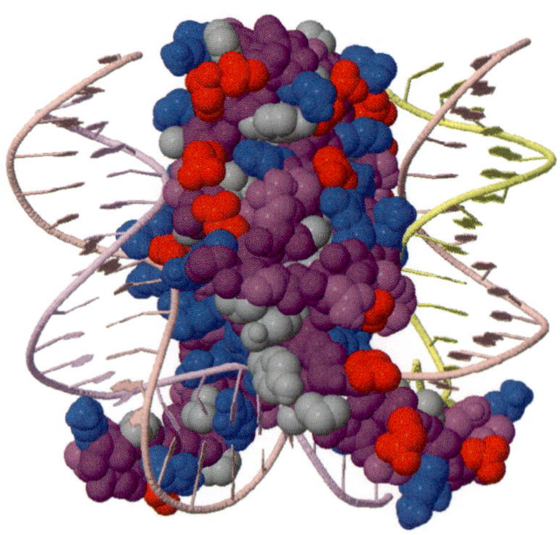

Figure 7-7. The IHF–DNA complex, illustrating both the large contact surface of ~ 4600 Å^2 between the protein and DNA and the massive, nearly180°, bend in the DNA induced by the protein. The DNA is represented as a skeletal model, while the protein atoms are represented as van der Waals spheres; acidic groups are *red*, basic groups *blue* and the polypeptide backbone *mauve*. Generated from PDB data file *1ihf* using the program Jmol.

DNA bending should be an energetically costly process, even for sequences of enhanced bendability. Some of the binding free energy must be used to bend a DNA molecule that, in the absence of protein, would not be bent. Therefore, **binding at the more readily bendable sites can contribute to the overall specificity of a DNA-binding protein**. Numerous observations indicate that **proteins recognize DNA not only through specific contacts with the nucleobases but also through the inherent conformational properties of the DNA that vary with the sequence**.

Proteins use primarily two mechanisms to bend DNA. In the first, **kinks are introduced through the intercalation of amino acid side-chains between adjacent base pairs in the minor groove**. In the IHF–DNA complex (Figure 7-7), two intercalating Pro residues projecting from the tip of each β-hairpin arm introduce large kinks at symmetrically displaced ApA nucleotide steps. Other proteins use nonaromatic, hydrophobic amino acids to intercalate into base steps from the minor groove side of the double helix, creating a widened minor groove and dramatically bent DNA. TATA box-binding protein induces strong bends of approximately 45° at either end of its specific sequence, the TATA box, through the intercalation of two Phe rings between two adjacent base pairs (Figure 7-8). The hyperthermophile chromosomal protein Sac7d bends DNA by 72°, primarily through the intercalation of Val26 and Met29 at a single CpA base step. This causes the DNA to bend into the major groove and away from the protein molecule.

The second mechanism for the introduction of bends into DNA involves **asymmetric neutralization of the negative charges of the phosphodiester groups by cationic amino acid residues**. Unbalanced electrostatic repulsions between the negative charges on DNA cause it to collapse towards the bound protein (Figure 7-8). Asymmetric charge neutralization is at least partly responsible for the DNA bend angle of 72° in the target DNA bound to serum response factor (SRF). Several positively charged amino acid residues of SRF bind the phosphate groups on only one side of the SRF site (Figure 7-9). The three positively charged amino acids interact with and pull the distal ends of the DNA. In addition, Arg143 lies in an extended conformation along the floor of the minor groove and stabilizes the relatively large propeller twists at the central base steps that are characteristic of AT-rich regions and facilitate bending of the DNA. In some of these cases, the DNA sequences have a propensity to bend.

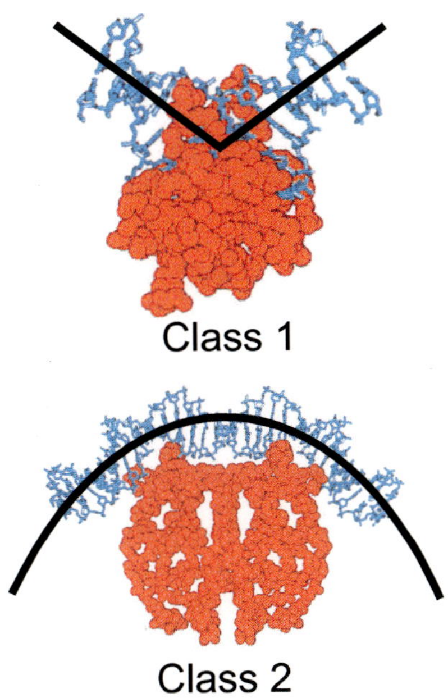

Figure 7-8. Two classes of DNA-bending proteins. Class 1-bending proteins, such as the TATA box-binding protein illustrated here, bind to the minor groove, unwind the DNA and induce bending away from the protein by intercalation of hydrophobic amino acid side-chains between base pairs. Class 2-bending proteins, such as CRP, form complexes in which the DNA bends towards the protein to maximize their interaction. Adapted from L. D. Williams & L. J. Maher (2000) *Ann. Rev. Biophys. Biomol. Structure* **29**, 497–521.

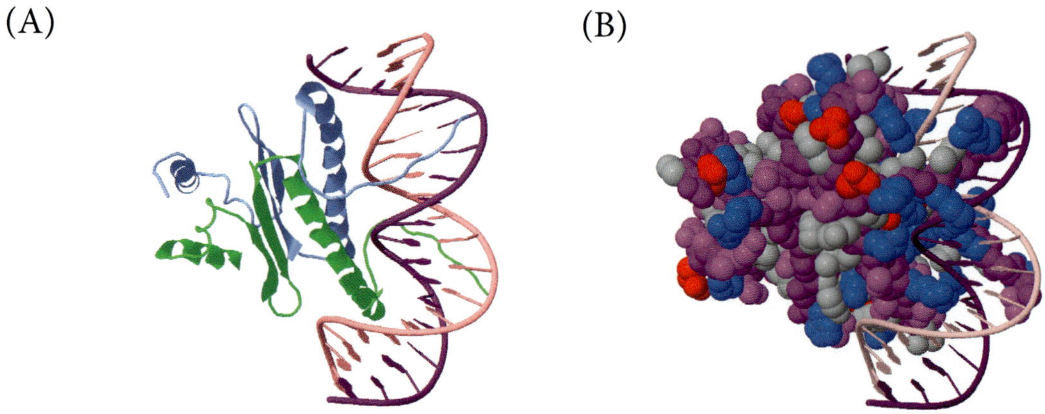

Figure 7-9. Interactions between positively charged amino acid side-chains and DNA phosphates in the serum response factor–DNA complex. (A) Skeletal model, illustrating the bend of the DNA and the polypeptide extensions that bind in the grooves of the DNA. (B) Space-filling model of the protein, emphasizing the polypeptide extensions. Basic groups are *blue*, acidic groups *red*. The contacts between the protein and DNA occur on only one side of the double helix, causing the DNA to bend around the protein. The His and Lys residues at either end serve as handles to bend the DNA further, although the DNA used is too short to allow formation of all the contacts involving the His residue. Generated from PDB file *1srs* using the program Jmol.

Bending is the most dramatic change observed upon DNA binding to proteins, but many more subtle conformational rearrangements of the DNA occur. A slight bending of the DNA of approximately 25° occurs in the 434 repressor–DNA complex (Figure 7-10) but more obvious is that the DNA in the region of the central four base pairs of the binding site is significantly overwound by approximately 20° (Section 3.1). These four base pairs are not in direct contact with the repressor, but sequences with A•T or T•A base pairs at these positions bind more strongly than do those with C•G or G•C. The intrinsic twist of a sequence affects its affinity; in this case, **sequences with lower affinity are more underwound**.

The role of intercalating residues in chromosomal high-mobility-group protein DNA binding, bending and specificity. J. Klass *et al.* (2003) *Nucleic Acids Res.* **31**, 2852–2864.

Integration host factor: putting a twist on protein–DNA recognition. T. W. Lynch *et al.* (2003) *J. Mol. Biol.* **330**, 493–502.

Predicting indirect readout effects in protein–DNA interactions. Y. Zhang *et al.* (2004) *Proc. Natl. Acad. Sci. USA* **101**, 8337–8341.

DNA twisting flexibility and the formation of sharply looped protein–DNA complexes. T. E. Cloutier & J. Widom (2005) *Proc. Natl. Acad. Sci. USA* **102**, 3645–3650.

7.3. DNA-BINDING STRUCTURAL MOTIFS

The structures of many complexes of protein and nucleic acid have been determined in great detail by X-ray crystallography and nuclear magnetic resonance (NMR). To promote crystallization of the complex and to keep it a practical size, the protein is bound to a nucleic acid of the smallest possible size, ideally containing only the specific sequence of nucleotides recognized.

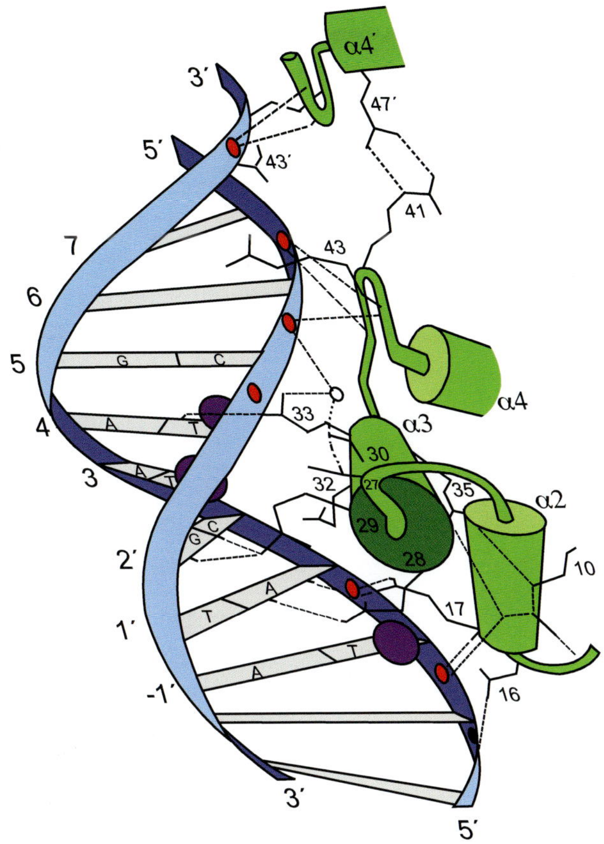

Figure 7-10. Interactions between one helix–turn–helix motif of the 434 repressor and its DNA binding site. Only part of the repressor protein is shown, with the helices depicted as *cylinders*. The DNA double helix is shown schematically, with the backbone of the individual chains depicted as *ribbons* and the positions of the phosphate groups by *circles*. The base pairs are indicated as the *flat segments* connecting the ribbons; the *dark spheres* on the thymidine nucleotides are the methyl groups. *Dashed lines* represent likely hydrogen bonds. Solvent molecules are required as bridges in some of these interactions, as illustrated by the interaction of the side-chain of residue 33 with a DNA backbone phosphate group and with the side-chain of residue 30. Adapted from A. K. Aggarwal *et al.* (1988) *Science* **242**, 899–907.

DNA binding has been found to be specified by a large number of disparate protein motifs: **many different protein structures can bind DNA specifically**. All these have one characteristic in common: **the protein needs to have interacting groups that protrude substantially from the protein surface**, to be able to contact the nucleotides at the base of the major groove and discriminate between different DNA base pairs by interacting with their edges (Figure 7-1). The most commonly encountered types of motif are listed in Table 7-2. The degree of sequence specificity can vary enormously within any particular class of motif. For example proteins with the helix–turn–helix–motif (Section 7.3.A) range from the *lac* repressor, with a high degree of sequence specificity, to the FIS protein, with little. Frequently more than one type of binding motif is found in a particular protein or even in a conserved sequence motif.

The best-studied protein structures will be described here briefly. A database of the structures and stabilities of DNA•protein complexes is available at www.rtc.riken.go.jp/jouhou/pronit/pronit.html.

Structural basis of DNA–protein recognition. R. G. Brennan & B. W. Matthews (1989) *Trends Biochem. Sci.* **14**, 286–290.

Structural studies of protein–nucleic acid interaction: the sources of sequence-specific binding. T. A. Steitz (1990) *Quart. Rev. Biophys.* **23**, 205–280.

A structural taxonomy of DNA-binding domains. S. C. Harrison (1991) *Nature* **353**, 715–719.

Transcription factors: structural families and principles for DNA recognition. C. O. Pabo & R. T. Sauer (1992) *Ann. Rev. Biochem.* **61**, 1053–1098.

7.3.A. Helix–turn–helix Motif

The best-characterized structural motif of DNA-binding proteins is **the helix–turn–helix**; it is present in proteins from both prokaryotes and eukaryotes. **It consists of two short α-helices separated in the sequence by a Gly residue** (Figure 7-11) that, in concert with its neighbors, acts as a flexible hinge allowing the polypeptide chain to bend between the two α-helices so that they can make hydrophobic contacts with each other. It is observed in a number of different proteins that have no other structural similarities (Figure 7-10). This motif **protrudes from the protein surface to allow one of the helices, the recognition helix, to enter the major groove of the DNA double helix** (Figure 7-11). The second helix stabilizes the structural motif by packing against the recognition helix, primarily through interactions between nonpolar side-chains. This is especially important for the structural integrity of the motif, as it can make few other stabilizing interactions with the rest of the protein. For this reason, the residues participating in this packing interaction are highly conserved in motifs that otherwise have little sequence or structural similarity. This conserved pattern of residues often makes it possible to detect the helix–turn–helix motif from just the primary structures of suspected DNA-binding proteins.

The distal recognition helix binds in the DNA major groove and recognizes the DNA sequence, primarily by hydrogen bonds from amino acid side-chains, especially those of Arg, Asn, Asp, Gln and Glu residues, which can make multiple hydrogen bonds. Water molecules are sometimes involved as bridges in those hydrogen bonds. Van der Waals contacts between protein and DNA are also involved.

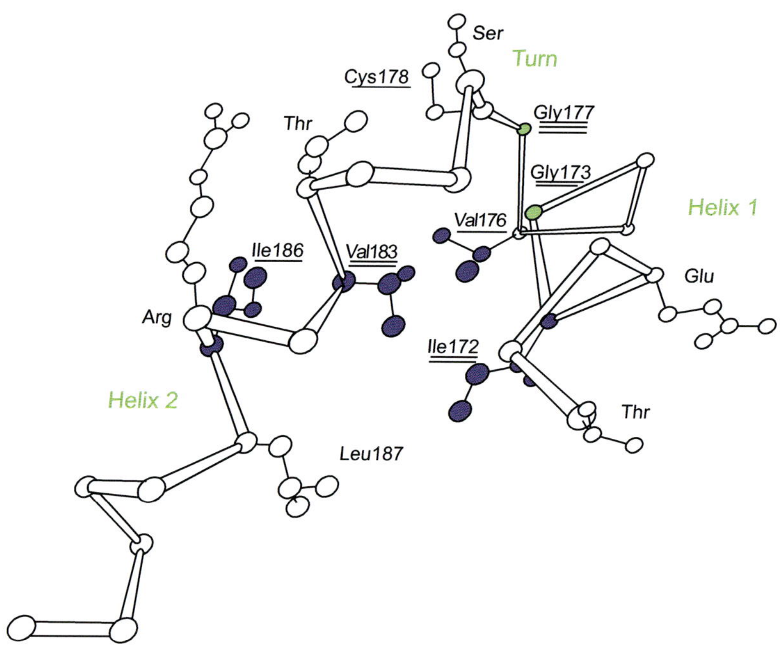

Figure 7-11. Structure of the helix–turn–helix motif of CRP. The first α-helix is to the *right*, the second is the recognition helix and is to the *left*. The degree of conservation of the various residues in other motifs is indicated by the number of *underlines*. The Gly177 residue is generally conserved in different proteins because it is important for the bend. Residues Ile172, Val176, Val183 and Ile186 are usually hydrophobic in other such motifs because they pack between the two helices. Data from T. A. Steitz (1990) *Quart. Rev. Biophys.* **23**, 205–280.

There is no simple code relating the amino acid sequence of the motif to the sequence of the DNA it recognizes; several side-chains can interact with each base pair, and more than one base pair can be in contact with a side-chain. The second, proximal helix interacts with the recognition helix and with the sugar–phosphate backbone of the DNA. The structure of this domain is highly conserved, but the orientation of the recognition helix within the major groove varies (Figure 7-10). The two DNA-binding helices are generally stabilized by interactions with other protein segments, such as two additional helices in the lambda CI repressor or a single additional helix in the homeobox domain.

Table 7-2. Protein motifs Involved in DNA binding

Motif	Target	Sequence specificity	Examples
β-strand	Major groove		Met repressor, Arc repressor
bZip	Major groove	High	GCN4, Fos-Jun
GATA	Major + minor grooves		GATA-1
GRP motif	Minor groove	Low preference	Hin recombinase, HMG14/17
Helix–loop–helix	Major groove	High	Myc, Achaete
Histone-fold	Backbone	None	Core histones, TAFs
HMG domain	Minor groove	Variable, but low	HMG1, SRY, TCF-1, UBF
HU class	Minor groove	Variable, but low	HU, IHF, TF1
Smad		Major groove	Smad MH-1
TBP domain	Minor groove	High	TBP
Helix–turn–helix	Major groove	Highly variable	FIS, C_I repressor, CRP
Homeodomain			Fushi tarazu
Winged helix			HNF3, histone H5
ETS domain			Fli-1, Elk-1
POU 'domain'			Oct1
Zinc-containing motifs			
Zinc finger	Major groove	Generally high	TFIIIA, Tramtrack, GAGA
Receptor DBD	Major groove		Glucocorticoid receptor
Gal4 DBD	Major groove		GAL4

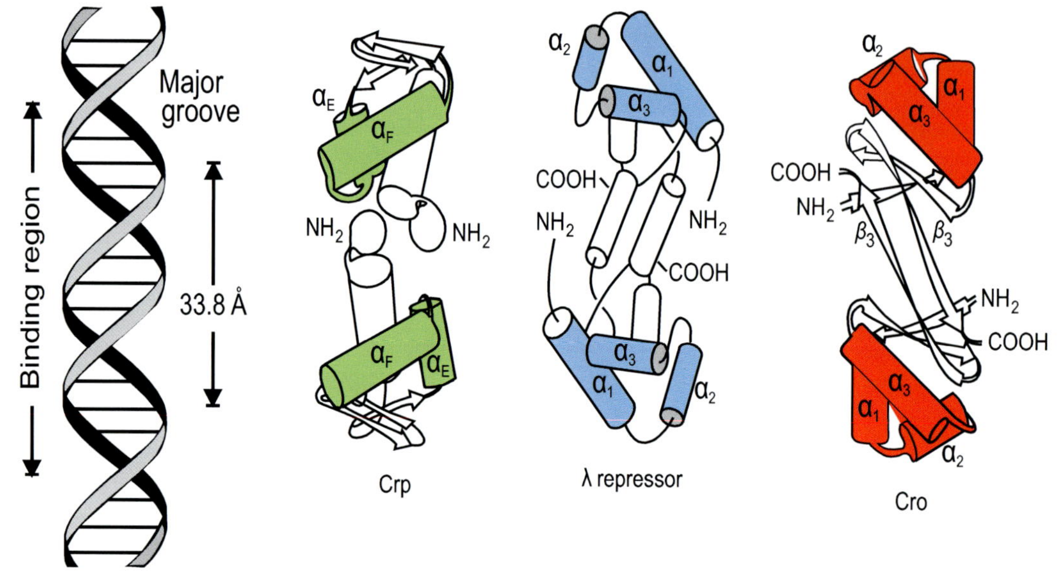

Figure 7-12. Comparisons of the dimeric structures of three DNA-binding proteins with the helix–turn–helix motif, relative to the structure of dsDNA (*left*). The three proteins are the Cro repressor, the amino-terminal domain of λ repressor and the carboxyl-terminal domain of CRP. The α-helices of each helix–turn–helix motif are *colored*. That in the foreground of each protein, and labeled as α_3 or α_F, is the recognition helix, which fits most closely into the major groove of the DNA. Note that the spacing between the two recognition helices of the dimer is very close to the 33.8 Å spacing between the DNA major grooves, but that the orientations of the helices relative to the major groove differ. Original figure kindly provided by B. W. Matthews (1993) in *Proteins: structures and molecular properties*, 2nd edn (T. E. Creighton, ed.), W. H. Freeman, NY, p. 356.

Eukaryotic helix–turn–helix proteins have essentially the same structure but contain longer turns or loops. These eukaryotic variants include the **homeodomain** fold, which contains both a helix–turn–helix element and a short extended minor groove-tracking sequence, the **POU**-specific 'domain', and the winged-helix fold. The last fold contains an extended β-sheet immediately following the recognition helix.

The eukaryotic proteins usually bind as monomers, but their prokaryotic counterparts normally bind as homodimers. The latter have both equivalent binding sites, making the same interactions to DNA with the same sequence, which requires that the sequence be **palindromic** or have **inverted repeats** (Section 2.2.C.6). For example, the binding site for the *lac* repressor has the following sequence of its two complementary DNA strands:

$$\begin{array}{l}5'\text{-}\mathbf{TGTGTGGAATTGT}X_9\mathbf{ACAATT}T\mathbf{CACACA}\\ 3'\text{-}\mathbf{ACACACCTTAACA}Y_9\mathbf{TGTTAA}A\mathbf{GTGTGT}\end{array} \quad (7.7)$$

All but one of 13 base pairs of each half obeys the palindromic nature of this binding site, indicated in bold. The nine nucleotides between the palindromic sequences are indicated by X_9 and Y_9. The palindromic sequences define a local two-fold symmetry axis at right angles to the DNA helix that relates the two halves of the DNA binding site. **The dimeric protein binds with its two-fold axis coinciding with that of the DNA sequence, and the distance between the two halves of the protein is the same as that of one turn of B-form DNA** (Figure 7-12). Having a double binding site increases markedly the strength and specificity of the binding interaction.

Although similar in many ways, the various helix–turn–helix motifs may not have evolved from a single evolutionary ancestor; instead, there may have been at least six to 11 different such domains in the last universal common ancestor of all life forms.

The many faces of the helix–turn–helix domain: transcription regulation and beyond. L. Aravind *et al.* (2005) *FEMS Microbiol. Rev.* **29**, 231–262.

Detecting DNA-binding helix–turn–helix structural motifs using sequence and structure information. M. Pellegrini-Calace & J. M. Thornton (2005) *Nucleic Acids Res.* **33**, 2129–2140.

1. Lac Repressor

The *lac* repressor is a member of a family of almost a dozen similar, mostly dimeric, repressors in *E. coli.* It represses the production of β-galactosidase in the absence of an inducer, such as lactose, which inactivates the *lac* repressor by binding to it.

The *lac* repressor is a tetramer of identical subunits of 360 amino acid residues each. The tetramer has a V-shape, in which each arm comprises two polypeptide chains; the two dimers are held together at the base of the V (Figure 7-5). Each polypeptide chain consists of five domains.

(1) The C-terminal tails of the four subunits of one *lac* repressor molecule, residues 330–360, form a four-helix bundle composed of four coiled coils, each carrying two leucine heptad repeats. This domain links the four polypeptide chains and is responsible for the tetrameric structure.

(2 & 3) Two globular domains, residues 60–329, form the core of *lac* repressor and form the dimeric part of *lac* repressor that binds the inducer.

(4) A helix–turn–helix motif at the N-terminus of each polypeptide chain binds directly to the DNA.

(5) A hinge α-helix preceding the helix–turn–helix motif in the sequence binds to the minor groove of the DNA.

Recognition of the *lac* operator by the *lac* repressor involves specific interactions between residues in the repressor's recognition helix and bases in the DNA major groove. Each dimer arm binds a separate copy of a palindromic DNA sequence (Equation 7.7). The two recognition α-helices of each dimer bind to successive major grooves of the DNA, while the two hinge helices of each dimer are close together in the structure and interact with the DNA minor groove between the two major-groove binding sites; both types of binding are specific for the DNA sequence. The hinge helices distort the structure of the DNA and open the minor groove, bending the DNA away from the protein.

Tetrameric *lac* repressor can bind to two palindromic sequences (Equation 7.7) if they are on different pieces of DNA or the same piece at an appropriate distance, corresponding to integral turns of the DNA double helix with a turn every 10.5 base pairs. The intervening DNA sequence is looped out from the tetramer. A single symmetrical substitution in each 10-base pair half-site of a symmetric *lac* binding sequence reduces the observed equilibrium association constant by three to four orders of magnitude; double symmetrical substitutions of the neighboring nucleotide reduce the binding by nearly six orders of magnitude, to approach the affinity for nonspecific sequences. The free energy effects of multiple substitutions at adjacent sites are nonadditive: the first reduces the free energy of binding by 3–5 kcal/mol, approximately halfway to the nonspecific level, whereas the second is less deleterious, reducing it by less than 3 kcal/mol.

If the C-terminal residues are deleted, *lac* repressor cannot form the tetramer, only a dimer. The dimer binds only weakly to a single palindromic sequence and in an asymmetric manner, with a different pattern of specific contacts between the two halves of the palindromic sequence. This involves a combination of elongation and twist by 48° of one subunit relative to the other, significant rearrangement of many side-chains, as well as sequence-dependent deformation of the DNA.

Plasticity in protein–DNA recognition: *lac* repressor interacts with its natural operator 01 through alternative conformations of its DNA-binding domain. C. G. Kalodimos *et al.* (2002) *EMBO J.* **21**, 2866–2876.

Role of hydration in the binding of *lac* repressor to DNA. M. G. Fried *et al.* (2002) *J. Biol. Chem.* **277**, 50676–50682.

Toward an integrated model of protein–DNA recognition as inferred from NMR studies on the *lac* repressor system. C. G. Kalodimos *et al.* (2004) *Chem. Rev.* **104**, 3567–3586.

The *lac* repressor. M. Lewis (2005) *C. R. Biol.* **328**, 521–48.

2. Lambda CI and Cro Repressors

The lambda bacteriophages encode two helix–turn–helix repressor proteins, CI and Cro, containing helix–turn–helix motifs. **The structures of these two repressor proteins differ substantially** (Figure 7-13). Whereas the Cro protein contains 56 amino acid residues and exists in solution as a monomer or dimer, the CI repressor contains 236 residues and can exist as a monomer, dimer, tetramer or octamer in solution. Each CI monomer contains two domains: an N-terminal domain of some 90 residues that contains the helix–turn–helix motif and binds specifically to the operator sites, and a C-terminal domain that does not interact with DNA. Between these two domains is a flexible hinge. Dimers of Cro are required to recognize operator DNA, but dimerization is weak, with dissociation constants in the micromolar concentration range. At nanomolar concentrations of subunits, assembly to dimers occurs only upon binding to DNA. It is slow, and Cro binding to DNA *in vivo* may be under kinetic rather than thermodynamic control. Binding of the Cro repressor is accompanied by large changes in both the protein and DNA: the subunits of Cro rotate 53° with respect to each other, and 19 base pairs of the DNA are bent by 40°. Recognition is achieved almost entirely by direct hydrogen bonding and van der Waals contacts between the protein and the exposed bases within the major groove of the DNA.

Both of these repressors bind to a set of six similar, but nonidentical, sites or operators in the phage genome. These sites are grouped in sets of three on each side of the structural gene for the CI protein. The affinities of the CI and Cro repressors for these six sites differ. When bound to DNA, the CI repressor dimers have the potential to form tetrameric complexes involving their C-terminal domains. The organization of the tripartite leftward and rightward operators of phage λ allows cooperative interactions between dimers bound to two adjacent sites. Because the occupancy of adjacent binding sites by CI repressor dimers depends on an interaction between two molecules, the binding curve is

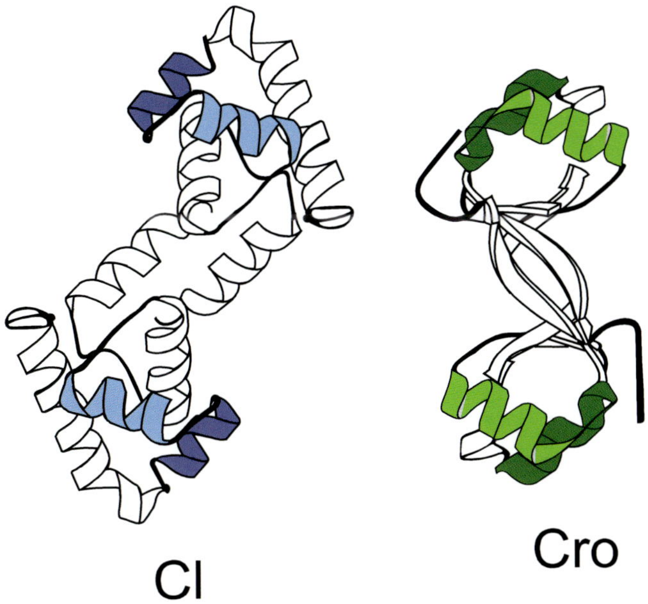

Figure 7-13. Structures of the CI and Cro repressors of lambda bacteriophage. Both have helix–turn–helix motifs (*colored*) with a spacing corresponding to that of successive turns of the major groove of B-DNA (Figure 7-12). In each case, a bound DNA double helix would lie with its helical axis approximately vertical. Adapted from a figure supplied by S. E. V. Phillips.

no longer hyperbolic, as it would be for independent binding, but instead is sigmoidal. An important consequence of this cooperativity is that occupancy changes rapidly over a small range of CI repressor concentrations and attains a level of >99% at concentrations significantly lower than those that would be required for the same level of independent binding. This cooperative protein•DNA binding implies that any modification of the protein–protein interactions required for cooperativity will result in a substantial change in occupancy.

Cooperativity in long-range gene regulation by the lambda CI repressor. I. B. Dodd *et al.* (2004) *Genes Dev.* **18**, 344–354.

Slow assembly and disassembly of lambda Cro repressor dimers. H. Jia *et al.* (2005) *J. Mol. Biol.* **350**, 919–929.

3. Homeodomains

Homeodomain-containing proteins are extremely important transcription regulators that control the coordinated expression of genes involved in development, differentiation and cellular transformation. The highly conserved 60-amino acid residue homeodomain allows these proteins to bind to DNA and modulate the expression of numerous target genes.

The homeodomain structure is built up from three α-helices, connected by short loop regions (Figure 7-14). α-Helices 2 and 3 and the loop region between them comprise the helix–turn–helix motif, but the recognition helix 3 is longer than normal. Homeodomains frequently bind to DNA as a monomer. The recognition helix 3 is positioned in the major groove of B-form DNA, and the ninth residue of this helix plays a crucial role in the specificity. The N-terminal residues, preceding helix 1, interact with the minor groove on the other side of the DNA double helix (Figure 7-14). Many other residues, including some from helix 2, interact nonspecifically with the DNA double helix.

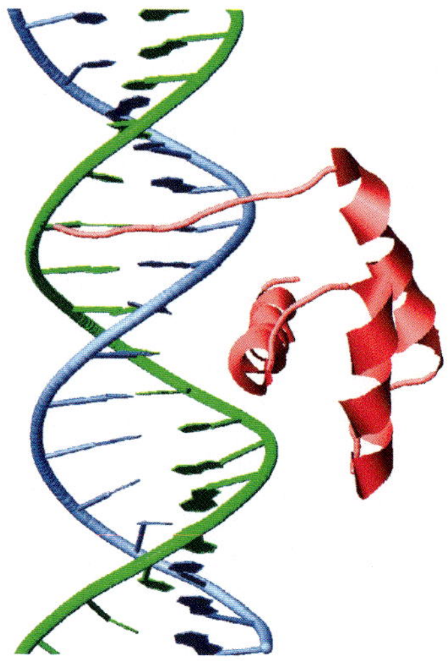

Figure 7-14. The structure of a homeodomain bound to dsDNA. The N-terminus of the polypeptide chain is at the *upper left*, whereas the C-terminus is behind, at the end of the α-helix. The three α-helices are depicted as *coils*. The helix at the C-terminus lies in the major groove of the DNA molecule. Generated from PDB file *1hdd* using the program Jmol.

The homeodomain binds specifically to the DNA sequence ATTA with a dissociation constant of about 1 nM. It also binds nonspecifically to other DNA, but with an affinity 10^2 lower. The homeodomain-containing proteins exhibit specificity *in vivo* but very similar DNA-binding affinities *in vitro*. Consequently, other mechanisms such as protein–protein interactions are probably involved in modulating their function.

Homeodomain revisited: a lesson from disease-causing mutations. Y. I. Chi (2005) *Human Gen.* **116**, 433–444.

NMR structural and kinetic characterization of a homeodomain diffusing and hopping on nonspecific DNA. J. Iwahara *et al.* (2006) *Proc. Natl. Acad. Sci. USA* **103**, 15062–15067.

4. POU Domains

POU proteins play a prominent role in the differentiation of the nervous system of several organisms, but also in more general cellular housekeeping. The POU region is a conserved sequence motif defining a family of eukaryotic transcription factors that contain **two structurally independent DNA-binding domains connected by a flexible linker of variable length and sequence** (Figure 7-15). The N-terminal POU-specific domain is a four-helix bundle containing the helix–turn–helix motif, while the C-terminal domain is structurally homologous to the homeobox domain. Both domains make sequence-specific contacts with DNA and together bind with high affinity to a conserved eight-base pair DNA binding site (ATGCAAAT) that is found in the promoter regions of a variety of genes. The N-terminal POU-specific domain binds to the ATGC part of the site, whereas the homeodomain recognizes the second half of the sequence.

The two domains are intrinsically flexible, which appears to confer functional diversity; **they encircle the DNA and are able to assume a variety of conformations that depend upon the DNA sequence.** In various members of this family, the orientations of the POU-specific domains relative to the POU homeodomain can be flipped, so that they interact with either the same or opposite sides of the DNA double helix. There is remarkable flexibility of the POU-specific domain in adapting to variations in sequence within the site.

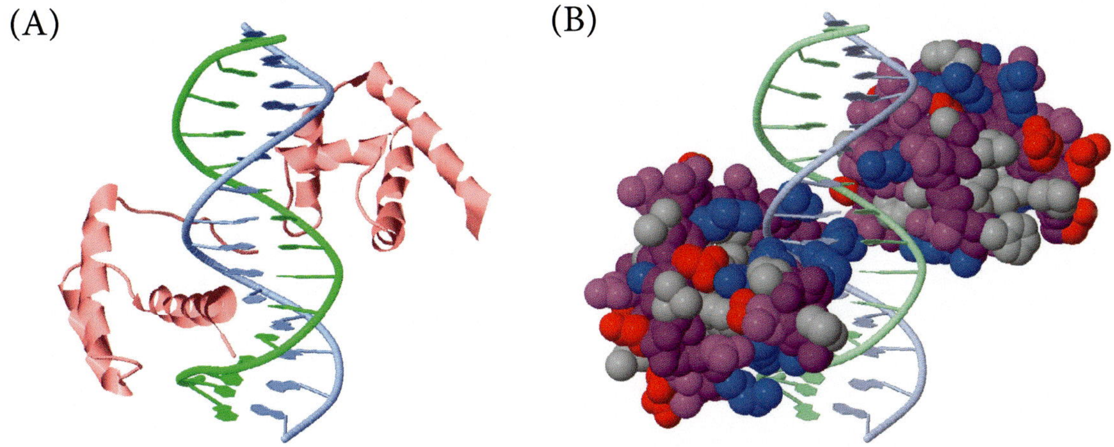

Figure 7-15. The structure of the POU domain and its binding to DNA. (A) Both the DNA and the protein are shown as simplified models; α-helices of the protein are depicted as *coils*. (B) The protein shown as a space-filling model. The domain like the homeodomain is at the *lower left*, while the four-helix bundle domain is at the *top right*. The polypeptide chain linking the two domains is flexible and only part is shown. Generated from PDB file *1oct* using the program Jmol.

The POU•protein complex provides various types of structures that can recruit specific regulator proteins to control transcription of the gene, providing a mechanism for assembling different combinations of proteins, as well as allosteric molecular recognition. One POU domain transcription factor, Oct4, interacts with another type of transcription factor, with which it controls gene expression, by dimerizing through interaction domains onto DNA in distinct conformational arrangements that are dictated by the nucleotide sequence of the binding site. Structures of the Oct1 transcription factor in the presence of two different DNA sites demonstrate how its POU DNA-binding domains can form two unrelated dimer arrangements, depending upon the DNA sequence. One arrangement allows binding of the Oct1-specific coactivator OBF-1, but binding of this coactivator is blocked in the second arrangement because the binding site is involved in its own dimer assembly. Conversely, the same overall assembly is maintained in the presence of two different DNA sequences by another POU transcription factor, Pit-1. The distances of the two Pit-1 half-binding sites on these two sequences differ by two base pairs, so the overall dimensions of the two complexes vary, allowing binding of a specific repressor in one conformation but not in the other.

Despite the various DNA-mediated molecular mechanisms, the net result is the same: conformation-dependent binding of further regulator proteins. **The DNA motif serves not only as a binding site for specific transcription factors, but also regulates their function by mediating specific transcription factor assemblies**, which determine binding to conformation-dependent coregulators.

Differential activity by DNA-induced quaternary structures of POU transcription factors. A. Remenyi *et al.* (2002) *Biochem. Pharmacol.* **64**, 979–984.

Crystal structure of a POU/HMG/DNA ternary complex suggests differential assembly of Oct4 and Sox2 on two enhancers. A. Remenyi *et al.* (2003) *Genes Dev.* **17**, 2048–2059.

5. *Trp Repressor*

The *trp* repressor from bacteria is a dimer of a 107-residue polypeptide chain. Each polypeptide chain comprises six α-helices, but three of these helices from each chain are entwined about each other (Figure 7-16); **rather than each chain making a globular structure and then dimerizing, the globular structure requires the dimer of these pairs of three N-terminal α-helices**. This central core supports structurally two 'heads' that interact with the palindromic binding site on DNA, one at each end of the molecule. Helices 4 and 5 of each polypeptide chain comprise a helix–turn–helix motif that interacts with the DNA. Recognition in this system is controlled by the binding of the corepressor ligand, L-tryptophan, as well as by the conformational and dynamic properties of the DNA target sequences, DNA sequence-dependent control of the oligomerization properties of the repressor, water-mediated interactions and specific interactions involving the peptide backbone and DNA phosphate moieties.

The *trp* repressor binds to DNA specifically only when two molecules of the amino acid L-tryptophan are bound, one to each half of the dimer (Figure 7-16). This binding causes the two 'heads' to tilt outwards and changes the orientation of the recognition helices. These two helices are now in the right position, 34 Å apart, to bind to the appropriate palindromic DNA sequence within the major groove. Without the bound tryptophan, the distance between the two heads is too short by 5–6 Å to permit such binding. **This is the best example of how binding of another ligand, a corepressor, can control binding of the protein to a DNA molecule**. The relative affinities for different sequences are conserved in the absence of the corepressor, and it is in all cases significantly higher than that observed for holo repressor binding to nonspecific DNA; therefore, the apo repressor does not participate only in nonspecific binding.

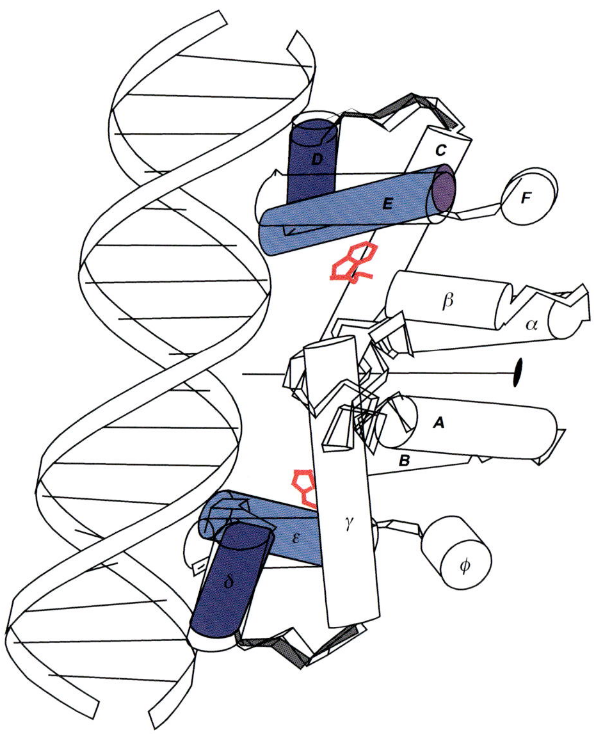

Figure 7-16. Structural change in the *trp* repressor induced by binding of the corepressor, L-tryptophan. Where the structure is altered by binding of L-tryptophan, the *shaded* model is that without the ligand. α-Helices A through F are from one polypeptide chain of the dimer, α through ϕ are from the other. The tryptophan corepressor bound to each of the two sites of the dimer is shown as a red skeletal model. The corepressor binds between helices B, C and E of each half of the dimeric protein. In doing so, it pushes helices D and E of the helix–turn–helix motif away from the center of the molecule and places the two equivalent recognition helices E and ε the correct distance apart to bind in adjacent major grooves of normal double helical DNA. The binding affinity for DNA is increased 10^3-fold by the presence of the corepressor. Adapted from R.-G. Zhange *et al.* (1987) *Nature* **327**, 591–597.

The *trp* repressor binds specific sequences even though there are no direct contacts between the protein and the bases. In the complex of the *trp* repressor with a symmetrical, palindromic, 18-base pair sequence, TGT**ACTAG**TT•AA**CTAGT**AC, the two ACTAG sequences centered five base pairs from the dyad (indicated by the dot) are those most important for binding. Remarkably, these sequences make no direct hydrogen bond or nonpolar contacts with the protein that might account for the specificity of binding; the only direct contact lies outside the specific sequence. Direct contacts between the protein and DNA are principally with the phosphate groups of the DNA backbone, which maintain the conformation of the backbone. The contacts between protein and DNA are mediated by three well-ordered water molecules in the major groove. Replacing the amino acid residues that interact with these water molecules reduces the affinity and, in some examples, alters the specificity of binding. The *trp* repressor illustrates how **some sequence specificity can require water molecules to mediate direct recognition of the operator sequence.**

Even on very small (25-base pair) targets, up to three molecules of repressor can bind simultaneously and slightly cooperatively. A 2:1 dimer–DNA complex forms first; at higher concentrations a third dimer binds with significantly lower affinity. There appears to be an inherent asymmetry in the normal target sequence and in interactions between neighboring repressor molecules. The *trp* repressor tends to aggregate even in the absence of DNA, but the interactions when on DNA depend upon the DNA sequence.

Probing the role of water in the tryptophan repressor–operator complex. M. P. Brown *et al.* (1999) *Protein Sci.* **8**, 1276–1285.

Probing the physical basis for trp repressor–operator recognition. A. O. Grillo *et al.* (1999) *J. Mol. Biol.* **287**, 539–554.

Water-mediated contacts in the trp-repressor operator complex recognition process. F. R. Wibowo *et al.* (2004) *Biopolymers* 73, 668–681.

6. *Cyclic AMP Receptor Protein/Catabolite Gene Activator Protein*

The **cyclic AMP receptor protein** (CRP) of *E. coli* is also known as the **catabolite gene activator protein** (CAP). It binds cyclic AMP (cAMP) and stimulates expression of the *lac* operon. It is the **paradigm for gene activator proteins**. CRP, in conjunction with cAMP, participates as a global transcription factor in a wide regulatory network, both activating and repressing a large number of operons. One mechanism of **catabolite repression**, which is a mechanism for modulating the expression of specific genes in response to the overall nutritional state of the organisms, is due to the reduction in the intracellular concentrations of both cAMP and CRP.

The CRP protein is a dimer composed of two identical subunits of 209 amino acid residues each (Figure 7-17). It binds to DNA when complexed with cAMP, which behaves as an allosteric effector; there are two cyclic AMP-binding sites in each monomer. CRP binds to specific DNA sites through a helix–turn–helix motif. When bound, it interacts with RNA polymerase and other regulatory proteins to regulate transcription of the target operons.

Each subunit has two domains. The larger N-terminal domain is responsible for dimerization and binding of cAMP binding. The second domain contains the helix–turn–helix motif that is responsible for DNA binding. Without cAMP, CRP binds to DNA nonspecifically. The binding of cAMP causes a change in conformation, which may involve realignment of the subunits and rearrangement of the domains, to one that binds to specific sequences with a dyad symmetry. The DNA sequences of many CRP binding sites are variations of the palindromic sequence 5′-AAA**TGTGA**TCT•AGA**TCACA**TTT-3′. The two TGTGA motifs highlighted tend to be conserved in various promoters. The cAMP–CRP complex binds to target DNA by inserting the helix–turn–helix motifs of the two subunits into successive major grooves of the DNA. The DNA is bent by 90° (Figure 7-8) but the significance of this bending is not clear.

The basic role of the cAMP–CRP complex is to enhance binding of RNA polymerase and facilitate its transcription of the adjacent gene. The complex of cAMP–CRP and RNA polymerase with DNA is cooperative, the binding of one stimulating binding of the other. Both CRP and RNA polymerase must bind to the same face of the DNA, and cAMP–CRP directly contacts the RNA polymerase. CRP and RNA polymerase holo enzyme associate cooperatively to form a 2:2 complex *in vitro*, with or without cAMP. A surface-exposed loop in the C-terminal domain of CRP and the C-terminal part of the α-subunit of RNA polymerase are primarily responsible for the contact between two proteins. An additional contact between the N-terminal domain of CRP and the N-terminal domain of α-subunit is involved in some cases.

CRP and the *lac* repressor (Section 7.3.A.1) associate cooperatively to form a 2:1 complex *in vitro*. Forming the complex is stimulated by cAMP, with a net uptake of one cAMP per molecule of CRP. CRP probably binds in the cleft between dimeric units in the *lac* repressor tetramer (Figure 7-5). These CRP–*lac* repressor interactions may play important roles in regulatory events that take place at overlapping CRP and repressor binding sites in the lactose promoter.

The *Escherichia coli* cyclic AMP receptor protein forms a 2:2 complex with RNA polymerase holoenzyme, *in vitro*. D. Dyckman & M. G. Fried (2002) *J. Biol. Chem.* **277**, 19064–19070.

Catabolite activator protein: DNA binding and transcription activation. C. L. Lawson *et al.* (2004) *Curr. Opinion Struct. Biol.* **14**, 10–20.

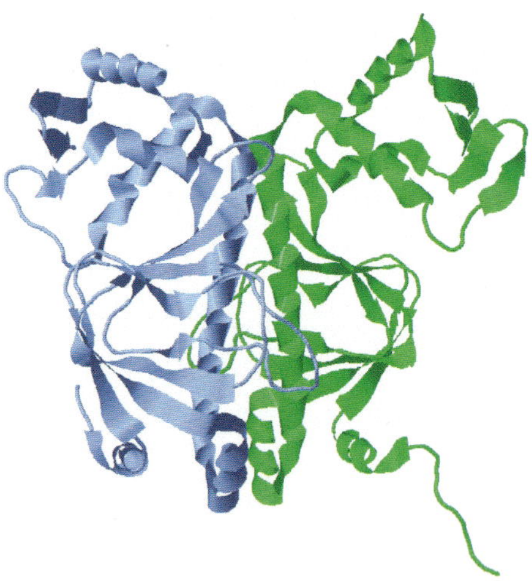

Figure 7-17. Structure of the CRP dimer. The helix–turn–helix motifs that interact with the major groove of DNA are at the top of the molecule; the bound DNA would lie horizontally along the top of the molecule. The spacing between the two sets in the dimer corresponds to the spacing between successive turns of the major groove of B-DNA. Generated from PDB file *1i5z* using the program Jmol.

7.3.B. TATA-binding Protein

The DNA sequence TATA occurs approximately 30 base pairs upstream of the start site of transcription of eukaryotic genes and helps to define the initiation site for transcription of the gene. The association of monomeric TATA-binding protein (TBP) with promoter DNA is an essential first step in many current models of eukaryotic transcription initiation. This step is followed by others in which additional transcription factors, and finally RNA polymerase, assemble at the promoter.

The C-terminal domain of TBP is composed of two direct repeats of about 80 amino acid residues; it resembles a saddle that binds with high affinity ($K_d \leq 5$ nM) and sits astride the DNA, in contact with its minor groove (Figure 7-18). Two Phe residues of TBP are inserted between the two last base pairs of the TATA element, causing the DNA structure to bend by about 80° and widening the minor groove without interfering with base pairing (Figure 7-8). Sequences at these sites that favor flexibility increase the affinity for TBP up to more than 100-fold. The rate of dissociation of the TBP–DNA complex is decreased, rather than increasing the rate of its formation. The importance of the DNA kinking varies, depending upon the nature of the TATA box and the surrounding sequences.

Little information is available concerning the structure of full-length TBP containing both the conserved C-terminal and the more variable N-terminal domains. The latter is involved in a monomer–dimer equilibrium with an apparent dissociation constant of approximately 8 μM.

Interactions with other components of the transcription machinery are vital for the function of TBP. Bending the DNA around the TATA box allows contacts between regulatory factors bound to their specific sequence, on one side, and the basal transcription machinery on the other side. One example of the mechanisms that regulate the interactions of TBP with promoter DNA is autorepression, where TBP sequesters its DNA-binding surface through dimerization. Once TBP is bound to DNA, some factors induce TBP to dissociate, while others convert the TBP–DNA complex into an inactive state. TBP *in vivo* is part of **transcription factor TFIID**, a multiprotein complex containing TBP and 14 TBP-associated factors (TAFs). The overall structure of this large complex resembles a molecular clamp formed by three major lobes connected by thin linking domains. The TAFs are dispensable for basal transcription *in vitro* but are required for the response to activator proteins and have selective effects on gene expression. The complexity of all these interactions of TBP with other proteins explains how TBP can be involved in the regulation of so many different genes, each with their own expression requirements.

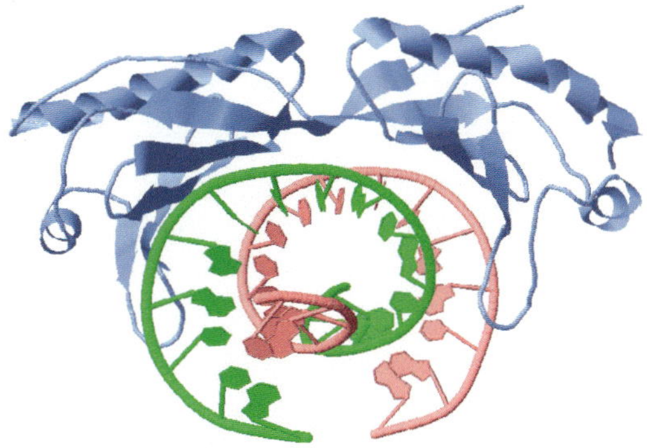

Figure 7-18. Structure of the complex of TATA-binding protein (TBP) with a TATA box. The protein binds the TATA sequence from the minor groove and bends the DNA into the major groove. The α-helices are represented as *coils* and the β-strands as *arrows*. Generated from PDB file *1cdw* using the program Jmol.

Self-association of the amino-terminal domain of the yeast TATA-binding protein. C. A. Adams *et al.* (2004) *J. Biol. Chem.* **279**, 1376–1382.

Comparison of the effect of water release on the interaction of the *Saccharomyces cerevisiae* TATA binding protein (TBP) with 'TATA Box' sequences composed of adenosine or inosine. S. Khrapunov & M. Brenowitz (2004) *Biophys. J.* **86**, 371–383.

7.3.C. Zinc-containing DNA-binding Motifs

A number of different DNA-binding proteins and motifs contain one or more Zn atoms that coordinate with amino acid residues, usually Cys or His, in the appropriate spatial orientation and thereby stabilize the structure of the domain. Six types of these domains have been identified, yet they are structurally diverse and the only common feature is the presence of zinc (Figure 7-19).

1. Zinc Fingers

Zinc fingers occur in many different versions in virtually every eukaryotic genome. They provide one more structural means for an α-helix to protrude from the protein surface sufficiently to interact with nucleotide bases in the major groove of DNA. **The most conspicuous feature of this structure is a Zn^{2+} ion chelated by two Cys and two His residues that occur in a characteristic sequence motif, $-X_3-$Cys$-X_{2-4}-$Cys$-X_{12}-$His$-X_{3-4}-$His$-X_4-$**, where X is any amino acid. Hundreds of such sequence motifs have been found in proteins involved in many aspects of gene regulation in eukaryotes. The four residues that chelate the zinc ion are held in the appropriate positions by the 3-D structure of the finger polypeptide chain, which consists of a two-stranded β-hairpin and a single α-helix (Figure 7-20). This fold is apparently stabilized by several hydrophobic side-chains that make up the core of the structure. **Each zinc finger seems to be a stable, autonomous, structural unit, even though it generally consists of only 28–31 residues**.

When bound to DNA, the proximal end of the α-helical region is directed into the major groove and makes contact with specific bases (Figure 7-4). Successive amino acid residues that make contact are separated by approximately one turn of the α-helix (Figure 7-20). One of the His residues that coordinate the zinc ion also contacts the sugar–phosphate backbone. A single zinc finger generally contacts three successive base pairs in DNA, usually in the same strand of the double helix (Figure 7-21).

Most zinc finger proteins contain multiple (up to 30) contiguous fingers separated by short linkers, frequently with the sequence –Thr–Gly–Glu–Lys– (Figure 7-20). An individual finger binds to DNA only weakly, but simultaneous interactions involving a number of zinc fingers increase the binding substantially. **Adjacent zinc fingers often individually specify four-base pair subsites that overlap by one base pair**. Different zinc fingers have different specificities for the triplet that they recognize. Proteins containing multiple zinc fingers are therefore of modular construction, and **nature seems to have devised a variety of such proteins with specificities for different DNA sequences by 'mixing and matching' various individual fingers with specificities for different triplets**. The simplicity of these direct interactions means that there is almost a direct code between the amino acids of the zinc finger and the bases of the DNA that they recognize, so it is possible to design zinc fingers to recognize specific DNA sequences. Adding domains with nuclease activity can be used to cleave specific DNA sequences.

There is considerable variation in the structures and functions of the zinc finger modules. The structure of the first finger is often especially variable. In the case of the SWI5 protein, for example, the first finger is preceded by a short α-helix, and an additional β-strand precedes the normal β-hairpin loop. Some of the multiple fingers may mediate interactions with other proteins and not be involved in DNA binding. In some proteins, zinc fingers are associated with conserved protein domains that help to define the functions of these regulators. Some zinc fingers bind specifically to RNA molecules.

In addition to the classical zinc fingers described here, there are at least 14 different classes of zinc fingers that differ in the nature and arrangement of their zinc-binding residues.

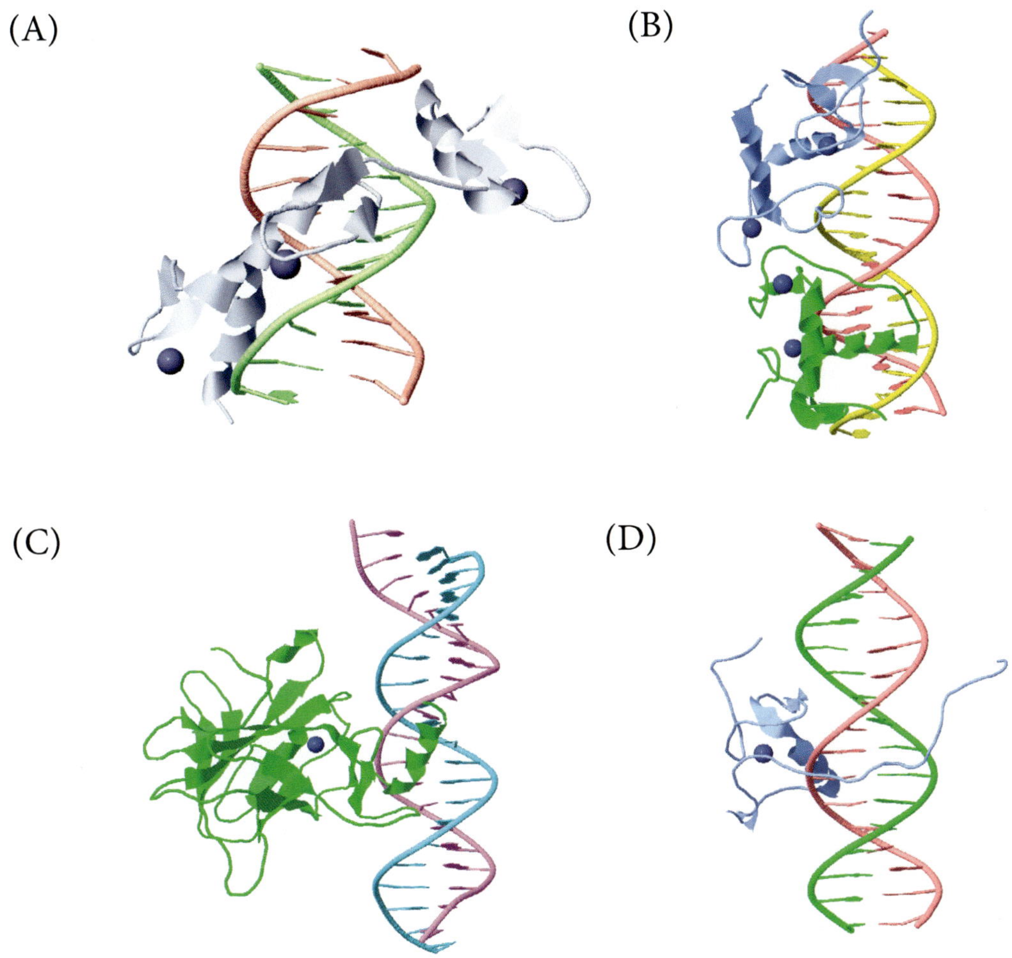

Figure 7-19. Four different zinc-containing transcription factors bound to their specific DNA sequences. The α-helices are depicted as *coils*, β-sheets as *arrows* and zinc ions as *spheres*. (A) Three zinc fingers of murine Zif268 (Section 7.3.C.1; generated from PDB file *1jk1*). (B) The dimeric DNA-binding domain of estrogen receptor (Section 7.3.C.2; generated from PDB file *1hcq*). (C) The core domain of the p53 tumor suppressor, a β-sandwich that serves as a scaffold to orient a loop–sheet–helix motif and two large loops. It is stabilized by a zinc ion that is coordinated by three Cys and one His residues. The zinc domain orients the side-chain of an Arg residue that contacts the minor groove of the DNA, while two Arg residues of the loop–sheet–loop motif contact the major groove (generated from PDB file *1tup*). (D) The DNA-binding domain of the chicken erythroid transcription factor GATA-1. A single zinc ion stabilizes the core of two irregular antiparallel β-strands and an α-helix. The α-helix and the loop connecting the two antiparallel strands interact with the major groove of the DNA, while the C-terminal tail wraps around the DNA to contact the minor groove (generated from PDB file *2gat*). Figure generated using the program Jmol.

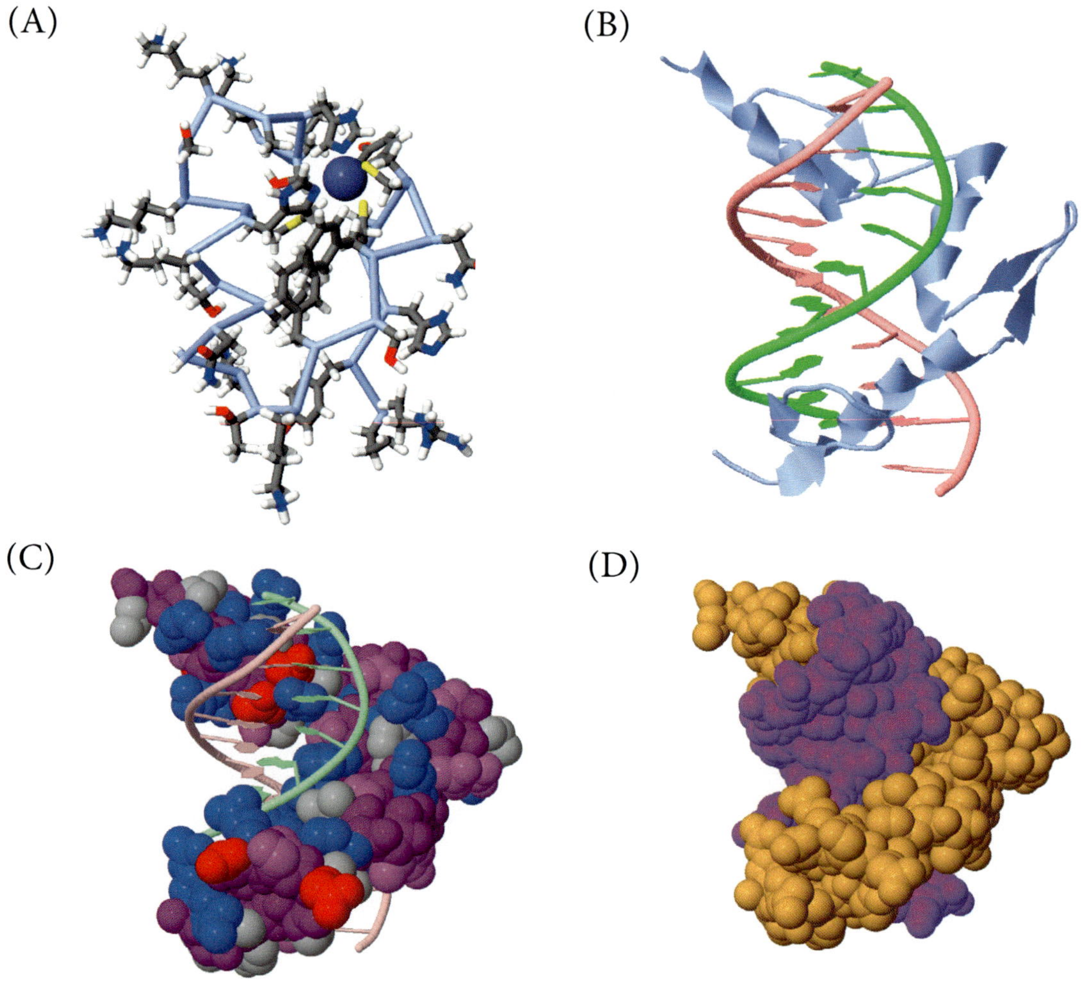

Figure 7-20. The structure of a single two-Cys, two-His zinc finger (A) and the interaction of a three-finger protein with DNA (B, C, D). (A) All the covalent bonds of the protein are depicted, except that a single bond links adjacent C^{α} atoms. The *sphere* is the Zn^{2+} ion, which is ligated by two His and two Cys side-chains. (B, C, D) The crystal structure of a polypeptide chain comprising three zinc fingers from protein Zif268. The protein is depicted in cartoon fashion in (B) and as space-filling in (C). In (D), both protein and DNA are depicted with their atoms as *spheres* of the appropriate van der Waals radii. Amino acid side-chains from residues at the amino end of each helix contact the edges of three sequential base pairs in the major groove of the DNA (Figure 7-21). Generated from PDB files *3znf* and *1a1f* using the program Jmol.

Designer zinc-finger proteins and their applications. M. Papworth *et al.* (2006) *Gene* **366**, 27–38.

Designer zinc finger proteins: tools for creating artificial DNA-binding functional proteins. M. Dhanesekaran *et al.* (2006) *Acc. Chem. Res.* **39**, 45–52.

Invariance of the zinc finger module: a comparison of the free structure with those in nucleic-acid complexes. D. Lu & A. Klug (2007) *Proteins* **67**, 508–512.

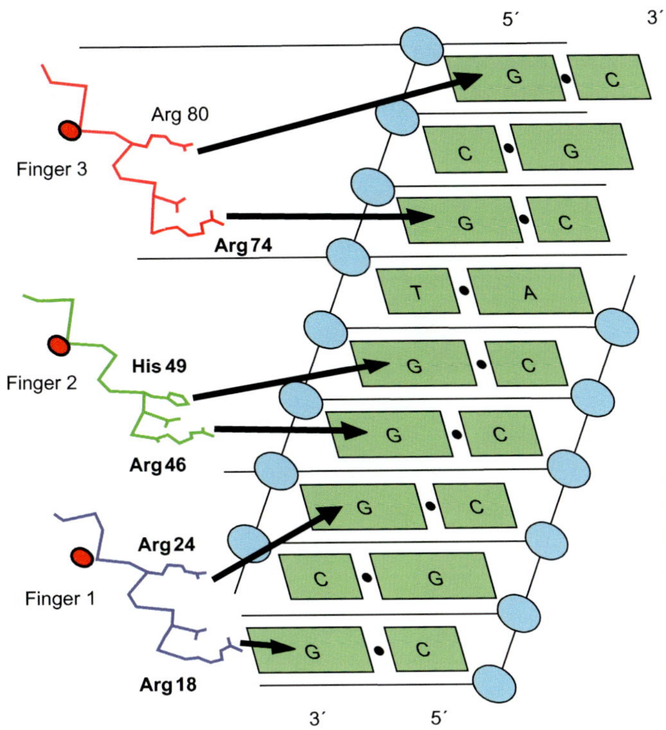

Figure 7-21. Hydrogen bonds from Arg and His residues of three zinc fingers to guanine bases in the DNA. The *arrows* indicate hydrogen bond interactions with the indicated bases of the DNA. The Zn atoms of the protein are *orange*; the phosphate groups of the DNA backbone are *blue*. Adapted from N. P. Pavletich & C. O. Pabo (1991) *Science* **252**, 809–817.

2. Steroid Hormone Receptors

The DNA-binding domains of the steroid hormone receptors, such as the **glucocorticoid** and **estrogen receptors**, consist of another type of DNA-binding motif that contains zinc and eight Cys residues that bind two zinc ions. **Each zinc ion is coordinated by four Cys residues in a loop that occurs in Cys–Cys ... Cys–Cys sequences.** Unlike the zinc fingers, the distal Cys residues are followed immediately by an α-helical region (Figure 7-22). The second α-helix is in direct hydrophobic contact with the first, forming a single structural domain known as the **double-loop zinc helix.** The α-helix immediately distal to the first Cys tetrad sequence contains the residues responsible for recognizing specific DNA sequences.

The DNA-recognition domains of each receptor exist as dimers and bind to a palindromic sequence that contains three conserved bases on each side. They differ in the separation between the conserved sequences in such a way that the individual components of the dimer must have different relative rotational orientations, as a result of differences in their dimerization.

What sets steroid hormone receptors (and other nuclear hormone receptors) apart from other families of sequence-specific transcriptional activators is the presence of **a ligand-binding domain that acts as a molecular switch to turn on transcriptional activity when a hormonal ligand induces a conformational change in the receptor.** The ligand-binding domains possess a conserved structure with 12 α-helices surrounding a central hydrophobic core. Upon binding hormone, helix 12 changes position and generates a pocket with helix 3 and helix 5, where transcriptional coactivators bind. The novel coactivator protein complex has an essential role in receptor-mediated transcriptional activation. Coactivators function as adaptors in a signaling pathway that transmits transcriptional responses from the DNA-bound receptor to the basal transcription machinery. Hormone **agonists** induce a conformational change in the C-terminal transcriptional activation domain, AF-2, that creates a new protein interaction site on the surface of the ligand-binding domain that is recognized

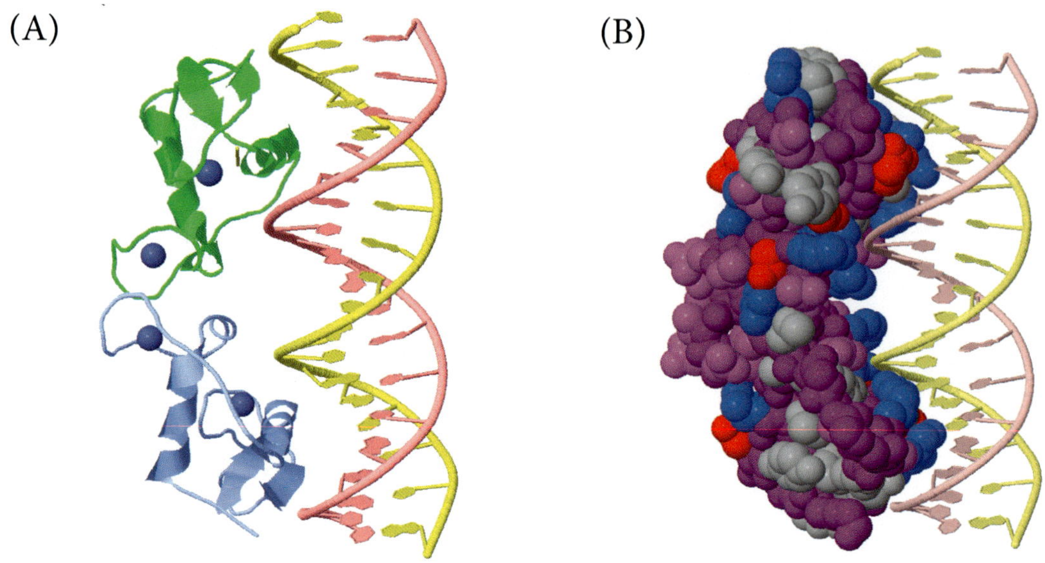

Figure 7-22. A dimer of the DNA-binding domain of the estrogen receptor binding to its specific DNA sequence. (A) The *spheres* are the four zinc ions, *coils* are α-helices and *arrows* β-strands. (B) The protein is shown in space-filling form. Generated from PDB file *1hcq* using the program Jmol.

by Leu–X–X–Leu–Leu sequence motifs in the p160 family of coactivators. In contrast, steroid **antagonists** for the estrogen receptor induce an alternate conformation in AF-2 that occludes the coactivator binding site and recruits corepressors that can actively silence steroid responsive genes. Thus **the cellular availability of coactivators and corepressors is an important determinant in the biological response to both steroid hormone agonists and antagonists.**

To affect transcription, the steroid hormone receptors must be able to gain access to their binding sites on the genomic DNA within the compacted chromosomes (Section 7.3.F). They do this by interacting directly with higher order chromatin structures and perturbing them using enzymatic multi-complexes that can either remodel or modify the chromosomal structure.

Structure and function of the glucocorticoid receptor ligand binding domain. R. K. Bledsoe *et al.* (2004) *Vitam. Horm.* **68**, 49–91.

Gene regulation by the glucocorticoid receptor: structure–function relationship. R. Kumar & E. B. Thompson (2005) *J. Steroid Biochem. Mol. Biol.* **94**, 383–394.

3. GAL4 Type

The GAL4 protein is typical of a third type of zinc-containing domain that binds DNA. **Two Zn atoms in each monomer are coordinated by six Cys residues; two of them interact with both Zn atoms.** This structure is connected by a flexible region of polypeptide chain to a short amphipathic α-helix that interacts with another monomer to form a short coiled coil of two parallel α-helices. The GAL4 protein functions as a symmetrical dimer and binds to a palindromic DNA recognition site. Direct contact with the bases in the DNA major groove is made by amino acid residues from a short α-helical region (Figure 7-23). Most of the contacts to the sugar–phosphate backbone involve amino acid residues in the more extended polypeptide chain.

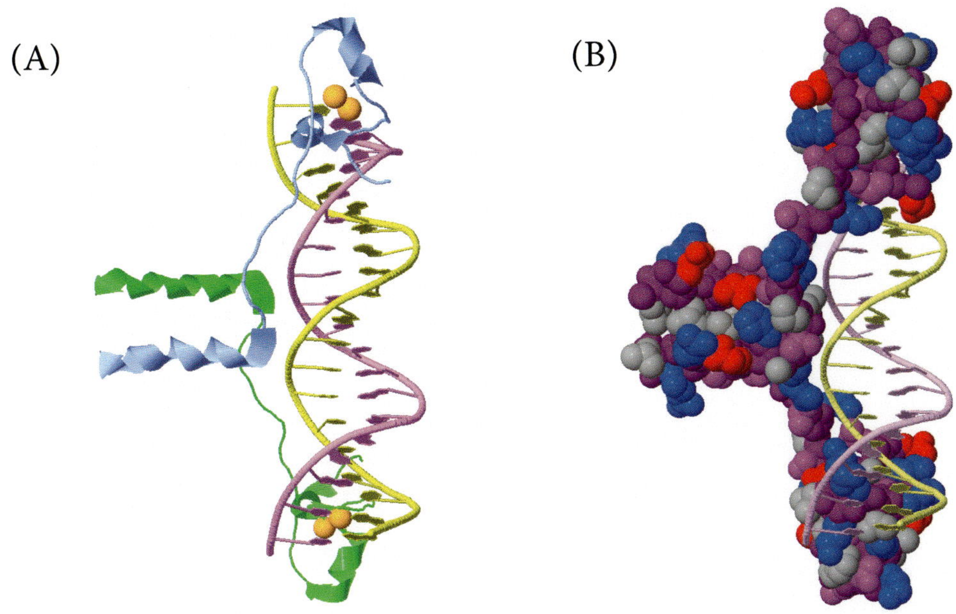

Figure 7-23. Structure of the GAL4 dimer–DNA complex. The protein is shown as a cartoon in (A) and with space-filling representation in (B). The two Zn atoms of each monomer are shown as *spheres*; they are coordinated by six Cys residues. Residues 8–40 of GAL4 comprise the α-helical DNA-recognition element, which interacts with the major groove of the DNA at each end of the elongated molecule. Residues 49–64 form the coiled-coil dimerization element, at the center, while residues 41–48 are in an extended conformation, interacting with the DNA backbone by its basic residues. Generated from PDB file *1d66* using the program Jmol.

Regions of GAL4 critical for binding to a promoter *in vivo* revealed by a visual DNA-binding analysis. A. Mizutani & M. Tanaka (2003) *EMBO J.* **22**, 2178–2187.

Yeast Gal4: a transcriptional paradigm revisited. A. Traven *et al.* (2006) *EMBO Rep.* **7**, 496–499.

7.3.D. bZip and Helix–Loop–Helix Domains

The **basic leucine zipper (bZip)** DNA-binding domain, and the closely related **helix–loop–helix (HLH)** domain, are found in many eukaryotic transcription factors (Figure 7-24). Both bind in the major groove of DNA. They also are able to form homo- or heterodimers with proteins of the same class. The simplest of these domains is the bZip domain, which forms a continuous α-helix when bound to DNA. The lower basic part of this helix (the b of bZip) forms sequence-specific contacts with DNA, while the upper part can dimerize with an appropriate partner through the formation of a parallel coiled coil, also known as a **leucine zipper**. The simplicity of the coiled-coil interaction means that the two polypeptide chains that interact need not be identical. **The dimer has the appearance of a pair of scissors binding to a palindromic DNA sequence in which each half-site is exposed on opposite faces of DNA.**

bZIP proteins can bend DNA by asymmetric charge neutralization, in which one or two basic side-chains interact with a symmetry-related pair of phosphate groups of the DNA. The nucleotide sequences flanking the target site can affect its ability to bind the bZip peptide specifically, probably by varying the overall DNA conformation.

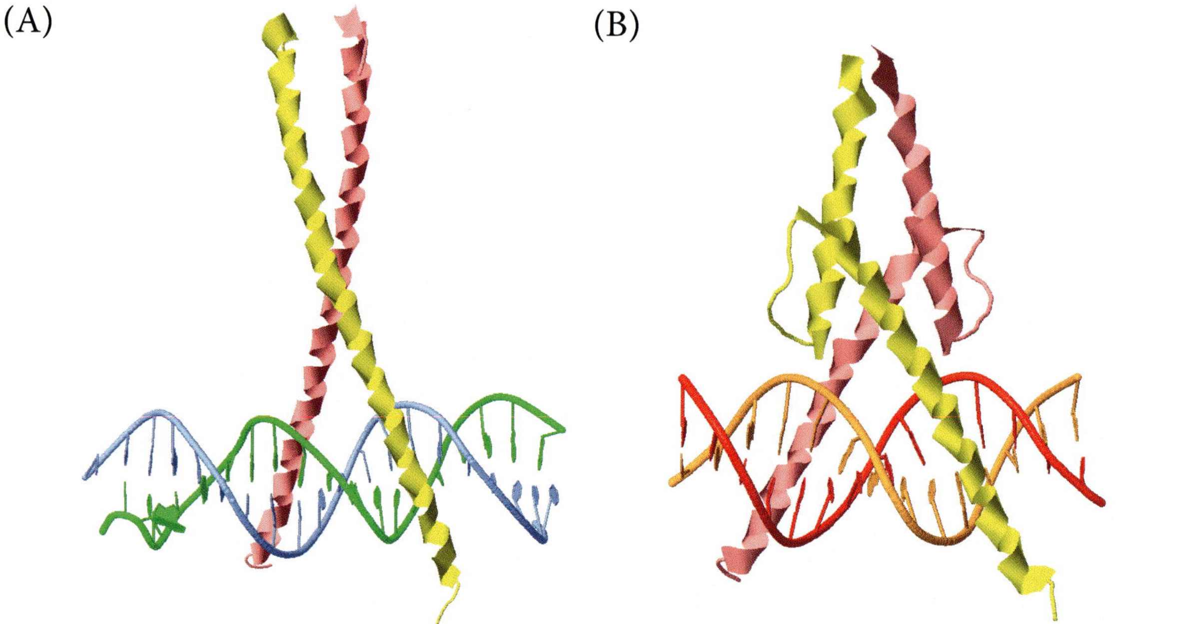

Figure 7-24. The structures of the DNA-bound forms of the bZIP protein GCN4 (A) the basic helix–loop–helix protein MyoD (B). Generated from PDB files *1ysa* and *1mdy* using the program Jmol.

The helix–loop–helix domain binds to DNA in a similar manner, with the difference that the α-helix involved in dimerization is separated from the DNA binding α-helix by a short loop. The four helices in the dimer form a parallel left-handed bundle.

Comprehensive identification of human bZIP interactions with coiled-coil arrays. J. R. Newman & A. E. Keating (2003) *Science* **300**, 2097–2101.

Reflections on apparent DNA bending by charge variants of bZIP proteins. P. R. Hardwidge *et al.* (2003) *Biopolymers* **69**, 110–117.

7.3.E. β-Sheets: Methionine Repressor

The **Met repressor recognizes DNA through β-strands**, rather than the α-helices of most other DNA-recognition proteins. It is a polypeptide chain of 104 amino acid residues that exists as a homodimer (Figure 7-25). Similar to the *trp* repressor (Section 7.3.A.5), **the two subunits of the Met repressor intertwine to form the symmetrical dimer**. The dimer noncooperatively binds two molecules of S-adenosyl-methionine (S-AdoMet):

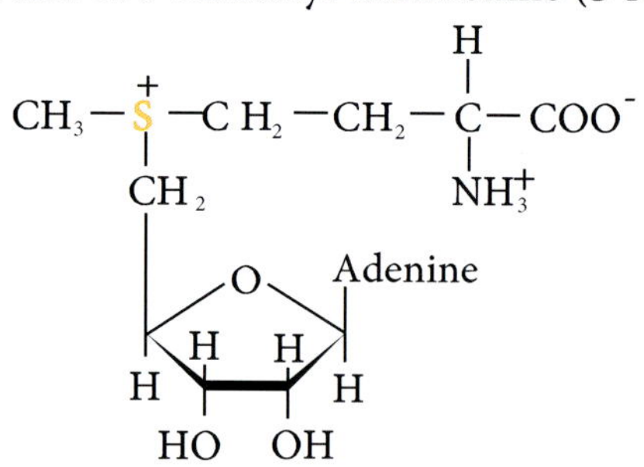

(7.8)

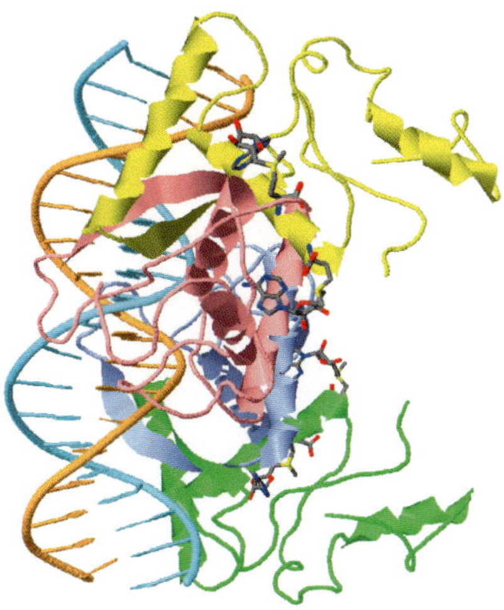

Figure 7-25. The structure of two dimeric methionine repressors bound to and bending dsDNA. The backbone of the protein is depicted, with β-strands depicted as *arrows*; two such strands from each dimer interact with the major groove of the DNA. The S-adenosylmethionine corepressor molecules bound to each monomer are depicted in skeletal form. Interactions between the DNA and the opposite ends of the nearly vertical α-helices (depicted as *coils*) cause bending of the DNA. Generated from PDB file *1mj2* using the program Jmol.

Each monomer of the Met repressor contains one β-strand and three α-helices, which account for about half of the sequence (Figure 7-25). The two β-strands of the dimer generate an antiparallel β-sheet, which packs against the α-helical regions.

This repressor recognizes an eight-nucleotide 'Met box', with the palindromic consensus sequence AGACGTCT, that is present in an altered form two to five times in each DNA region. **Two repressors having bound S-AdoMet bind tightly and cooperatively to DNA fragments containing two consecutive consensus Met boxes, while binding is negligible to nucleotide sequences containing only one Met box.** A DNA fragment containing two adjacent eight-base pair Met boxes binds two dimers. One dimer binds to each half-site, and the two-fold symmetry axis of each binding site coincides with the two-fold axis of the protein dimer. Two neighboring dimers interact through α-helix 1 to generate a tetrameric protein structure,.

The antiparallel protein β-sheet is inserted into the major groove of B-form DNA containing Met box sequences. Amino acid side-chains on the exposed face of the β-sheet form hydrogen bonds directly with specific base pairs. Thr25 from each β-strand contacts an adenine base, and the neighboring residue Lys23 of each strand contacts a guanine. The backbone from the N-terminus of α-helix 2 also makes contacts. The repressor senses the conformation and flexibility of the DNA backbone, so binding can also depend upon the sequences of base pairs that are not in contact with the protein.

S-AdoMet binds to the dimer at two independent and symmetrical sites, one on each monomer. The trivalent S atom of S-AdoMet is positively charged and situated near the partial negative charge at the C-terminus of a helix. The adenine moiety of S-AdoMet inserts into a hydrophobic pocket, replacing the side-chain of Phe65, while the methionine moiety lies at the protein surface. Consequently, methionine alone does not bind to the aporepressor, while S-adenosylhomocysteine (which lacks the terminal $-CH_3$ group and the positive charge on the S atom) binds with about half the affinity of S-AdoMet.

How the binding of S-AdoMet to the repressor regulates its binding to DNA is not clear, because it neither causes conformational changes in the protein nor contacts the DNA in the repressor–operator complex. A long-range electrostatic effect involving the positive charge on the S atom seems the most plausible explanation.

Direct and indirect readout in mutant Met repressor–operator complexes. C. W. Garvie & S. E. Phillips (2000) *Structure* **8**, 905–914.

Structural basis for the differential regulation of DNA by the methionine repressor MetJ. A. M. Augustus *et al.* (2006) *J. Biol. Chem.* **281**, 34269–34276.

7.3.F. Histone-Fold

Histones are small basic proteins that are involved in organizing the chromosomal DNA of eukaryotes (Section 3.7). There are normally four distinct histones (H2A, H2B, H3 and H4), but with related sequences, and two of each assemble into an octameric structure. This structure interacts with about 200 base pairs of double-stranded DNA, to form the **nucleosome** (Figure 3-12). It interacts with nearly all the DNA in a chromosome and is not very specific for the nucleotide sequence. **The histone-fold motif is a feature of all four histones.** It consists of a long central α-helix connected by loops to two shorter α-helices (Figure 7-26). The histone-fold is found in other proteins and it might have arisen by gene duplication of a helix–strand–helix motif, with the long central α-helix being produced by fusion of two helices.

Extensive interactions between two histone-folds in a **handshake motif**, with extensive hydrophobic interactions between the nonpolar faces of the two long amphipathic α-helices, produce the heterodimers that exist in **the nucleosome core particle** (Figure 7-26). Each dimer has three independent DNA-binding sites: two contributed by the loops at each end of the dimer, and one formed by the two N-termini of α-helices of each histone-fold at the dimer apex. **Histone–DNA contacts include extensive salt bridges and hydrogen bonds from both main chains and side-chains of the histones to the phosphate backbone, together with nonpolar contacts with DNA sugar groups and electrostatic interactions of the positively charged N-termini of α-helices in the histone-folds with DNA phosphate groups.** Such interactions might be expected for proteins that can bind to a variety of DNA sequences; the number of specific base contacts in the core particle structure is very small (one of them being a nonpolar contact of the 5-methyl group of a thymidine

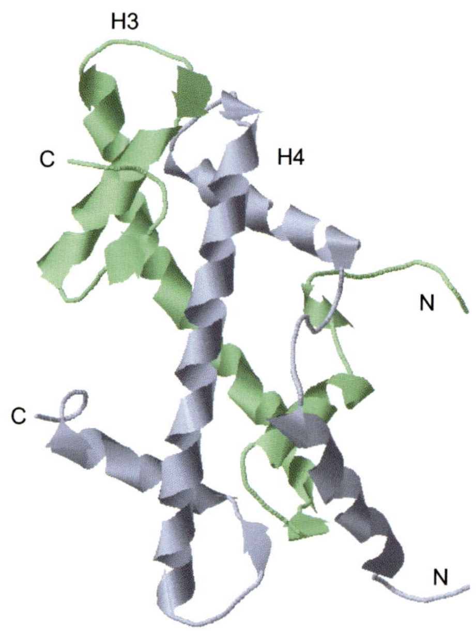

Figure 7-26. The histone-folds in histones H3 and H4 when part of the nucleosome bound to DNA. Histone H3 is *green*, H4 *mauve*. Generated from PDB file *1aoi* using the program Jmol.

in the major groove). Strikingly, Arg side-chains contact the minor groove each of the 14 times it faces inwards, probably to screen electrostatic repulsions between the DNA phosphate groups. **Each histone-fold dimer in the octamer organizes 27–28 base pairs of DNA**, with stretches of four base pairs between contacts.

The histone-fold has also been found to occur in other proteins, with little or no sequence homology with the histones, that are also involved in DNA/protein recognition and protein–protein interactions. For example, nine of the 14 **TAFs** (Section 7.3.B) contain histone-fold domains. The histone-fold motif thus appears to be widespread, and it may primarily be a robust structural motif particularly well-suited to protein dimerization in large complexes.

A database for histone and proteins containing the histone-fold is available at http://research.nhgri.nih.gov/histones/.

Crystal structures of histone S in mutant nucleosomes reveal altered protein–DNA interactions. U. M. Muthurajan *et al.* (2004) *EMBO J.* **23**, 260–271.

Histones and histone modifications. C. L. Peterson & M. A. Laniel (2004) *Curr. Biol.* **14**, R546–551.

Assembly of variant histones into chromatin. S. Henikoff & K. Ahmad (2005) *Ann. Rev. Cell. Dev. Biol.* **21**, 133–153.

Histones in functional diversification. Core histone variants. R. H. Pusarla & P. Bhargava (2005) *FEBS J.* **272**, 5149–5168.

7.3.G. Bacterial Type II DNA-binding Proteins: Heat-unstable (HU) and Integration Host Factor (IHF)

The bacterial type II DNA-binding proteins condense bacterial DNA genomes by binding at multiple sites and inducing coherent bends. They include the abundant **HU** (**heat-unstable**) and **integration host factor** (**IHF**) proteins, plus protein TF1 that is produced by bacteriophages. HU and TF1 bind with little, if any, sequence specificity, but IHF binds at specific sites that have some sequence similarities. All of these proteins bind to DNA as dimers. IHF is normally a heterodimer, while HU can be either a homo- or heterodimer.

These proteins bend the DNA very substantially (Figure 7-7). For example, within the 2.5 turns of double helix that comprise its binding site, IHF causes a bend of at least 160°, perhaps even more than 180°. In the cases of both IHF and HU, the two subunits intertwine to form a compact core, from which are extended two long β-sheet ribbons (Figure 7-7). These arms lie in the minor groove, extending from the inside to the outside of the wrapped DNA; there are two substantial kinks in the DNA where they terminate. IHF also clamps the hairpin by contacts in the minor groove using α-helices from both subunits of the dimer and extensive electrostatic interactions with the backbone phosphate groups.

A Pro side-chain located at the tips of the β-arms is absolutely conserved and produces these two kinks by intercalating partially between adjacent base pairs. This opens the minor groove on the outside of the bend. The narrowed groove on the inside should increase the repulsions between the phosphate groups on opposite strands, but this is diminished by interactions that neutralize their charges.

The conformation of the DNA dictates the sequence specificity of IHF binding, not specific contacts with the bases. Two short sequences separated by approximately half a turn provide the sequence specificity. The conserved sequence CAA is favored over other sequences at the kink site because it can best accommodate the severe distortion induced by the protein. **IHF binds rapidly to straight or partially bent DNA, followed by slower DNA bending** on the 10^{-4} s time scale. The rate of bending is similar to that for opening a single A•T base pair inside duplex DNA. Spontaneous thermal disruption of base pairing at an A•T site may be sufficient to overcome the free energy barrier needed to bend or kink the DNA before forming a tight complex with IHF.

DNA–protein interactions and bacterial chromosome architecture. J. Stavans & A. Oppenheim (2006) *Phys. Biol.* **3**, R1–R10.

Direct observation of DNA bending/unbending kinetics in complex with DNA-bending protein IHF. S. V. Kuznetsov *et al.* (2006) *Proc. Natl. Acad. Sci. USA* **103**, 18515–18520.

Bacterial protein HU dictates the morphology of DNA condensates produced by crowding agents and polyamines. T. Sarkar *et al.* (2007) *Nucleic Acids Res.* **35**, 951–961.

7.3.H. Single-strand DNA-binding Proteins

DNA does not normally exist naturally as a single-strand (ssDNA), except in a few bacterial viruses. It normally occurs only transiently during the replication, repair and recombination of double-stranded DNA molecules, so there are only a few **ssDNA binding proteins (SSBs) but all organisms have at least one**; some of them also bind single-stranded RNA (Section 7.4.C). **These proteins are involved in the replication and repair of DNA and in genetic recombination, when they protect ssDNA from damage by nucleases, prevent hairpin formation and block reannealing of the DNA until the process is completed**. Many ssDNA-binding proteins interact physically and functionally with a variety of other DNA processing proteins. These interactions are thought to order and guide the parade of proteins that 'trade places' on the ssDNA, a model known as **hand-off**, as the processing pathway progresses.

Minimalist protein design: a β-hairpin peptide that binds ssDNA. S. M. Butterfield *et al.* (2005) *J. Am. Chem. Soc.* **127**, 24–25.

1. Prokaryotic Single-strand DNA Binding Proteins

SSB proteins from prokaryotes (and mitochondria of eukaryotes) have a tetrameric structure of identical 19-kDa polypeptide chains that binds specifically to ssDNA with high affinity and moderate cooperativity. As a result, **many molecules of SSB coat ssDNA, and they can induce denaturation of double-stranded DNA**. Binding is virtually independent of the nucleotide sequence.

The N-terminal domain of each monomer in the tetrameric structure (Figure 7-27-C) contains a single α-helical segment and two β-pleated sheets that fold to create a barrel from which β-hairpin loops protrude, the oligonucleotide/oligosaccharide binding fold (OB-fold) (Section 7.3.H.3). In contrast, the C-terminal domain, which includes the last third of the polypeptide chain, is disordered in both instances, but it adopts a fixed structure upon interacting with RNA polymerase and other proteins. The monomers form dimers that can then further dimerize, forming the tetramer. The structure has external patches of positive charge that, together with the flexible loops, guide the wrapping of ssDNA around the tetramer. Aromatic residues on the external surface of the assembly are essential for ssDNA binding and interact directly with the nucleotides of the DNA. SSB binds ssDNA with virtually no change in structure. Each subunit of the tetrameric protein contains the same potential ssDNA binding site, so ssDNA can bind to the protein in several ways. Normally, a 65-nucleotide stretch of ssDNA interacts with and wraps around all four subunits of the tetramer; the wrapping occurs on a time scale of tens of microseconds. Binding to ssDNA is accompanied by a substantial negative enthalpy change and negative heat capacity change, due in part to linkage of temperature-dependent protonation and DNA base unstacking equilibria.

An SSB tetramer can be transferred from one ssDNA molecule to another without proceeding through a free protein intermediate, and the rate of transfer is determined by the availability of free DNA binding sites within the initial SSB–ssDNA donor complex. It occurs by a 'direct transfer' mechanism in which an intermediate composed of two DNA molecules bound to one SSB tetramer forms transiently prior to the release of the acceptor DNA.

Microsecond dynamics of protein–DNA interactions: direct observation of the wrapping/unwrapping kinetics of single-stranded DNA around the *E. coli* SSB tetramer. S. V. Kuznetsov *et al.* (2006) *J. Mol. Biol.* **359**, 55–65.

3D structure of *Thermus aquaticus* single-stranded DNA-binding protein gives insight into the functioning of SSB proteins. R. Fedorov *et al.* (2006) *Nucleic Acids Res.* **34**, 6708–6717.

2. Eukaryotic Replication Protein A

Replication protein A (RPA) was originally identified as a eukaryotic SSB protein essential for the *in vitro* replication of simian virus 40 DNA, but subsequently it was found to be **indispensable in almost all DNA metabolic pathways**, including DNA replication, repair and recombination, plus cell-cycle and DNA-damage checkpoints. It is a more complex protein than the prokaryotic SSB, consisting of three different polypeptide chains of molecular weights 70, 32 and 14 kDa. The 70-kDa subunit of RPA is composed of multiple domains: an N-terminal domain involved in protein interactions, a central DNA-binding domain, a putative ($C–X_2–C–X_{13}–C–X_2–C$) zinc finger, and a C-terminal domain involved in interacting with the other subunits.

RPA binds tightly to ssDNA, with a dissociation constant of about 10^{-10} M, and it prefers ssDNA over RNA or double-stranded DNA by a factor of about 10^3. The three subunits of human RPA contain six conserved **DNA-binding domains** (DBDs), of which four bind ssDNA (DBD-A, -B and –C on the largest subunit, DBD-D on the 32-kDa subunit) with dissociation constants smaller than 10^{-9} M. Each DBD has an OB-fold (Figure 7-27-D). DBDs A and B are very similar and individually have very low intrinsic affinities for ssDNA, but tandem DBDs (AA, AB, BA and BB) bind ssDNA with about 100-fold greater affinity. DBD-A has the highest intrinsic affinity for ssDNA, while DBD-D has the lowest; it binds ssDNA only after the other three binding sites are occupied. Binding of ssDNA occurs in at least three different stages, with initially eight to 10 nucleotides interacting, then 13–14, and finally 27, when it uses all four of its DBDs. At low salt concentrations, however, RPA tends to bind to only 18 nucleotides of ssDNA, using only three of its DBDs. Binding is not markedly cooperative.

A dynamic model for replication protein A (RPA) function in DNA processing pathways. E. Fanning *et al.* (2006) *Nucleic Acids Res.* **34**, 4126–4137.

Functions of human replication protein A (RPA): from DNA replication to DNA damage and stress responses. Y. Zou *et al.* (2006) *J. Cell. Physiol.* **208**, 267–273.

3. Oligonucleotide/Oligosaccharide Binding Fold

The SSBs share a common core ssDNA-binding domain with a conserved oligonucleotide/ oligosaccharide binding fold (OB-fold). The OB-fold is a small β-barrel formed from five β-strands connected by modulating loops (Figure 7-27); its topology is that of the Greek key. Although the OB-fold is most prominent in proteins that bind nucleic acids, it also occurs in many other types of proteins, where it can interact with oligosaccharides or other proteins. In all OB-folds, two or three loops at the same end of the β-barrel act as clamps to bind to their ligands. The loops change in length and sequence to accommodate various ligands. Those that bind nucleic acids usually have little specificity for the nucleotide sequence and bind nonhelical structures, although a few recognize specific single-stranded regions of nucleic acids.

OB-fold domains range between 70 and 150 amino acid residues in length, depending on the lengths of the loops between the β-strands. Strands 3 and 5 can close the β-barrel by hydrogen-bonding in a parallel manner, but in some cases they are not in close contact and the barrel is not completely closed. An α-helix often occurs between strands 3 and 4 and packs against the bottom of the barrel (Figure 7-27). Ligands are usually bound by the loop connecting strands 2 and 3, designated L_{23}, assisted by loops L_{12}, $L_{3\alpha}$, $L_{\alpha 4}$ and L_{45} (the α refers to the α-helix) The loops define a cleft that runs across the surface of the OB-fold perpendicular to the axis of the β-barrel. The majority of nucleic acid ligands bind within this cleft, usually perpendicular to the antiparallel β-strands and with a polarity running 5′ to 3′ in the nucleic acid from strands 4 and 5 to strand 2. The prokaryotic SSB is, however, unusual in binding ssDNA with the opposite polarity. Interactions also occur with the surface of the barrel defined by strands 2 and 3.

It might be expected that SSBs that are not sequence-specific would interact primarily with the backbone of ssDNA, and not with the nucleotides. In fact, **the bases of the ssDNA are usually in close contact with the protein, while the phosphodiester groups are primarily exposed to solvent**; similar binding interactions occur in other proteins that bind single-stranded nucleic acids and loop structures. The bases interact with the protein primarily through stacking interactions with aromatic amino acid side-chains and packing interactions with hydrophobic parts of side-chains, which can also involve the deoxyribose rings. Deleting these side-chains decreases the binding affinity substantially. The nonspecific nature of these nonpolar interactions means that the binding sites recognize single-stranded nucleic acid structures but not the identities of the bases. In addition, these binding sites seem to be intrinsically flexible, so that they can mold themselves to be complementary to nucleic acids with different sequences. Some OB-folds bind their ligands with no change in conformation, whereas others undergo folding from a very disordered structure in the absence of the ligand. The nucleic acid ligand can undergo changes in conformation in order to bind. For example, the local secondary structure can be disrupted in order to bind, with nucleotides bulged out of helices or removed from stacking interactions.

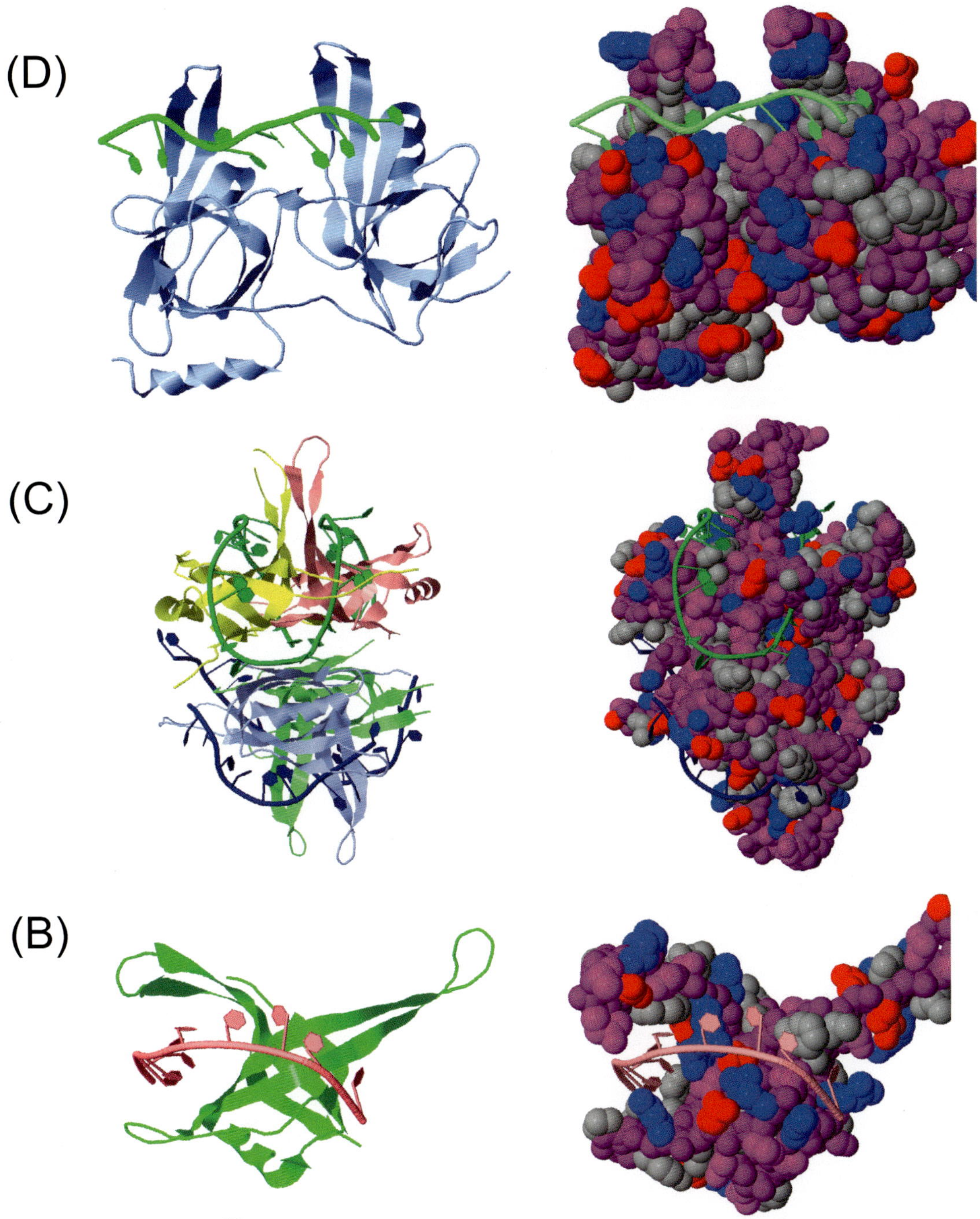

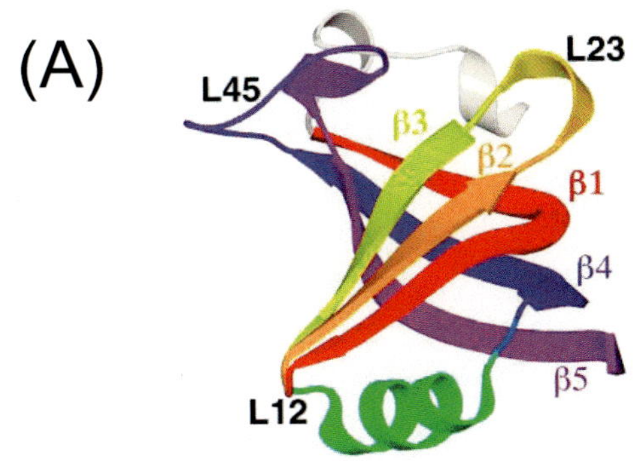

Figure 7-27. The OB-fold in binding ssDNA molecules. (A) The canonical OB-fold. The β-strands are numbered by their order in the primary structure; L are the variable loops. (B) The OB-fold of the SSB from *Helicobacter pylori* binding poly(dT) (PDB data file *2vw9*); only the polypeptide backbone is shown on the *left*, whereas the space-filling model is on the *right*. (C) The tetrameric SSB of *E. coli* bound to two 35-mer ssDNAs (PDB data file *1eyg*). (D) Two domains of the human replication protein A bound to poly(dC) (PDB data file *1jmc*). Figure generated using the program Jmol.

Individual OB domains have low affinity for their ligands, but tethering four of them on the same molecule produces high affinity. Yet the DNA molecules dissociate at the much faster rates associated with weak binding than the slow rates that would be expected from a single high-affinity binding site. The reason is that a bound ssDNA molecule can dissociate from each single binding site relatively independently of the other sites. **This type of linkage of multiple weak interactions on a dynamic binding scaffold appears to be a highly effective mechanism for multi-step processing in biology.**

The four ssDNA-binding domains of RPA and each of the monomers of the prokaryotic SSB protein have the characteristic OB-fold (Figure 7-27). Each monomer of SSB makes extensive contacts with ssDNA, using a large interaction surface that includes both sides of a protracted two-stranded β-sheet that extends from loop L_{23}. In RPA, binding of ssDNA causes a major reorientation between, and significant conformational changes within, the OB-folds that comprise the DNA-binding domain. Two OB-folds have a tandem orientation in the presence of DNA but adopt multiple orientations in its absence. Within the OB-folds, extended loops implicated in DNA binding change conformation significantly when the DNA dissociates.

Many other proteins use the OB-fold to interact with nucleic acids, including RNA. For example, *E. coli* ribosomal protein S1 is composed of six repeating homologous OB-folds. The second OB-fold in protein S1 is essential for the recognition of transfer messenger RNA, while the four C-terminal OB-folds play a role in stabilizing the complex of S1 and the transfer messenger RNA. Other ribosomal proteins (S17, S12 and IF1) consist of single OB-folds. Telomere end-binding proteins contain four OB-fold domains used for recognition of single-stranded telomeric DNA.

The OB proteins, even those restricted to binding nucleic acids, are renowned for **the absence of any discernible sequence similarities**. Yet the close similarities between their structures and interactions with ligands suggest that they have evolved from a common ancestor. That ancestor must have existed early in evolution, as it is present in all extant life forms.

OB-fold domains: a snapshot of the evolution of sequence, structure and function. V. Arcus (2002) *Curr. Opin. Struct. Biol.* **12**, 794–801.

Insights into ssDNA recognition by the OB-fold from a structural and thermodynamic study of Sulfolobus SSB protein. I. D. Kerr *et al.* (2003) *EMBO J.* **22**, 2561–2570.

Nucleic acid recognition by OB-fold proteins. D. L. Theobald *et al.* (2003) *Ann. Rev. Biophys. Biomol. Structure* **32**, 115–133.

7.4. RNA-BINDING PROTEINS

RNA differs from DNA covalently in only two seemingly minor ways: the 2′-hydroxyl group replaces an H atom and the methyl group of the thymine base is missing (the base is then known as uracil) (Chapter 4). Yet these two small covalent differences have major consequences for the structures and functions of RNA. The structures of RNA are much more diverse than those of DNA, and they even have catalytic activities (Section 14.6). RNA serves as the genetic material in some viruses, but in most cases its main function is, as messenger RNA, to convey the genetic information from the DNA genetic material to the ribosomes for the synthesis of proteins. Other major RNAs are the transfer

RNAs, which serve as adaptors between the genetic information of nucleic acids and the amino acids of proteins (Section 7.4.E). All of these functions of RNA require that they interact with a wide variety of proteins. RNA is also an essential and functional component of many **ribonucleoproteins**, usually abbreviated as RNP. The most prominent example is the **ribosome** (Section 7.4.F).

Not many RNA-binding proteins have been characterized structurally, but sequence homologies detected in a variety of proteins that bind RNA have recognized a number of structural motifs that are believed to participate in binding RNA (Table 7-3). RNA and DNA are recognized in fundamentally similar ways, but **the structural motifs that bind RNA tend to be distinct from those that bind DNA**. Exceptions are the zinc finger (Section 7.3.C.1), OB-fold (Section 7.3.H.3), homeodomain (Section 7.3.A.3) and KH domains (Section 7.4.C), which bind both RNA and DNA.

RNA is most readily distinguished from DNA by the presence of the 2′-hydroxyl group and the A-type double helix that it adopts in place of the B-type helix of DNA. On the other hand, A-form RNA has a deep and narrow major groove and a shallow minor groove, which make the bases relatively inaccessible, so it is more difficult for proteins to recognize specific sequences in double-stranded RNA. Most RNAs are single-stranded, however, and consequently adopt double helices only locally, with hairpins, internal loops, bulges and pseudoknots in the remainder of the RNA molecule. Many proteins recognize irregular RNA structures in which the bases are more exposed for

Table 7-3. Examples of RNA-binding domains and sequence motifs

Sequence motif	Representative proteins	Target RNA
Alu binding module	SRP 14/9 heterodimer	*Alu* domain of SRP RNA
dsRNA binding motif	*E. coli* RNase III	RNA transcripts
	Drosophila Staufen	maternal bicoid mRNA
KH domain	hnRNP protein K	mRNA precursor
OB domain	Class II aminoacyl tRNA synthetase	tRNA
RGG box	hnRNP proteins	mRNA precursor
RNP domain	U1A spliceosomal protein	U1 snRNA
	hnRNP protein C	mRNA precursor
	hnRNP protein A1	mRNA precursor
	Poly(A) binding protein	mRNA poly(A) tail
S1 domain	Ribosomal protein S1	mRNA
	Initiation factor 1	
Sm domain	Spliceosomal core proteins	snRNAs
Zn-finger	TFIIIA	5S rRNA

hnRNP, heterogeneous nuclear RNP; mRNA, messenger RNA; rRNA, ribosomal RNA; RNP, ribonucleoprotein; snRNA, small nuclear RNA; SRP, signal recognition particle.

distinguishing by hydrogen bonds or stacking interactions. Binding to the protein may actually make the RNA structure less regular. Some RNA molecules fold into complex tertiary structures that may be recognized on the basis of their unique shapes.

A **groove binder** class of RNA-binding proteins places an α-helix, 3_{10}-helix, β-ribbon or irregular loop into the groove of an RNA helix, recognizing both the specific sequence of bases and the shape or dimensions of the groove, which are sometimes distorted from the normal A-form structure. Many known RNA-binding proteins are all β- or αβ-proteins; the αβ-folds are comparable to those present in some ribosomal protein subunits and may have evolved from them. These proteins often contain an exposed β-sheet, which is a good RNA-binding surface because aromatic and hydrophobic side-chains emanating from it can interact with the bases of the RNA molecule and create pockets that recognize the bases of single-stranded RNA. Some of these proteins recognize completely unstructured RNA; in others RNA secondary structure indirectly promotes binding by constraining bases in an appropriate orientation. **Binding specificity is generally a result of base-specific hydrogen bonds, nonpolar contacts and mutual accommodation of the protein and RNA surfaces.**

The amino acid residues most involved in contacts with the RNA are Arg, Asn, Ser and Lys; less preferred are Ala, Ile, Leu and Val; Trp and Cys are rarely observed. Of the total number of amino acids located at the interfaces, 22% are hydrophobic, 40% charged (32% positive, 8% negative), 30% polar and 8% Gly. Arg and Lys interact electrostatically with phosphate groups. Pro and Asn prefer to interact with bases more than ribose and phosphate groups; Met, Phe and Tyr prefer ribose over phosphate and bases. **No base predominates at protein–RNA interfaces**, but A bases are preferred by Ile, Pro and Ser residues, C is preferred by Leu, G by Asp and Gly, and U by Asn. Of all the contacts, 72% are van der Waals interactions, 23% involve hydrogen bonds and 5% are very short contacts. Of the potential H bonds, 54% are standard, 33% are of the C–H … O type, and 13% are between ionized groups; the protein main chain is involved in 32% of the hydrogen bonds. Amino acid residues that are not frequently in direct contact with RNA components usually interact with RNA atoms via their polypeptide backbone atoms in indirect hydrogen bonds involving water molecules.

The majority of RNA-binding proteins have modular structures in which an RNA-binding domain is combined with other domains with other functions.

The yeast three-hybrid system has become a useful tool in identifying interactions between RNA and proteins. An RNA sequence is tested in combination with an RNA-binding protein linked to a transcription activation domain for transcription of a reporter gene. A productive RNA–protein interaction activates that reporter gene *in vivo*.

Protein families and RNA recognition. Y. Chen & G. Varani (2005) *FEBS J.* **272**, 2088–2097.

Extending the size of protein RNP RNA complexes studied by nuclear magnetic resonance spectroscopy. C. D. Mackereth *et al.* (2005) *Chembiochem.* **6**, 1578–1584.

RNA RNP protein interactions in the yeast three-hybrid system: affinity, sensitivity and enhanced library screening. B. Hook *et al.* (2005) *RNA* **11**, 227–233.

Crystallization of RNA RNP protein complexes. E. Obayashi *et al.* (2007) *Methods Mol. Biol.* **363**, 259–276.

7.4.A. Ribonucleoprotein (RNP) Domain

The sequence of the ribonucleoprotein (RNP) domain has been found in well over 200 distinct RNA-binding proteins from diverse species, so it probably appeared early in evolution. It is one of the most common protein sequence motifs found in entire genomes. It is also known as **RRM** (RNA recognition motif) or **RNP-CS** (RNP-consensus sequence)-type RNA-binding domain. Some proteins contain multiple copies of the RNP domain, but others contain only one. For example, there are four tandem copies of the RNP domain in the protein that binds to the polyadenylate (poly A) tail of messenger RNA in eukaryotes.

RNP motifs can recognize a disparate range of RNA structures and sequences, with greatly varying affinities and specificities. In some cases they bind RNA very tightly, with dissociation constants of less than 10^{-9} M, and with exquisite sequence specificity, to form stable RNP particles. In other cases, binding to RNA is much weaker, with dissociation constants of roughly 10^{-6} M and poor discrimination between different RNA sequences. Weak binding by a single RNP motif, however, is augmented if there are several such domains in a protein that can bind to the same RNA molecule simultaneously. In many cases, affinity for an RNA is also determined by parts of the protein outside the RNP motif. The diversity of RNA structures recognized by the RNP motif is remarkable, in part because the structures of both the protein and the RNA are malleable and undergo substantial alterations in forming a complex.

An RNP domain generally consists of about 80 amino acid residues. It is characterized by two short sequence motifs called RNP1 (eight amino acid residues) and RNP2 (six residues) that are highly conserved. Structurally, the RNP domain is constructed of a four-stranded antiparallel β-sheet with two α-helices lying on one face (Figure 7-28). The two middle β-strands of the sheet contain the

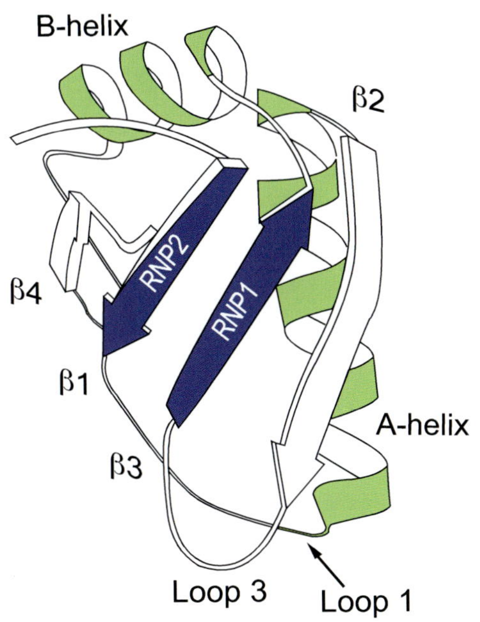

Figure 7-28. The structure of the RNP domain. The amino acid sequences that can identify RNP domains RNP1 and RNP2 comprise two of the β-strands. Adapted from K. Nagai *et al.* (1990) *Nature* **348**, 515–520.

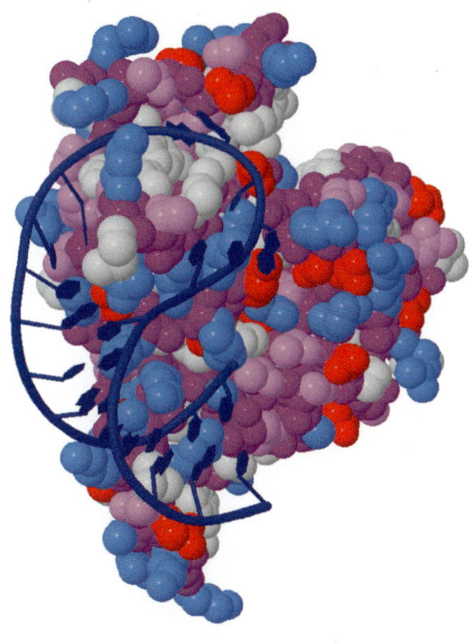

Figure 7-29. Structure of the complex between the U1A spliceosomal protein and its hairpin RNA binding site in U1 snRNA. The RNA is depicted as a skeletal model, whereas the van der Waals surface of the protein is depicted; basic groups are *blue*, acidic groups are *red*. Generated from PDB file *1a9n* using the program Jmol.

RNP1 and RNP2 sequences, and the side-chains of the three highly conserved aromatic residues within RNP1 and RNP2 project from the surface of the β-sheet. In the case of the U1A protein (Figure 7-29), an RNA hairpin loop of 10 nucleotides binds to the surface of the β-sheet as an open structure; the polypeptide loop 3 between the β2 and β3 strands protrudes through the RNA loop. The first seven nucleotides of the loop fit into a groove formed on the surface of the β-sheet. Their bases are splayed out and make nearly all the contacts between the RNA and protein. These seven bases stack onto adjacent bases and protein side-chains; they are also involved in an intricate hydrogen bond network with the side-chains and polypeptide backbone of protein. **These stacking interactions are comparable to those observed in the binding of ssDNA to proteins** (Section 7.3.H).

RNA recognition by RNP proteins during RNA processing. G. Varani & K. Nagai (1998) *Ann. Rev. Biophys. Biomol. Structure* **27**, 407–445.

7.4.B. Double-stranded RNA-binding Domain

The double-stranded RNA-binding domain (dsRBD) is a sequence motif of 65–75 amino acid residues that is found in multiple copies in RNA-binding proteins from diverse origins. These proteins are involved in diverse functions, such as localizing messenger RNA in the cell, editing messenger RNA and processing ribosomal and transfer RNAs. **The dsRBD module consists of a three-stranded antiparallel β-sheet and two α-helices packed on one side** (Figure 7-30-A); their order is α–β–β–β–α in the primary structure. The dsRBD is homologous to the N-terminal domain of ribosomal protein S5, suggesting that dsRBD evolved from a ribosomal protein.

A dsRBD usually recognizes 12 base pairs of double-stranded RNA (dsRNA), interacting with two successive minor grooves and across the intervening major groove on one face of a primarily A-form RNA helix (Figure 7-30-B). The nature of these interactions explains why dsRBD is specific for dsRNA, rather than single-stranded RNA (ssRNA) or double-stranded DNA (dsDNA). A dsRBD of Staufen protein recognizes the shape of the A-form double helix of dsRNA through interactions between conserved residues within loop 2 and the RNA minor groove, and between loop 4 and the phosphodiester backbone across the adjacent major groove. In addition, the A helix interacts with the single-stranded loop that caps the RNA helix. The dsRBD does not bind to dsDNA because it would have the recognizably different B-form and because of contacts with the 2′-OH groups of the ribose moiety in the minor groove. The interface between the protein and RNA is not tightly packed but is relatively flexible, with the protein interacting primarily with the sugar–phosphate backbone, not the bases, so it is not apparent how there can be sequence specificity.

The dsRBD of endonuclease Rnt1p targets this enzyme to its RNA substrates by recognizing hairpins closed by AGNN tetraloops (Section 4.1.B). The tetraloop is retained in the RNA–dsRBD complex (Figure 7-30-B). The dsRBD contacts the RNA at successive minor, major and tetraloop minor grooves on one face of the N-terminal α-helix. Neither the universally conserved G nor the highly conserved A bases of the AGNN tetraloop are recognized by specific hydrogen bonds, but an α-helix fits snugly into the minor groove of the RNA tetraloop and top of the stem, interacting in a nonsequence-specific manner with the sugar–phosphate backbone and the two nonconserved tetraloop bases.

Double-stranded RNA is most common in some animal viruses, and cells often use its presence to detect a viral infection. The dsRBD is usually involved and activates the enzyme dsRNA-dependent

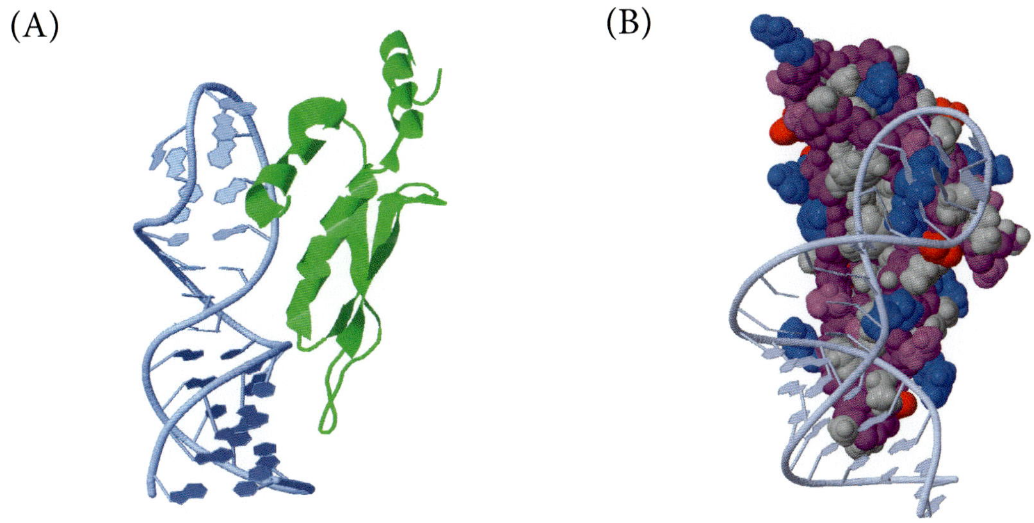

Figure 7-30. Structure of the dsRBD domain from the ribonuclease III Rnt1p complexed with the 5′-terminal hairpin of one of its substrates closed by an AGNN tetraloop. (A) The topology of the protein domain is depicted schematically. (B) The protein domain is shown in space-filling form, with the RNA bound on it surface. This view is rotated from that shown in (A). Generated from PDB data file *1t4l* using the program Jmol.

protein kinase, which can lead to a shutdown in gene translation, thereby inhibiting protein synthesis and replication of the virus.

The double-stranded-RNA-binding motif: interference and much more. B. Tian *et al.* (2004) *Nature Rev. Mol. Cell. Biol.* **5**, 1013–1023.

The double-stranded RNA-binding motif, a versatile macromolecular docking platform. K. Y. Chang & A. Ramos (2005) *FEBS J.* **272**, 2109–2117.

Structure and specific RNA binding of ADAR2 double-stranded RNA binding motifs. R. Stefl *et al.* (2006) *Structure* **14**, 345–355.

7.4.C. KH Domain

Three copies of a sequence motif within hnRNP K, one of the nuclear proteins that bind to precursors of messenger RNA (mRNA) and homologous sequences, are known as **KH domain**s.

The KH domain interacts with ssRNA and ssDNA. It has an αβ-structure containing a three-stranded β-sheet and three α-helices (Figure 7-31). The binding groove is an αβ-structure formed by the juxtaposition of two α-helices, one β-strand and two flanking loops. A groove with a hydrophobic floor accommodates a stretch of three C nucleotides, but specificity for flanking residues is provided by hydrogen-bonding groups in the KH domain.

KH domain: one motif, two folds. N. V. Grishin (2001) *Nucleic Acids Res.* **29**, 638–643.

X-ray crystallographic and NMR studies of the third KH domain of hnRNP K in complex with single-stranded nucleic acids. P. H. Backe *et al.* (2005) *Structure* **13**, 1055–1067.

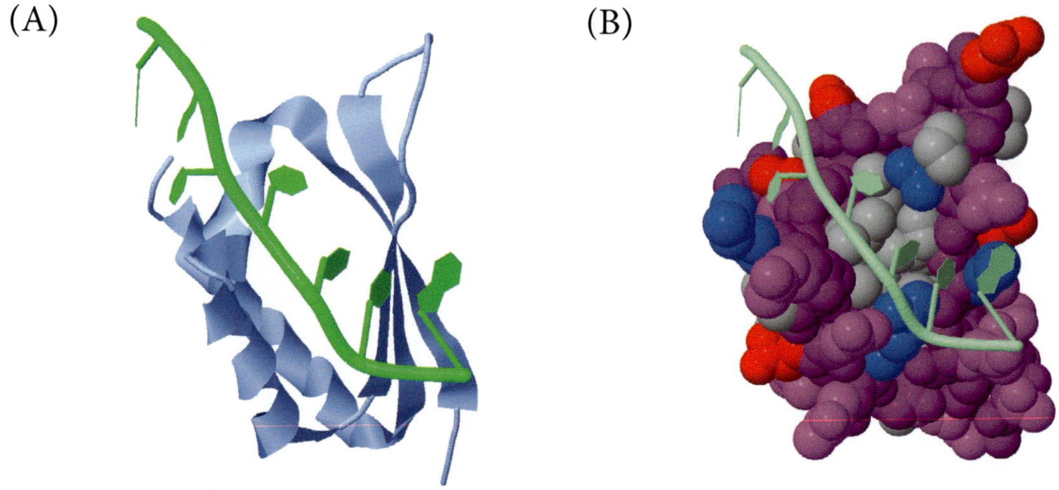

Figure 7-31. The structure of the complex between a KH domain from a human poly(C)-binding protein and a 7-nucleotide segment of ssDNA, with the sequence 5′-AACCCCTA-3′. (A) The course of the polypeptide chain of the KH domain is shown schematically. (B) The protein is depicted in space-filling form. Generated from PDB file *2p2r* using the program Jmol.

7.4.D. MS2 Bacteriophage Coat Protein

The coat protein of MS2 bacteriophage is one of the best-characterized RNA-binding proteins. The bacteriophage contains a genomic ssRNA of 3569 nucleotides that is packaged into an icosahedral protein shell made of 180 copies of the coat protein. A dimer of the coat protein binds to a hairpin on the RNA that is near the ribosome-binding site of the virally encoded replicase gene; this inhibits translation of the replicase mRNA and triggers assembly of the coat protein and packaging of the genomic RNA. The RNA hairpin binding site consists of a tetraloop and an adenosine bulge. It binds to the surface of a continuous β-sheet across the dyad axis of the coat protein dimer (Figure 7-32). Consequently, only a single RNA molecule binds asymmetrically to the symmetric protein dimer:

Figure 7-32. Two copies of the RNA hairpin of MS2 bacteriophage binding to 1.5 dimers of the coat protein. Only one subunit of the lower dimer is depicted. The protein is illustrated as *ribbons*, the RNA as a skeletal model. Note that each dimer is symmetrical but can bind only one RNA molecule because it lies on the two-fold axis. Generated from PDB file *2izm* using the program Jmol.

the bulged adenosine and an adenosine in the tetraloop bind to equivalent sets of residues from each subunit. The amino acid residues that are energetically most important for binding the RNA are clustered in the middle of the RNA binding interface and are surrounded by residues that are less important.

RNA–protein interactions in spherical viruses. H. H. Bink & C.W. Pleij (2002) *Arch. Virol.* **147**, 2261–2279.

Investigating the structural basis of purine specificity in the structures of MS2 coat protein RNA translational operator hairpins. C. Halgstrand *et al.* (2002) *Nucleic Acids Res.* **30**, 2678–2685.

Alanine scanning of MS2 coat protein reveals protein–phosphate contacts involved in thermodynamic hot spots. D. Hobson & O. C. Uhlenbeck (2006) *J. Mol. Biol.* **356**, 613–624.

7.4.E. Recognizing Transfer RNAs

Many aspects of the nature of protein–RNA recognition can be illustrated with **transfer RNAs** (tRNA). They are the adaptor molecules that specify the amino acid residues that correspond to each of the triplet codons (Figure 7-21). Their recognition involves two extremes of specificity. Each cell contains 20 aminoacyl-tRNA synthetases, one for each amino acid, that are responsible for attaching the correct amino acid to each of the tRNAs. Each of these enzymes must specifically recognize and aminoacylate *only* their cognate tRNA, using the correct amino acid. In contrast, all the tRNAs interact with the certain proteins. For example, the prokaryotic elongation factor Tu (EF-Tu) introduces aminoacylated tRNAs (aa-tRNA) into the A-site of the ribosome and binds to nearly all the aa-tRNAs, with the exception of the initiator Met tRNA and selenocysteinyl-tRNA. Consequently, it must recognize features that are common to all the elongator aa-tRNAs. Intermediate and varying degrees of specificity are involved in the processing of the tRNA precursors. The various tRNAs are synthesized initially as precursor RNA molecules, which must be processed correctly by specific nucleases and many different enzymes that modify the bases (Figure 4-16). The modification enzymes have varying specificities; some recognize almost all tRNAs, others recognize a subset, while others are specific for a unique tRNA.

Bacteria such as *E. coli* have at least 46 different tRNA molecules, with anticodons that correspond to the various amino acids. There are, however, only 20 aminoacyl-tRNA synthetases, one for each of the normal amino acids, so some of the synthetases must recognize more than one tRNA. For example, the seryl-tRNA synthetase recognizes the six tRNAs that accept serine and must ignore all the others, even though all the tRNAs have very similar structures (Figure 4-17). The aminoacyl-tRNA synthetases can be grouped into two classes, I and II, based upon some conserved sequence motifs and the structural architecture and action of their catalytic domains. Yet they are remarkably diverse, with polypeptide chains that vary from 303 to 951 amino acid residues and α, α_2, α_4 and $\alpha_2\beta_2$ quaternary structures. Given that tRNAs have similar secondary and tertiary structures, the molecular basis for the specific recognition between aminoacyl-tRNA synthetases and tRNA resides in the so-called **tRNA identity elements**.

The class I glutaminyl-system and the class II aspartyl- and seryl-systems are best understood. They interact with their tRNA in very different ways but are similar in having a large interface that binds to the tRNA. The large interface probably increases the binding affinity and ensures the correct positioning and orientation of the tRNA. The interactions include nonspecific backbone contacts, often involving basic residues. Those interactions that are specific for the bases and responsible for discriminating between the various tRNAs are largely restricted to the anticodon and the acceptor stem. The second general feature is that protein–RNA contacts involve mutual induced fit via conformational changes in either or both of the protein and the tRNA. In the process, flexible protein loops can become ordered and protein domains can be reoriented and stabilized. In the case of the tRNA, base pairs in the acceptor stem can be broken, the bases in the anticodon loop can be unstacked, and the 3′-end can be distorted.

On the evolution of structure in aminoacyl-tRNA synthetases. P. O'Donoghue & Z. Luthey-Schulten (2003) *Microbiol. Mol. Biol. Rev.* **67**, 550–573.

Two conformations of a crystalline human tRNA synthetase–tRNA complex: implications for protein synthesis. X. L. Yang *et al.* (2006) *EMBO J.* **25**, 2919–2929.

The crystal structure of the ternary complex of phenylalanyl-tRNA synthetase with $tRNA^{Phe}$ and a phenylalanyl–adenylate analogue reveals a conformational switch of the CCA end. N. Moor *et al.* (2006) *Biochemistry* **45**, 10572–10583.

1. Class I Glutaminyl-tRNA Synthetase

Glutaminyl-tRNA synthetase (GlnRS) is a monomeric protein and binds a single $tRNA^{Gln}$. The specificity of GlnRS for $tRNA^{Gln}$ is largely determined by interactions with identity elements in the tRNA acceptor stem and anticodon stem loop (Figure 7-33). Both have very different conformations in the free tRNA. In the complex, the anticodon stem of the tRNA is extended from five to seven base pairs; the two additional base pairs are not the usual Watson–Crick pairs. The interface between the distal two β-barrel domains of the synthetase forms three separate recognition pockets into which the three anticodon bases (CUG) fit by being splayed out. In the active site of the synthetase, the tRNA is oriented in the active site of the synthetase so that base pairs 2 and 3 of the anticodon can be recognized. The 3′-end of the tRNA that accepts the amino acid reaches the catalytic center only by forming an unusual hairpin turn, in which the first base pair is broken but compensated at least to some extent by formation of a new hydrogen bond.

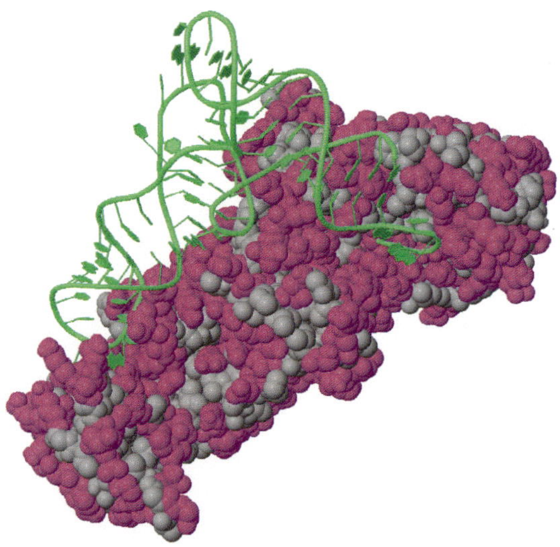

Figure 7-33. The monomeric glutaminyl tRNA synthetase from *E. coli* binding one molecule of $tRNA^{Gln}$. The tRNA is shown in skeletal form, whereas the protein is illustrated with a space-filling model. Polar amino acid residues are *blue*, nonpolar residues are *gray*. Generated from PDB file *1qrs* using the program Jmol.

Many bacteria lack a glutaminyl-tRNA synthetase, and $tRNA^{Gln}$ is initially mischarged with Glu by a nondiscriminating glutamyl-tRNA synthetase. This enzyme thus charges both $tRNA^{Glu}$ and $tRNA^{Gln}$ with glutamate. The Glu charged to $tRNA^{Gln}$ is subsequently converted to Gln. All eukaryotes and some bacteria contain a discriminating GluRS that only generates Glu–$tRNA^{Glu}$. The difference between the two types has been tracked down to one amino acid residue: an Arg residue required for discrimination is changed to a Gly that permits both tRNAs to serve as substrate.

A number of archaeal organisms generate Cys–tRNA(Cys) in a two-step pathway, first charging phosphoserine onto $tRNA^{Cys}$ and subsequently converting it to Cys-$tRNA^{Cys}$.

Crystal structure of a non-discriminating glutamyl-tRNA synthetase. J. O. Schulze *et al.* (2006) *J. Mol. Biol.* **361**, 888–897.

---

2. Class IIb Aspartyl-tRNA Synthetase

Aspartyl-tRNA synthetase (AspRS) is a dimer that can bind two tRNAs simultaneously and symmetrically, one to each subunit (Figure 7-34). In contrast to the class I case, class II synthetases interact specifically with the major groove side of the acceptor stem of their cognate tRNA, so that the 3′-end of the tRNA can enter the synthetase active site without significantly distorting its structure. Recognition of the anticodon by AspRS recognizes the anticodon using an OB-fold (Section 7.3.H.3) at the N-terminus. The normally compact anticodon loop undergoes a large conformational change upon binding, exposing five bases of the anticodon loop. The three anticodon bases, GUC, lie on the surface of the β-sheet and are recognized by specific hydrogen bonds. The central U stacks with a conserved Phe residue and hydrogen bonds to Gln and Arg residues that are also conserved. The tRNA synthetase recognizes primarily the shape of the tRNA and interacts with the tRNA backbone. Many of the normal bases of the tRNA can be replaced by alternatives, so long as the new nucleotides preserve the original backbone structure of the tRNA.

Human aspartyl-tRNA synthetase contains an extension at its N-terminus that is involved in the transfer of Asp–tRNA to elongation factor 1α. This extension may reduce the rate of dissociation of

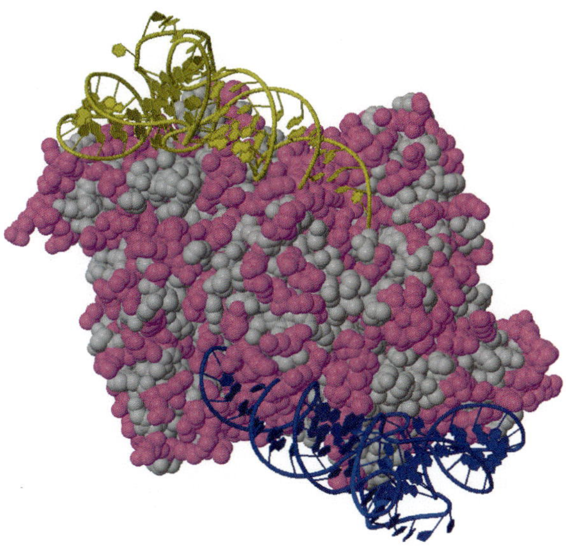

Figure 7-34. Two molecules of $tRNA^{Asp}$ bound to the dimeric aspartyl tRNA synthetase from yeast. The view is along the two-fold symmetry axis. The tRNAs are illustrated as skeletal models, whereas the protein is a space-filling model. Polar amino acid residues are *blue*, nonpolar residues are *mauve*. Generated from PDB file *1asy* using the program Jmol.

Asp–tRNA from human aspartyl-tRNA synthetase and provide enough time for elongation factor 1α to interact with the Asp–tRNA.

Expanding tRNA recognition of a tRNA synthetase by a single amino acid change. L. Feng *et al.* (2003) *Proc. Natl. Acad. Sci. USA* **100**, 5676–5681.

Recognition of acceptor-stem structure of $tRNA^{Asp}$ by *Escherichia coli* aspartyl-tRNA synthetase. H. Choi *et al.* (2003) *RNA* **9**, 386–393.

Structure of the N-terminal extension of human aspartyl-tRNA synthetase: implications for its biological function. H. K. Cheong *et al.* (2003) *Int. J. Biochem. Cell Biol.* **35**, 1548–1557.

3. Class II Seryl-tRNA Synthetase

Another dimeric class II synthetase is Seryl-tRNA synthetase (SerRS), but it recognizes tRNA significantly differently from AspRS (Figure 7-35), perhaps because it must recognize six different tRNAs for serine. The six serine anticodons are part of two distinct groups, so the serine tRNAs share no common anticodon base and the anticodon is not an identity element for $tRNA^{Ser}$. One identity element is the long variable arm of $tRNA^{Ser}$, which is shared only by two other tRNAs. Only one molecule of $tRNA^{Ser}$ can bind to the dimer, because it binds nonsymmetrically across both subunits. SerRS is unique in having a 100-residue N-terminal domain that is a flexible antiparallel coiled coil, 60 Å long and exposed to the solvent, known as the *helical arm*. The helical arm changes its orientation and interacts with the TΨC loop and the long variable arm of the tRNA. The tRNA variable arm has a minimum length in this case because the backbone of six base pairs contacts the protein. The synthetase contacts primarily the tRNA backbone, not the bases, and recognizes primarily the unique shape of $tRNA^{Ser}$. The shape is largely determined by two bases inserted into the D-loop that make novel interactions in the core of the tRNA. One of these bases is stacked against the first base pair of the long variable arm and determines its spatial orientation.

The synthetase does not contact the anticodon stem loop, so it can recognize any of the six anticodons for serine.

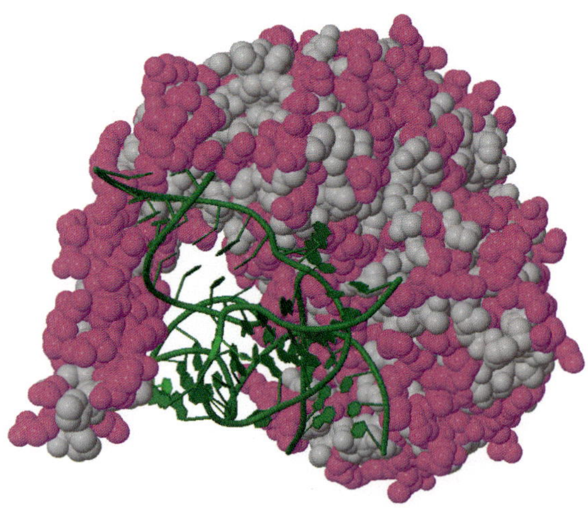

Figure 7-35. Structure of the ternary complex of seryl-tRNA synthetase, $tRNA^{Ser}$ (GGA), and a nonhydrolyzable analog of seryl-adenylate. A space-filling model of the protein is represented, with polar groups *mauve*. The tRNA is depicted as a skeletal model; if it is depicted as a space-filling model, the complex is seen to be closely packed. The tRNA synthetase is a dimer of identical subunits and there are two active sites, each with seryl-adenylate bound, but only a single tRNA molecule binds because it lies on the two-fold axis. Generated from PDB file *1ser* using the program Jmol.

4. *Elongation factor EF-Tu*

EF-Tu is involved in polypeptide biosynthesis, delivering charged tRNAs to the ribosome. It is a member of the important class of G proteins, which bind and hydrolyze GTP to drive reactions that would otherwise be unfavorable energetically. In the activated GTP-bound state, EF-Tu binds all aminoacylated tRNAs and more tightly than their uncharged forms.

EF-Tu needs to recognize and bind to all tRNAs, so activated EF-Tu makes contacts only to relatively limited and conserved regions of aminoacylated tRNAs: to the aminoacylated CCA end, the 5′-phosphate and the T-stem helix (Figure 7-36). The CCA 3′-end binds in a cleft between domains 1 and 2 of EF-Tu, and conserved residues from domain 2 interact with the base and phosphate of nucleotide A76. The backbone of EF-Tu makes hydrogen bonds to the ester group linking the carboxyl group of the amino acid to the 3′-OH of the ribose and to the free amino group of the amino acid, thereby being specific for charged tRNAs. The amino acid itself projects into a pocket formed between domains 1 and 2 that must be not very specific. The complex is extremely elongated.

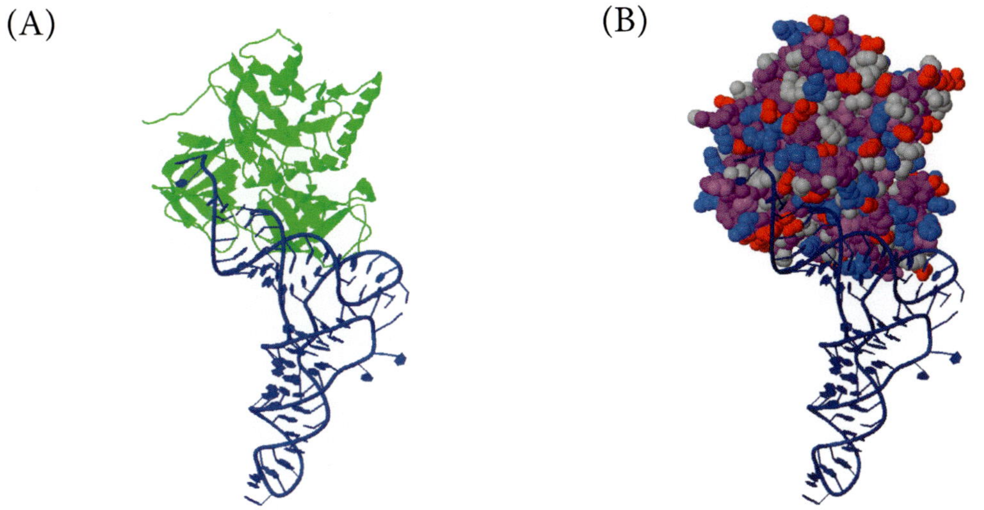

Figure 7-36. Structure of the complex of EF-Tu and a tRNA. The backbone and bases of the tRNA are depicted in skeletal form. (A) The polypeptide backbone of the EF-Tu structure is depicted with *coils* for α-helices and *arrows* for β-strands. (B) The protein is depicted with van der Waals *spheres* for each atom. Generated from PDB file *1b23* using the program Jmol.

Crystal structure of the bovine mitochondrial elongation factor Tu.Ts complex. M. G. Jeppesen *et al.* (2005) *J. Biol. Chem.* **280**, 5071–5081.

A protein extension to shorten RNA: elongated elongation factor-Tu recognizes the D-arm of T-armless tRNAs in nematode mitochondria. M. Sakurai *et al.* (2006) *Biochem. J.* **399**, 249–256.

Directed mutagenesis identifies amino acid residues involved in elongation factor Tu binding to yeast Phe-tRNAPhe. L. E. Sanderson & O. C. Uhlenbeck (2007) *J. Mol. Biol.* **368**, 119–130.

7.4.F. The Ribosome

Ribosomes are the ultimate RNPs, having molecular weights of roughly 2.5 million in prokaryotes and 4 million in eukaryotes. They are so large that they and their component RNA molecules are usually designated by their sedimentation coefficients. A complete 70S prokaryotic ribosome has an irregular structure with a maximum dimension of about 250 Å, and it dissociates into two parts, the 50S and 30S subunits. In *E. coli*, the 50S subunit consists of two RNA molecules, one 23S (2904 nucleotides) and the other 5S (120 nucleotides) plus 33 protein molecules. The 30S subunit consists of one 16S RNA molecule (1542 nucleotides) and 21 different protein molecules. The RNAs make up 66% of the mass. The larger eukaryotic 80S ribosome dissociates into 40S and 60S subunits. The 40S subunit consists of an 18S, 1874-nucleotide RNA and 33 polypeptide chains, while the 60S subunit consists of 28S (4718 nucleotides), 5.8S (160 nucleotides) and 5S (120 nucleotides) RNA molecules, plus 49 polypeptide chains. RNA molecules make up 60% of the total mass.

Ribosomes are the cellular factories where polypeptide chains are assembled using the genetic information in mRNA, so they have very complex functions and interact with a number of different molecules. They also have very complex structures that are known in great detail by X-ray crystallography (Figure 7-37). Much information about the overall architecture had been gained earlier by electron microscopy.

The structures of the individual subunits of the ribosome are dictated primarily by the structures of their large RNA molecules, which have extensive secondary structures (Figure 7-37). The proteins are generally in peripheral positions, but with unstructured loops enmeshed with the RNA. Nearly all of the known types of RNA tertiary structures have been found in ribosomal RNA. Most of the base pairs involved are standard Watson–Crick type (Figure 2-13) and proteins tend to interact most with RNA stems with classical double-helical structures (Figure 4-5). Some of the proteins are homologous to homeodomains, zinc fingers, leucine zippers and other motifs found in proteins that bind nucleic acids. Based on their topology, the protein structures fall into six groups. Many share similar α + β sandwich folds, but the topology of this domain varies considerably, as do the ways in which the proteins interact with RNA.

An extraordinary variety of protein–RNA interactions is observed. The proteins function primarily to stabilize interactions between the RNA domains that maintain the complex's structural integrity. Many electrostatic interactions occur between the phosphate groups of the RNA backbone and numerous Arg and Lys residues, particularly those in flexible extensions at the ends of the polypeptide chain. Bases of the RNA are recognized in both the minor groove and the widened major groove of RNA helices, as well as through hydrophobic binding pockets that capture bulged nucleotides and through insertion of amino acid residues into hydrophobic crevices in the RNA. Many of the proteins have, in addition to globular domains, long extended regions, either in the termini or internal loops, which make extensive contact to the RNA component and are involved in stabilizing the RNA tertiary structure.

Remarkably, the ribosomal RNA is primarily responsible for the major catalytic activities of the ribosome, including the step that forms the peptide bond between the new amino acid and the nascent polypeptide chain, plus the hydrolysis of GTP to fuel the steps that involve moving the nascent chain from one site to another (Chapter 8). The proteins appear to be involved in more peripheral roles and are important for stabilizing the overall structure and interacting with other molecules, such as the tRNAs (Section 7.4.E).

(A)

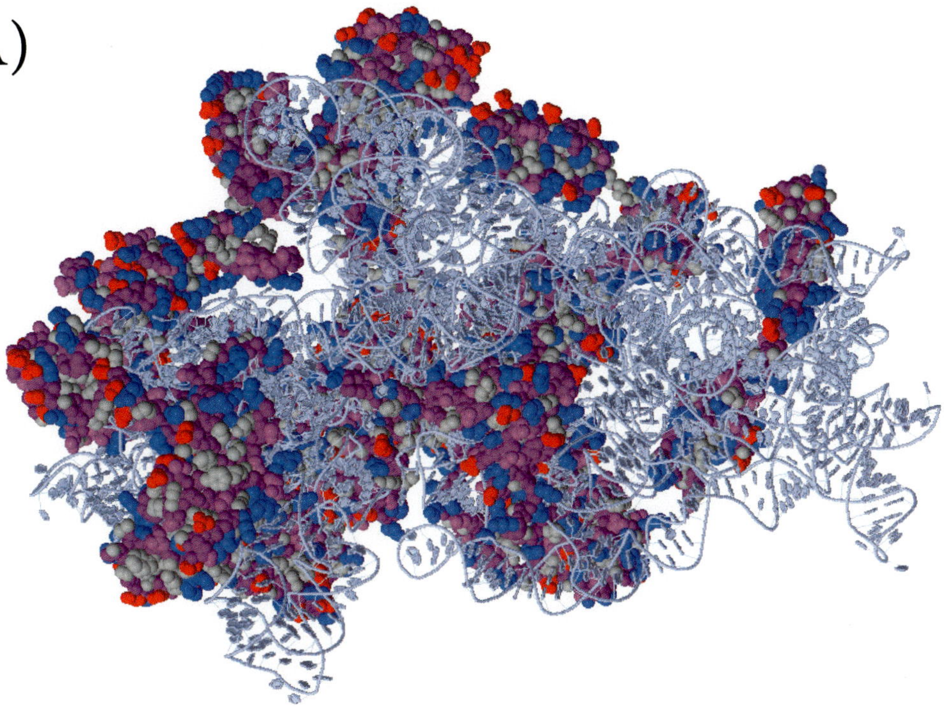

(B)

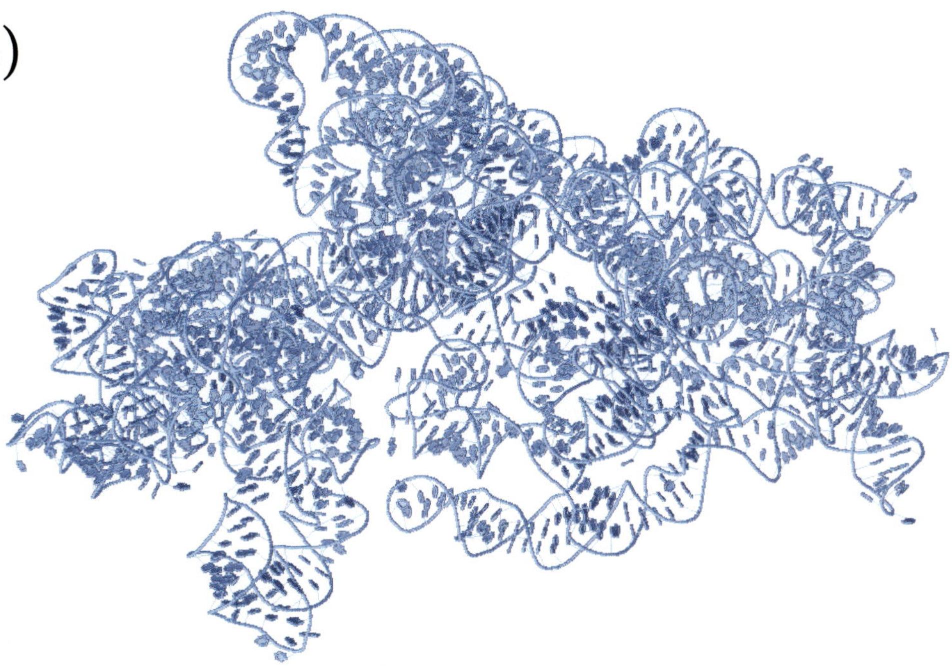

Figure 7-37. Structure of the small (30S) subunit of the ribosome from *Thermus thermophilus*. (A) The entire structure of the ribosomal subunit, with both RNA and proteins. The RNA is shown as a skeletal model, whereas the proteins are shown as *spheres* of the van der Waals radii of the atoms; acidic groups are *red*, basic groups are *blue*, and atoms of the backbone are *mauve*. (B) The structure of the RNA in the ribosome, with the proteins deleted. The secondary structure is described in Figure 4-6-B. Generated from PDB file *2avy* using the program Jmol.

RNA structure: reading the ribosome. H. F. Noller (2005) *Science* **309**, 1508–1514.

Structure of the 70S ribosome complexed with mRNA and tRNA. M. Selmer *et al.* (2006) *Science* **313**, 1935–1942.

Ribosome dynamics: insights from atomic structure modeling into cryo-electron microscopy maps. K. Mitra & J. Frank (2006) *Ann. Rev. Biophys. Biomol. Structure* **35**, 299–317.

~ CHAPTER 8 ~

CATALYTIC NUCLEIC ACIDS

RNA molecules, in the form of ribozymes, are capable of catalyzing chemical reactions in the absence of proteins; they occur naturally in nature and their identification was the first indication that nucleic acids could have catalytic capabilities. Ribozymes can increase the rates of their reactions by as much as 10^{11}, with specificities of better than 1000:1. **The catalytic RNA folds into a compact three-dimensional structure that is directed towards chemical reaction of a single specific phosphate linkage** (Section 4.2.C). All of the known natural ribozymes participate in RNA maturation or processing and catalyze transesterification reactions of the phosphodiester groups of the RNA backbone. The majority of natural catalytic RNAs catalyze only intramolecular reactions but the ribozyme and 'substrate' can be separated onto individual RNA molecules, when the ribozyme becomes a true catalyst that is capable of multiple turnovers.

The existence of catalytic RNAs supports the idea that life started in a **primeval RNA world** in which RNA molecules alone, without DNA or proteins, were responsible for storing and replicating the genetic information, as well as for catalyzing biochemical reactions. Subsequently, DNA or proteins would have been found to be more efficient at most of these functions and thus replaced the RNA. Present-day catalytic RNAs may be vestiges of the original functions of RNA.

Nucleic acid enzymes. R. Fiammengo & A. Jaschke (2005) *Curr. Opinion Biotechnol.* **16**, 614–621.

Ribozyme catalysis: not different, just worse. J. A. Doudna & J. R. Lorsch (2005) *Nature Struct. Biol.* **12**, 395–402.

Structure, folding and mechanisms of ribozymes. D. M. Lilley (2005) *Curr. Opinion Struct. Biol.* **15**, 313–323.

HDV ribozymes. M. D. Been (2006) *Curr. Top. Microbiol. Immunol.* **307**, 47–65.

Bacterial RNase P: a new view of an ancient enzyme. A. V. Kazantsev & N. R. Pace (2006) *Nature Rev. Microbiol.* **4**, 729–740.

8.1. Natural Reactions

Catalytic RNAs are involved in diverse functions but they have similar mechanisms of action. Each natural ribozyme uses an activated nucleophile to attack a phosphodiester linkage in-line, which

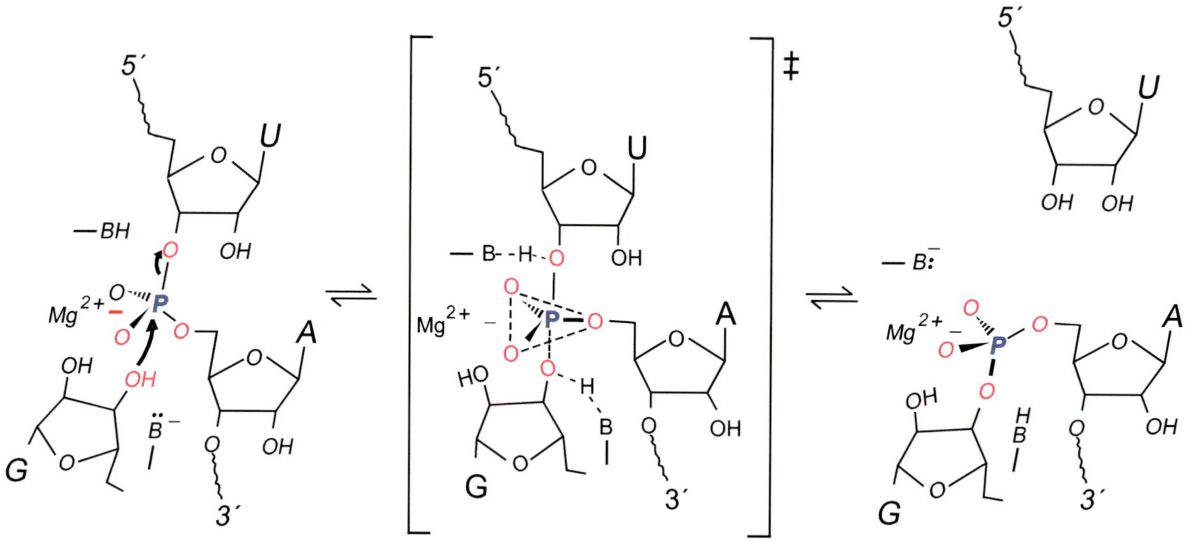

Figure 8-1. The basic transesterification reaction catalyzed by ribozymes. On the *left*, the 3′-hydroxyl group of guanosine acts as the nucleophile in attacking the P atom of the phosphate diester to be cleaved. The reaction is imagined to be assisted by two basic groups, B, and an Mg^{2+} ion complexed to the phosphate. In the transition state (*middle*) the phosphate group is now planar, and it is inverted in the product (*right*), which has the guanosine replacing the chain that has been expelled. Adapted from T. Cech (1987) *Science* **236**, 1532–1539.

produces inversion of the stereochemistry of the phosphate group (Figure 8-1). The various catalytic RNA reactions differ primarily in the identities of the nucleophile and leaving group.

Four small catalytic RNAs have been identified in **satellite RNAs** from plants and animals: the **hammerhead**, **hairpin**, **hepatitis delta virus** (HDV) and **Varkud satellite** (VS) ribozymes. These catalytic RNAs differ substantially in structure but perform the same cleavage reaction. The four satellite RNAs replicate via a complementary double-stranded RNA intermediate using a rolling circle mechanism, in which replication produces a continuous RNA molecule composed of tandem repeats of the satellite RNA. The function of the ribozymes is to self-cleave the satellite RNA repeats into monomers. All use as internal nucleophile the 2′-OH group that is located immediately adjacent to the phosphate group to be cleaved (Figure 8-1). In a one-step reaction, the 5′-oxygen of the phosphodiester is displaced and the RNA backbone is cleaved, producing RNAs with 2′–3′-cyclic phosphate and 5′-OH groups at their termini. Chemical alkaline hydrolysis of unstructured RNA produces the same products (Figure 4-2) but the ribozyme reaction is directed towards a single phosphodiester linkage.

A second, more common, type of catalytic RNA comprises the **self-splicing introns**. These RNAs catalyze excision of the intron of which they are a part from a messenger RNA precursor; that includes linking together the flanking exons (Figure 8-2). They can also catalyze the reverse splicing reaction, which permits them to insert into RNA molecules and act as mobile genetic elements and transfer genetic information between organisms. There are two types of self-splicing introns, group I and group II, which differ in their structures and reaction mechanisms (Figure 8-3). **Group I introns** use external nucleophiles to catalyze two consecutive transesterification reactions. In the first, the 3′-OH of an external, bound guanosine attacks the phosphate group at the 5′-end of the intron. This reaction adds the guanosine covalently to the 5′-end of the intron and releases the 5′-end of the exon. The

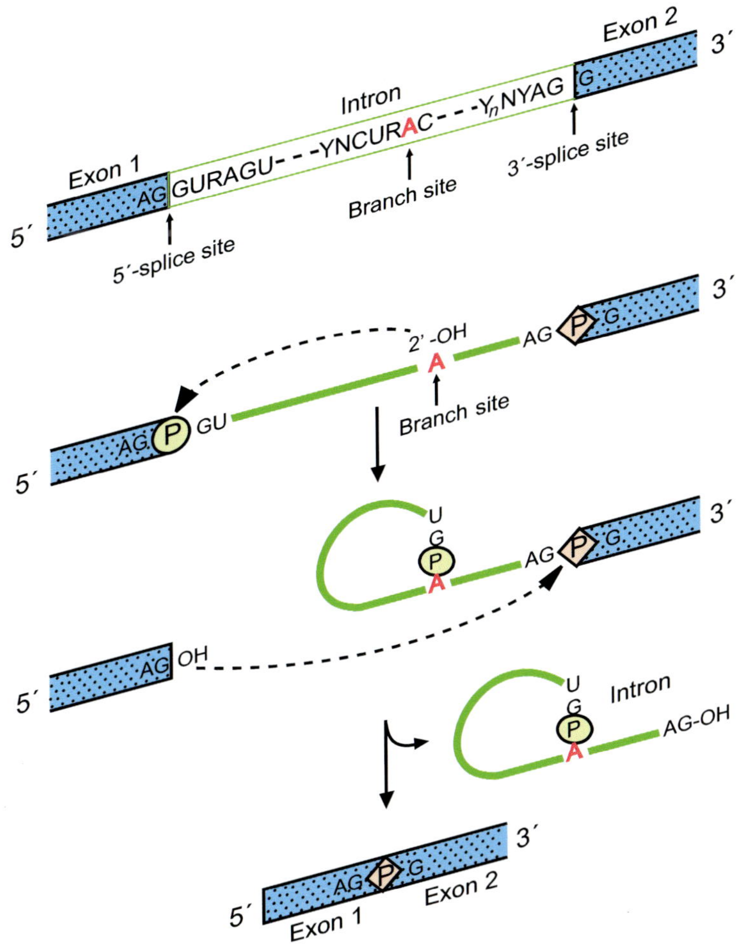

Figure 8-2. The two-step splicing pathway of nuclear messenger RNA precursors. Exon and intron sequences are indicated by *stippled* and *open boxes*, respectively. The sequences given are consensus sequences found at the mammalian 5′- and 3′-splice sites the branch site of U2-dependent introns, where N is any base, Y is a pyrimidine and R is a purine. The *dashed arrows* indicate the nucleophilic attacks on the splice sites by the 2′-OH of the branch site in the first step and the 3′-OH of the cleaved 5′-exon in the second step. The phosphate groups at the 3′- and 5′-splice sites are indicated indicate by P in a *circle* or a *triangle*.

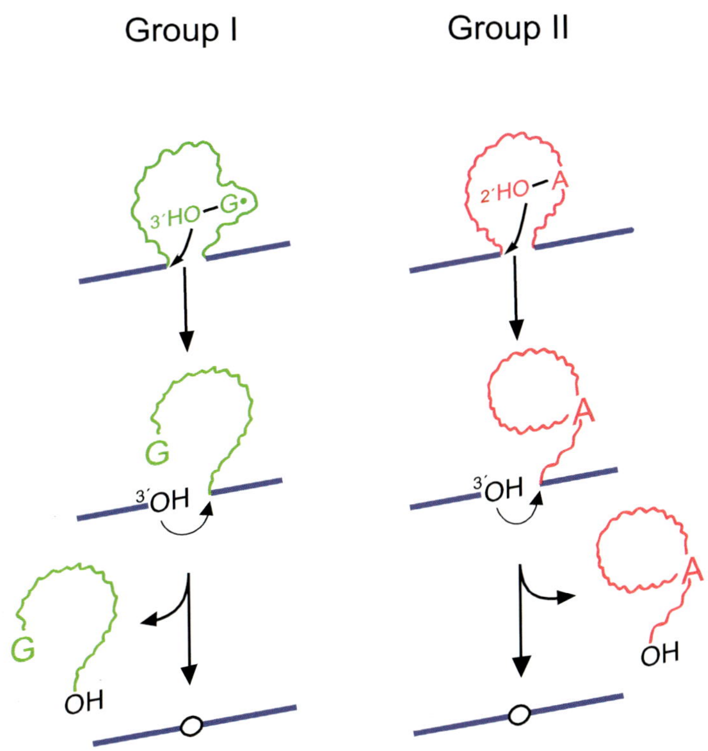

Figure 8-3. Comparison of the mechanisms of action of group I and group II self-splicing introns. The group I ribozymes use as nucleophile a guanosine cofactor that is not part of the polynucleotide chain, which then becomes incorporated covalently into the RNA. The group II ribozymes use a bulged A residue of the RNA molecule (Figure 8-2).

second step is essentially the reverse of the first step, but the 3′-OH of a guanine nucleotide at the 3′-end of the intron is the leaving group. The result is that the intron has been excised, leaving an intact RNA molecule.

The reaction mechanism of **group II introns** is intermediate between the previous two. They use an internal nucleophile, the 2′-OH of a bulged A nucleotide that it is not adjacent to the phosphate group to be cleaved but much further away, at the 5′-exon–intron boundary (Figure 8-2). As in the case of the group I introns, the 3′-OH group of the 5′-exon is the leaving group for the first step and the nucleophile for the second step. The released intron has a **lariat-like structure**, with the 5′-end of the intron covalently attached to the 2′-OH of the bulged A nucleotide. The same reaction mechanism is employed in the more complex process of messenger RNA precursor splicing that is catalyzed by the **spliceosome**.

A third type of ribozyme is **ribonuclease P**. It participates in the biosynthesis of tRNAs (Section 4.2.B) by catalyzing the removal of extensions from the 5′-ends of tRNA precursors. It consists naturally of both an RNA and a protein component. Both are essential *in vivo* but the RNA molecule alone is sufficient to carry out the tRNA processing reaction, although at a reduced rate. Unlike the other naturally occurring catalytic RNAs that use either the 2′- or 3′-OH groups as nucleophiles, RNaseP uses water, which hydrolyzes the RNA backbone. The original 5′-sequence is removed, producing a tRNA with a terminal 5′-phosphate group. Using an external nucleophile and acting on an external substrate, RNaseP is the only known natural RNA that catalyzes multiple turnovers and acts as a true enzyme *in vivo*.

The **Varkud satellite** ribozyme is the largest nucleolytic ribozyme. It consists of five helical sections, organized by two three-way junctions. The substrate stem–loop binds into a cleft formed between two helices, while making a loop–loop contact with another section of the ribozyme. The phosphate that is to be acted on makes a close contact with an internal loop (the A730 loop), the probable active site of the ribozyme. This loop contains a particularly critical nucleotide, A756, as most changes to this nucleotide lead to three-orders of magnitude slower cleavage. The groups that would participate in Watson–Crick base pairing are especially important.

Many of the functions of the **ribosome** (Section 7.4.F) are catalyzed by the RNA components, rather than the proteins, including the step in which a peptide bond is formed to extend the growing polypeptide chain. The 2×10^7-fold rate enhancement produced by the ribosome is achieved entirely by lowering the entropy of activation. The enthalpy of activation for generating a peptide bond is slightly less favorable on the ribosome than in solution, in contrast with protein enzymes. The ribosome enhances the rate of peptide bond formation mainly by positioning the substrates and/or by excluding water from the active site, rather than by conventional chemical catalysis.

Chemical models for ribozyme action. T. Lonnberg & H. Lonnberg (2005) *Curr. Opinion Chem. Biol.* **9**, 665–673.

The transition state for formation of the peptide bond in the ribosome. A. Gindulyte *et al.* (2006) *Proc. Natl. Acad. Sci. USA* **103**, 13327–13332.

Nucleobase catalysis in ribozyme mechanism. P. C. Becilacqua & R. Yajima (2006) *Curr. Opinion Chem. Biol.* **10**, 455–464.

Alternative roles for metal ions in enzyme catalysis and the implications for ribozyme chemistry. R. K. Sigel &

A. M. Pyle (2007) *Chem. Rev.* **107**, 97–113.

8.2. Ribozyme Structure and Catalysis

The three-dimensional arrangement of the nucleotides, metal ions, water molecules and cofactors that comprise the RNA structure are vital for catalysis. A binding site on an RNA that can recognize and bind a specific substrate requires that the structure be relatively complex (Section 4.2). All the ribozymes appear to have compact, folded structures reminiscent of proteins and tRNA (Section 4.2.C) in which an active site is pre-organized and located within crevices. Catalytic RNAs generally require divalent metal ions, usually Mg^{2+} or Mn^{2+}. These ions are important for both structure and function. Those involved in catalysis usually serve as an electrophile to activate the nucleophile or stabilize the leaving group.

Understanding how these structures produce catalysis has been hindered by the structural plasticity of RNA molecules, especially in the case of the hammerhead ribozyme. It is relatively simple, and a minimal model consists of only 42 nucleotides. To prevent catalysis and cleavage from taking place, the RNA molecules studied crystallographically consist in one case of a substrate that contains deoxyribose nucleotides; in a second case, a single 2′-O-methyl is substituted for the 2′-OH nucleophile. A third structure is of an all-ribose substrate but in the absence of divalent metal ions. The three structures are similar but they do not have an active site geometry that could explain catalysis; it appears that these structures must change significantly during catalysis.

The situation was resolved by the structure of a full-length hammerhead ribozyme consisting of 63 nucleotides, with a 2′-O-methyl substitution in place of the 2′-OH nucleophile to prevent cleavage of the phosphodiester bond. The additional nucleotides in this structure increase the rate of catalysis 1000-fold and produce very significant changes throughout the folded structure. The structure of the active site now can explain catalysis (Figure 8-4). The active site has the 2′-OH group of the adjacent ribose ring in position to attack the phosphate group to be cleaved, a guanine base positioned to act as the base, and the 2′-OH group of another guanine residue to act as the acid, as shown in Figure 8-1. No ions appear to be involved in catalysis in this case.

Ribozymes are now believed to function using the same principles identified in enzymes, by placing functional groups in the active site in correct proximity to assist the chemical reaction and by binding and stabilizing the transition state.

Structure of ribonuclease P: a universal ribozyme. A. Torres-Larios *et al.* (2006) *Curr. Opinion Struct. Biol.* **16**, 327–335.

Tertiary contacts distant from the active site prime a ribozyme for catalysis. M. Martick & W. G. Scott (2006) *Cell* **126**, 309–320.

Dissecting the multistep reaction pathway of an RNA enzyme by single-molecule kinetic ‘fingerprinting’. S. Liu *et al.* (2007) *Proc. Natl. Acad. Sci. USA* **104**, 12634–12639.

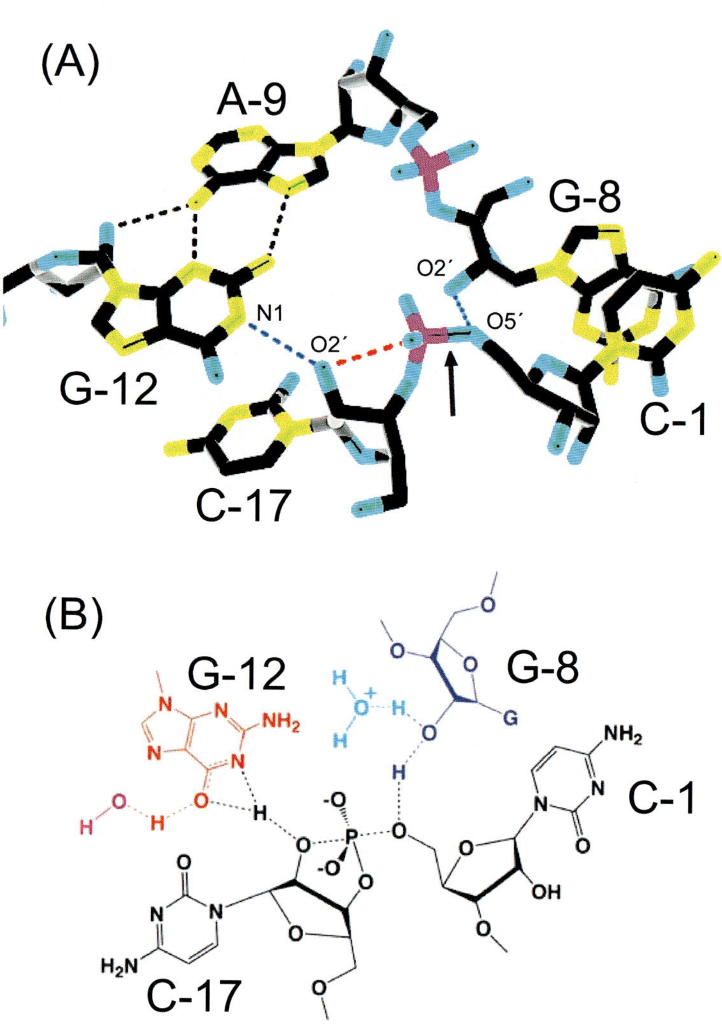

Figure 8-4. The structure of the active site and the inferred transition state of the full-length hammerhead ribozyme. The complete structure of the RNA is shown in Figure 4-18. (A) The structure of the active site determined crystallographically. C atoms are black; O, green; N, yellow; P, mauve. Hydrogen bonds are indicated by *black*, *red* or *blue dashed lines*. The *red* and *blue* ones are postulated to be involved in catalysis. The phosphodiester bond to be cleaved is indicated by the arrow. The 2′-OH of C-17 is in close proximity to the phosphate group, very nearly in-line with the P atom and the 5′-O leaving group, as expected if it is the nucleophile that attacks the P atom (Figure 8-1). The nucleobase of G-12 is in position to act as a base and accept the proton from the 2′-OH group. The 2′-OH of G-8 is in position to act as an electrophile in the bond cleavage. (B) The proposed transition state for the reaction. The general base, G-12, is *red*, the general acid, G--8 *blue* and the substrate *black*. The water molecules are believed to be necessary to accept or supply protons to the basic and acidic catalytic groups of G-12 and G-8, avoiding the necessity of them existing as fully charged ions. Adapted from figures kindly supplied by W. G. Scott.

8.3. Selection for Novel Ribozymes and Deoxyribozymes

Large libraries of RNA molecules with up to 10^{16} different sequences have generated novel RNA catalysts. Such libraries are typically used to isolate RNA or DNA molecules with catalytic activity in an *in vitro* selection procedure known as **SELEX** (**S**ystematic **E**volution of **L**igands by **EX**ponential enrichment). SELEX is a combinatorial amplification technique that exerts evolutionary pressure on a population of RNA molecules with random sequences to select and optimize for a particular catalytic function. The population is selected for those RNA molecules that have the desired property, then replicated and subjected to further selection. Only a handful of the original 10^{16} sequences survive many rounds of stringent selection steps. The range of reactions catalyzed by *in vitro*-selected ribozymes is now well beyond those known for natural RNAs. Some of the notable activities identified by this approach include:

(1) an RNA that functions like an RNA polymerase, using nucleoside triphosphates and releasing pyrophosphate as the leaving group (Section 6.2.B);

(2) an RNA that functions like a polynucleotide kinase, being able to transfer the γ-phosphate from ATP to the 5′- or internal 2′-OH of an RNA substrate;

(3) RNAs that use Ca^{2+} or Pb^{2+} instead of Mg^{2+} for catalysis.

RNAs have also been identified that catalyze chemical reactions other than simply transesterification at phosphate groups. One acts like an amino acyl RNA synthetase, in that it can transfer an amino acyl group from Phe–AMP onto the 3′-end of an RNA substrate. Other ribozymes have been selected that catalyze formation of carbon–nitrogen and carbon–sulfur bonds. A transition-state analog used as the bait selected an RNA able to catalyze the isomerization of a bridged biphenyl linkage.

This Darwinian approach to *in vitro* evolution has also demonstrated that nucleic acid-based catalysts are not restricted to RNA, as **DNA can act similarly**. Catalytic DNA molecules are known as **deoxyribozymes** or **DNAzymes**. They appear to act similarly to their RNA counterparts but have not been investigated as thoroughly.

Ribozymes have the potential to use their nucleobases directly in chemical catalysis in a variety of ways reminiscent of enzymes, including hydrogen bonding to the transition state, stabilizing charge development and transferring protons as general acid–base catalysts. Consequently, the chemical reactions found to be catalyzed by RNA and DNA are likely to continue to expand.

Functional gene-discovery systems based on libraries of hammerhead and hairpin ribozymes and short hairpin RNAs. M. Sano *et al.* (2005) *Mol. Biosyst.* **1**, 27–35.

Sensors made of RNA: tailored ribozymes for detection of small organic molecules, metals, nucleic acids and proteins. S. Muller *et al.* (2006) *IEE Proc. Nanobiotechnol.* **153**, 31–40.

Functional DNA nanotechnology: emerging applications of DNAzymes and aptamers. Y. Liu & J. Liu (2006) *Curr. Opinion Biotechnol.* **17**, 580–588.

Revitalization of six abandoned catalytic DNA species reveals a common three-way junction framework and diverse catalytic cores. W. Chiuman & Y. Li (2006) *J. Mol. Biol.* **357**, 748–754

8.4. Ligand-Binding Nucleic Acids: Aptamers

Aptamers are artificial small single-stranded RNA or DNA molecules that can be generated *in vitro* to bind with high affinity to a wide range of molecules, ranging from small chemical compounds to large proteins. The name is derived from *aptus*, the Latin word for 'to fit'. Aptamers are isolated from complex libraries of synthetic nucleic acids by a SELEX procedure that involves repetitively reducing the complexity of the library by partitioning on the basis of selective binding to the target molecule, followed by re-amplification. The selected RNA and DNA molecules bind their targets with similar affinities as antibodies and are able to distinguish between isotypes of an enzyme, so aptamers have been also been claimed to be **synthetic antibodies**. Moreover, aptamers can be easily stabilized by chemical modifications for *in vivo* applications. Numerous observations have shown that stabilized aptamers against extracellular targets such as growth factors, receptors, hormones and coagulation factors are very effective inhibitors of the corresponding protein function *in vivo*. RNA aptamers have also been expressed in living cells, where they have inhibited a protein implicated in intracellular signal transduction.

Peptide aptamers are small peptide sequences that have been selected in similar ways to recognize a predetermined target.

Aptamers come of age: at last. D. H. Bunka & P. G. Stockley (2006) *Nature Rev. Microbiol.* **4**, 588–596.

DNA and RNA aptamers: from tools for basic research towards therapeutic applications. H. Ulrich *et al.* (2006) *Comb. Chem. High Throughput Screen.* **9**, 619–632.

Structures of RNA switches: insight into molecular recognition and tertiary structure. H. Schwalbe *et al.* (2007) *Angew. Chem. Int. Ed. Engl.* **46**, 1212–1219.

INDEX

Page numbers for major entries, illustrations and tables are in **bold**.
Page numbers for items that are primarily a reference are in *italics*.